Effect of Dietary Supplementation on the Growth and Immunity of Fish and Shellfish—2nd Edition

Effect of Dietary Supplementation on the Growth and Immunity of Fish and Shellfish—2nd Edition

Guest Editors

Qiyou Xu
Jianhua Ming
Fei Song
Changle Qi
Chuanpeng Zhou

Basel • Beijing • Wuhan • Barcelona • Belgrade • Novi Sad • Cluj • Manchester

Guest Editors

Qiyou Xu	Jianhua Ming	Fei Song
Life Science Department	Life Science Department	Department of Biology
Huzhou University	Huzhou University	South China Normal
Huzhou	Huzhou	University
China	China	Guangzhou
		China

Changle Qi	Chuanpeng Zhou
School of Life Sciences	South China Sea Fisheries
Huzhou University	Research Institute
Huzhou	Chinese Academy of Fishery
China	Sciences
	Guangzhou
	China

Editorial Office
MDPI AG
Grosspeteranlage 5
4052 Basel, Switzerland

This is a reprint of the Special Issue, published open access by the journal *Fishes* (ISSN 2410-3888), freely accessible at: https://www.mdpi.com/journal/fishes/special_issues/8KZ26T1E0M.

For citation purposes, cite each article independently as indicated on the article page online and as indicated below:

Lastname, A.A.; Lastname, B.B. Article Title. *Journal Name* **Year**, *Volume Number*, Page Range.

ISBN 978-3-7258-3811-0 (Hbk)
ISBN 978-3-7258-3812-7 (PDF)
https://doi.org/10.3390/books978-3-7258-3812-7

Cover image courtesy of Zhihao Zhou

Contents

About the Editors

Qiyou Xu

Qiyou Xu received his Ph.D. in animal nutrition and feed science from Northeast Agricultural University in 2004. He worked as a research assistant and associate professor at the Animal Nutrition Institute of Northeast Agriculture University (1993–2005), before working as an associate professor and then professor at the Heilongjiang River Fisheries Research Institute of the Chinese Academy of Fishery Sciences (2005–2018). In 2018, he moved to Huzhou University as a professor. His research topics mainly include aquatic animal nutrition, feed, and aquaculture.

Jianhua Ming

Jianhua Ming received his Ph.D. in aquatic animal nutrition and feed science from Nanjing Agricultural University in 2011. He worked as an aquatic engineer in aquatic feed enterprises for 10 years (1996–2006). In 2011, he moved to Huzhou University as an associate professor and then professor. His research topics mainly include aquatic animal nutrition and feed science, aquatic animal physiology and molecular immunology, aquatic animal health aquaculture, and metabolic regulation.

Fei Song

Fei Song received her Ph.D. in nutrition and feeds for aquatic animals from the Ocean University of China in 2018. She works as an associate professor at Department of Life Science at South China Normal University (2018-present). Her research topic is the nutritional physiology of fish.

Changle Qi

Changle Qi received his Ph.D. degree (Zoology) from East China Normal University in 2020. He works at Huzhou University as an associate professor (2020–present). He is currently researching invertebrate nutrition. His research interests cover gonadal development, molting, and oxidative stress.

Chuanpeng Zhou

Chuanpeng Zhou received his Ph.D. in aquaculture from Nanjing Agricultural University in 2012, followed by a postdoctoral fellowship at the South China Sea Fisheries Research Institute, Chinese Academy of Fishery Sciences, and the South China Sea Institute of Oceanology, the Chinese Academy of Sciences. Since 2013, Dr. Zhou has served as an associate researcher at the South China Sea Fisheries Research Institute, the Chinese Academy of Fishery Sciences, focusing on studies in aquatic animal nutrition and feed science. His research areas include fish nutrition feed technology, carbohydrate metabolism and utilization in fish, and the interactions between fish nutrition and immunology.

Preface

With the development of the aquatic feed industry, new feedstuff and feed additives have been developed. Studies will be needed to ascertain their digestibility, along with the composite nutritional value of the formulation to the particular targeted species. This Reprint focuses on the application of new feedstuffs and additives in aquaculture species' growth and immunity, especially recently developed feedstuffs and additives.

The global aquaculture industry faces increasing challenges, including the need to improve feed efficiency, reduce reliance on fishmeal, and mitigate disease risks while ensuring sustainable production. Dietary supplementation has emerged as a promising strategy to address these challenges by optimizing nutrient utilization, promoting growth, and enhancing immune responses in fish and shellfish. This Reprint systematically investigate the effects of dietary supplements such as glutamate, soybean isoflavones, β-1,3-glucan, yeast derivatives, α-lipoic acid, bile acid, hormones, and copper on growth performance, antioxidant capacity, immune regulation, and intestinal health in aquatic species, providing an essential theoretical foundation for precision nutrition strategies. This applied information on a wide range of specific ingredients for use in commercial aquaculture.

We thank all of the authors and reviewers who have participated in this Special Issue. This work was supported by the "Pioneer" and "Leading Goose" R&D Program of Zhejiang (2023C02024) and the Zhejiang Provincial Natural Science Foundation of China under Grant No. LTGN23C190003.

This Reprint should be useful to students, researchers, fish nutritionists and feed formulators. We hope that this Reprint will fulfill its intended purpose of serving as an important and valuable reference resource.

Qiyou Xu, Jianhua Ming, Fei Song, Changle Qi, and Chuanpeng Zhou
Guest Editors

Editorial

Research Progress and Application Prospects of Dietary Supplements in Growth and Immune Regulation of Aquatic Animals

Jianhua Ming [1], Qiyou Xu [1,*], Changle Qi [1], Fei Song [2] and Chuanpeng Zhou [3]

[1] College of Life Sciences, Huzhou University, Huzhou 313000, China; mingjianhua686@163.com (J.M.); qichangle1989@163.com (C.Q.)

[2] Department of Biology, South China Normal University, Guangzhou 510631, China; sophioe@163.com

[3] South China Sea Fisheries Research Institute, Chinese Academy of Fishery Sciences, Guangzhou 510300, China; chpzhou@163.com

* Correspondence: 02655@zjhu.edu.cn

Received: 25 March 2025

Accepted: 28 March 2025

Published: 1 April 2025

Citation: Ming, J.; Xu, Q.; Qi, C.; Song, F.; Zhou, C. Research Progress and Application Prospects of Dietary Supplements in Growth and Immune Regulation of Aquatic Animals. *Fishes* 2025, 10, 152. https://doi.org/ 10.3390/fishes10040152

The global aquaculture industry faces increasing challenges, including the need to improve feed efficiency, reduce reliance on fishmeal, and mitigate disease risks while ensuring sustainable production. Dietary supplementation has emerged as a promising strategy to address these challenges by optimizing nutrient utilization, promoting growth, and enhancing immune responses in fish and shellfish. This Special Issue has compiled 11 research articles (article 1–11) that systematically investigate the effects of dietary supplements such as glutamate, soybean isoflavones, β-1,3-glucan, yeast derivatives, α-lipoic acid, bile acid, hormones, and copper on growth performance, antioxidant capacity, immune regulation, and intestinal health in aquatic species, providing essential theoretical foundations for precision nutrition strategies. This editorial synthesizes current research advancements and outlines future directions in the field.

Metabolic regulation of functional amino acids: Glutamate is a functional amino acid that plays critical roles in nutrition, metabolism, and signaling [1,2]. In this Special Issue, Zheng et al. (article 1) found that supplementing 2% glutamate in a low-protein diet (30%) significantly increased hepatopancreatic superoxide dismutase (SOD) activity in juvenile Chinese mitten crab (*Eriocheir sinensis*), though it did not affect weight gain. It revealed that glutamate downregulated mTOR pathway genes (mTOR, S6K1) to inhibit protein synthesis while promoting glutathione (GSH) biosynthesis, enhancing antioxidant defenses under low-protein conditions. However, the application effect of glutamate in fish showed species differences. Jiang et al. (article 2) showed that 0.4–0.6% L-glutamic acid promoted muscle development and improved intestinal digestive enzyme activity in largemouth bass (*Micropterus salmoides*) by activating mTOR and PI3K/Akt pathways, suggesting that its growth-promoting effect was closely related to the metabolic characteristics of the species. These results align with findings in Jian carp (*Cyprinus carpio* var. Jian) [3]. Similarly, Ding et al. [4] revealed that dietary arginine had no significant effect on the growth performance of Chinese perch (*Siniperca chuatsi*), but significantly affected its antioxidant capacity, intestinal digestion, and nutrient metabolism. Based on the antioxidant and intestinal health indicators, the optimal dietary arginine requirement for Chinese perch was 2.99–3.37% of the dry diet. These functional amino acids play an important role in the nutrition of aquatic animals, highlighting the necessity of precise nutrition strategies.

Immune regulation and intestinal health: As an immunostimulant, β-glucan is widely used in aquatic animals to enhance immunity and promote growth [5]. In this Special Issue, Xu et al. (article 3) studied the immunomodulatory effects of β-1,3-glucan in oriental river prawn (*Macrobrachium nipponense*). Adding 0.2% β-1,3-glucan significantly elevated hepatopancreatic acid phosphatase (ACP) and superoxide dismutase (SOD) activities and enhanced resistance to *Aeromonas hydrophila* by upregulating pattern recognition receptor (PRR)-related genes (*lgbp*, *lectin*, and *lbp*). Intestinal microbiota analysis revealed that β-1,3-glucan suppressed Cyanobacteria abundance and promoted *Rhodobacter* proliferation, improving gut microbial balance. Prebiotics and probiotics have immunomodulatory effects on aquatic animals and improve their health benefits [6,7]. Gao et al. (article 4) evaluated the effects of yeast prebiotics, probiotics, postbiotics (butyrate), and black soldier fly meal (BSFM) on hepatic immune gene expression in zebrafish. The results showed that butyrate significantly downregulated pro-inflammatory cytokine expression (TNF-α and IL-1β) post-pathogen challenge, highlighting its potent anti-inflammatory properties; BSFM attenuated inflammation by suppressing the NF-κB/p65 signaling pathway, likely mediated by immune-modulating components such as chitin and antimicrobial peptides [8]. These findings provide mechanistic support for the application of butyrate and BSFL as functional feed additives. Notably, while yeast prebiotics and probiotics showed limited immunomodulatory effects, the probiotic group exhibited a downward trend in *angptl4* expression following LPS challenge, suggesting indirect regulation of immune stress via lipid metabolism pathways [9]. This observation opens new avenues for exploring multi-target mechanisms of probiotics. Bile acids (BAs), the main components of bile, are initially synthesized by cholesterol in the liver and further into the intestine [10]. As one of the bile acids, taurochenodeoxycholic acid (TCDCA) is involved in nutritional regulation and has an adjuvant therapeutic effect on metabolic or immune disorders [11]. In this Special Issue, Xu et al. (article 5) demonstrated that TCDCA supplementation improved growth performance, muscle development, and nutrient quality of *Procambarus clarkii*, and ameliorated muscular autophagy and the gut microbiota relating to fatty acid and protein metabolism, as well as immunity. These data provide a theoretical basis for the application of TCDCA in the cultivation of crustaceans.

Antioxidant and metabolic regulation: Oxidative stress is a major constraint on aquatic animal health. α-Lipoic acid (LA) mitigated hepatic oxidative damage by scavenging free radicals (ROS) and enhancing antioxidant enzyme activities (SOD and CAT) [12]. In this Special Issue, Fang et al. (article 6) explored the effects of LA on largemouth bass fed high-carbohydrate diets. Supplementation with 0.5–1.0 g/kg LA significantly enhanced growth performance, reduced hepatic glycogen deposition, and upregulated insulin signaling molecules (*ira*, *irb*, and *atk1*) and glycolysis-related genes (*pfkl* and *pk*) expression through the activation of the insulin pathway, offering a novel strategy to improve carbohydrate tolerance in carnivorous fish. Soybean isoflavones are natural phytoestrogens that mainly exist in soybean and other leguminous plants. Shi et al. (article 7) investigated the effects of soybean isoflavones on growth and lipid metabolism in juvenile Chinese mitten crab. Results demonstrated that dietary supplementation with 0.004% or 0.008% soybean isoflavones significantly increased weight gain (WG) and specific growth rate (SGR) while reducing non-esterified fatty acids (NEFA) and triglyceride (TG) in the hepatopancreas. Mechanistic analysis revealed that soybean isoflavones suppressed lipid synthesis-related genes (*stebp-1* and *Δ9 fad*) and upregulated lipolysis genes (*caat*, *tgl*, *cpt-1a*, *cpt-1b*, and *cpt2*), thereby improving the utilization of high-fat diets. These findings provide a theoretical basis for soybean isoflavones as functional additives in crustacean aquaculture.

Dietary hormonal additives can directly regulate the physiological state of aquatic animals and play an important role in their growth, metabolism, and immunity [13]. In this Special Issue, Liang et al. (article 8) found that triiodothyronine (T3), thyroxine (T4), and propylthiouracil (PTU) altered gut microbiota and short-chain fatty acid (SCFA) metabolism in little yellow croaker (*Larimichthys polyactis*). T4 treatment enhanced gut microbiota α-diversity, reduced Proteobacteria abundance, and increased Bacteroidota abundance. Acetic acid (AA) levels rose in the T3/T4 groups, whereas isobutyric acid (IBA) decreased in the T4 group. Notably, *Vibrio* abundance negatively correlated with AA but positively with IBA content, suggesting SCFAs modulate pathogen colonization. This work pioneers the "thyroid-gut microbiota axis" model in aquatic species, offering novel strategies for metabolic disease prevention. Melatonin, an endogenous rhythm regulator, shows promise in crustacean health management. Chen et al. (article 9) reported that dietary melatonin (50 mg/kg) in crayfish (*Procambarus clarkii*) synchronously enhanced diurnal non-specific immunity (lysozyme and ALP) and antioxidant capacity (CAT and GPx), while significantly upregulating mRNA expression of circadian clock genes (*clock*, *bmal1*, and *per1*). This study is the first to elucidate melatonin's molecular mechanism in maintaining physiological rhythms via circadian gene regulation, providing a theoretical basis for precision feeding strategies in intensive aquaculture.

Physiological functions of trace elements: At present, the demand of largemouth bass for selenium (Se) [14], iron (Fe) [15], and zinc (Zn) [16] in feed has been reported, but the dietary appropriate level of copper (Cu) has not been reported. Cu, as an essential trace element, plays a pivotal role in antioxidant defense and immune regulation in aquatic animals. In this Special Issue, Kayiira et al. (article 10) found that dietary appropriate supplementation of Cu (3.00–3.66 mg/kg) could enhance antioxidant capacity and influence relative Nrf2 and NF-κB signaling pathways, hence improving the health and immunity of largemouth bass. Notably, excessive copper (>5.72 mg/kg) elevated plasma triglyceride (TG) content and disrupted glucose metabolism, underscoring the need for strict dosage control. This study provides essential data for standardizing copper requirements in carnivorous fish.

Dietary supplements have become an important tool for improving the health and productivity of aquatic animals by regulating growth metabolism, antioxidants, and immune pathways. This Special Issue presents a series of original research results, focusing on the latest findings in the application of aquatic animal additives, providing a theoretical basis for the precise formulation design of aquafeed, which will promote the healthy and sustainable development of the aquaculture industry.

Author Contributions: J.M., writing—original draft preparation and editing; Q.X., supervision; C.Q., writing—review and editing; F.S., writing—review and editing; C.Z., writing—review and editing. All authors have read and agreed to the published version of the manuscript.

Institutional Review Board Statement: Not applicable.

Data Availability Statement: In this editorial, no new data were created. All the results we cited in this manuscript are published articles, which can be found in the database.

Acknowledgments: We would like to thank all the authors and reviewers who participated in this Special Issue. This work was supported by the Research on Public Welfare Technology Application of Science and Technology Project of Huzhou in China (2024GZ30) and the Zhejiang Provincial Basic and Public Welfare Research Program in China (LGN20C190006).

Conflicts of Interest: The authors declare no conflicts of interest.

List of Contributions:

1. Zheng, J.; He, Y.; Shi, M.; Jia, L.; Xu, Y.; Tan, Y.; Qi, C.; Ye, J. Effects of Dietary Glutamate on the Growth Performance and Antioxidant Capacity of Juvenile Chinese Mitten Crab (*Eriocheir sinensis*). *Fishes* **2024**, *9*, 306.
2. Jiang, F.; Huang, W.; Zhou, M.; Gao, H.; Lu, X.; Yu, Z.; Sun, M.; Huang, Y. Effects of Dietary L-glutamic acid on the Growth Performance, Gene Expression Associated with Muscle Growth-

Related Gene Expression, and Intestinal Health of Juvenile Largemouth Bass (*Micropterus salmoides*). *Fishes* **2024**, *9*, 312.

3. Xu, T.; Wang, J.; Xu, H.; Wang, Z.; Liu, Y.; Bai, H.; Zhang, Y.; Kong, Y.; Liu, Y.; Ding, Z. Dietary β-1,3-Glucan Promotes Growth Performance and Enhances Non-specific Immunity by Modulating Pattern Recognition Receptors in Juvenile Oriental River Prawn (*Macrobrachium nipponense*). *Fishes* **2024**, *9*, 379.

4. Gao, N.; Zhang, J.; Shandilya, U.K.; Lumsden, J.S.; Barzugar, A.B.; Huyben, D.; Karrow, N.A. Hepatic Gene Expression Changes of Zebrafish Fed Yeast Prebiotic, Yeast Prebiotic, Black Soldier Fly Meal, and Butyrate. *Fishes* **2024**, *9*, 495.

5. Xu, X.; Zheng, X.; Song, C.; Liu, X.; Zhou, Q.; Sun, C.; Wang, A.; Zhu, A.; Liu, B. Taurochenodeoxycholic Acid Improves Growth, Physiology, Intestinal Microbiota, and Muscle Development in Red Swamp Crayfish (*Procambarus clarkii*). *Fishes* **2025**, *10*, 38.

6. Fang, Z.; Pan, X.; Gong, Y.; Zhang, N.; Chen, S.; Liu, N.; Chen, N.; Li, S. Dietary Alpha-Lipoic Acid Alleviated Hepatic Glycogen Deposition and Improved Inflammation Response of Largemouth Bass (*Micropterus salmoides*) Fed on High Dietary Carbohydrates. *Fishes* **2025**, *10*, 9.

7. Shi, M.; He, Y.; Zheng, J.; Xu, Y.; Tan, Y.; Jia, L.; Chen, L.; Ye, J.; Qi, C. Effects of Soybean Isoflavones on the Growth Performance and Lipid Metabolism of the Juvenile Chinese Mitten Crab *Eriocheir sinensis*. *Fishes* **2024**, *9*, 335.

8. Liang, X.; Zhang, Y.; Ye, T.; Liu, F.; Lou, B. Thyroid-Active Agents Triiodothyronine, Thyroxine and Propylthiouracil Differentially Affect Growth, Intestinal Short Chain Fatty Acids and Microbiota in Little Yellow Croaker *Larimichthys polyactis*. *Fishes* **2025**, *10*, 69.

9. Chen, J.; Du, Y.; Zhang, M.; Wang, J.; Ming, J.; Shao, X.;Wang, A.; Tian, H.; Zhang, W.; Xia, S.; et al. Effects of Melatonin on the Growth and Diurnal Variation of Non-Specific Immunity, Antioxidant Capacity, Digestive Enzyme Activity, and Circadian Clock-Related Gene Expression in Crayfish (*Procambarus clarkii*). *Fishes* **2025**, *10*, 114.

10. Kayiira, J.C.; Mi, H.; Liang, H.; Ren, M.; Huang, D.; Zhang, L.; Teng, T. Effect of Dietary Copper on Growth Performance, Antioxidant Capacity, and Immunity in Juvenile Largemouth Bass (*Micropterus salmoides*). *Fishes* **2024**, *9*, 369.

11. Zhang, W.; Xia, S.; Liu, B.; Tian, H.; Liu, F.; Yang, W.; Yu, Y.; Zhao, C.; Dewangan, N.K.; Wang, A.; et al. Effects of Dietary Protein Levels on Growth Performance, Plasma Parameters, and Digestive Enzyme Activities in Different Intestinal Segments of *Megalobrama amblycephala* at Two Growth Stages. *Fishes* **2025**, *10*, 60.

References

1. Brosnan, J.T.; Brosnan, M.E. Glutamate: A Truly Functional Amino Acid. *Amino Acids* **2013**, *45*, 413–418. [PubMed]
2. Andersen, S.M.; Waagbø, R.; Espe, M. Functional amino acids in fish nutrition, health and welfare. *Front. Biosci.* **2016**, *8*, 143–169.
3. Zhao, Y.; Li, J.; Yin, L.; Feng, L.; Liu, Y.; Jiang, W.; Wu, P.; Zhao, J.; Chen, D.; Zhou, X. Effects of Dietary Glutamate Supplementation on Flesh Quality, Antioxidant Defense and Gene Expression Related to Lipid Metabolism and Myogenic Regulation in Jian Carp (*Cyprinus carpio* var. Jian). *Aquaculture* **2019**, *502*, 212–222. [CrossRef]
4. Ding, L.; Chen, J.; He, F.; Chen, Q.; Li, Y.; Chen, W. Effects of Dietary Arginine Supplementation on Growth Performance, Antioxidant Capacity, Intestinal Digestive Enzyme Activity, Muscle Transcriptome, and Gut Health of *Siniperca chuatsi*. *Front. Mar. Sci.* **2024**, *10*, 1305192.
5. Tian, J.; Yang, Y.; Du, X.; Xu, W.; Zhu, B.; Huang, Y.; Ye, Y.; Zhao, Y.; Li, Y. Effects of Dietary Soluble β-1,3-Glucan on the Growth Performance, Antioxidant Status, and Immune Response of the River Prawn (*Macrobrachium nipponense*). *Fish Shellfish Immunol.* **2023**, *138*, 108848. [PubMed]
6. Akhter, N.; Wu, B.; Memon, A.M.; Mohsin, M. Probiotics and Prebiotics associated with Aquaculture: A Review. *Fish Shellfish Immunol.* **2015**, *45*, 733–741. [CrossRef] [PubMed]
7. Dawood, M.A.O.; Koshio, S. Recent Advances in the Role of Probiotics and Prebiotics in Carp Aquaculture: A Review. *Aquaculture* **2016**, *454*, 243–251.
8. Koutsos, E.; Modica, B.; Freed, T. Immunomodulatory Potential of Black Soldier Fly Larvae: Applications Beyond Nutrition in Animal Feeding Programs. *Transl. Anim. Sci.* **2022**, *6*, txac084. [CrossRef] [PubMed]
9. Alnassar, N.; Hillman, C.; Fontana, B.D.; Robson, S.C.; Norton, W.H.J.; Parker, M.O. *Angptl4* Gene Expression as a Marker of Adaptive Homeostatic Response to Social Isolation across the Lifespan in Zebrafish. *Neurobiol. Aging* **2023**, *131*, 209–221. [PubMed]

10. Di Ciaula, A.; Garruti, G.; Baccetto, R.L.; Molina-Molina, E.; Bonfrate, L.; Wang, D.Q.H.; Portincasa, P. Bile Acid Physiology. *Ann. Hepatol.* **2017**, *16*, S4–S14. [PubMed]

11. Bao, L.G.; Hao, D.C.; Wang, X.; He, X.L.; Mao, W.; Li, P.F. Transcriptome Investigation of Anti-inflammation and Immuno-Regulation Mechanism of Taurochenodeoxycholic Acid. *BMC Pharmacol. Toxicol.* **2021**, *22*, 23.

12. Packer, L.; Witt, E.H.; Tritschler, H.J. Alpha-lipoic Acid as a Biological Antioxidant. *Free Radic. Biol. Med.* **1995**, *19*, 227–250. [CrossRef] [PubMed]

13. Power, D.M.; Llewellyn, L.; Faustino, M.; Nowell, M.A.; Björnsson, B.T.; Einarsdottir, I.E.; Canario, A.V.; Sweeney, G.E. Thyroid Hormones in Growth and Development of Fish. *Comp. Biochem. Physiol. C Toxicol. Pharmacol.* **2001**, *130*, 447–459. [PubMed]

14. Zhu, Y.; Chen, Y.; Liu, Y.; Yang, H.; Liang, G.; Tian, L. Effect of Dietary Selenium Level on Growth Performance, Body Composition and Hepatic Glutathione Peroxidase Activities of Largemouth Bass *Micropterus salmoide*. *Aquac. Res.* **2012**, *43*, 1660–1668. [CrossRef]

15. Mao, X.; Chen, W.; Long, X.; Pan, X.; Liu, G.; Hu, W.; Gu, D.; Tan, Q. Effect of Dietary Iron (Fe) Level on Growth Performance and Health Status of Largemouth Bass (*Micropterus salmoides*). *Aquaculture* **2024**, *581*, 740446. [CrossRef]

16. Gu, D.; Mao, X.; Abouel Azm, F.R.; Zhu, W.; Huang, T.; Wang, X.; Ni, X.; Zhou, M.; Shen, J.; Tan, Q. Optimal Dietary Zinc Inclusion Improved Growth Performance, Serum Antioxidant Capacity, Immune Status, and Liver Lipid and Glucose Metabolism of Largemouth Bass (*Micropterus salmoides*). *Fish Shellfish Immunol.* **2024**, *144*, 109233. [CrossRef] [PubMed]

Article

Effects of Dietary Glutamate on the Growth Performance and Antioxidant Capacity of Juvenile Chinese Mitten Crab (*Eriocheir sinensis*)

Jiajun Zheng, Yisong He, Mengyu Shi, Li Jia, Yang Xu, Yue Tan, Changle Qi * and Jinyun Ye *

College of Life Science, Huzhou University, Huzhou 313000, China; 18987437460@163.com (J.Z.); 19816907180@163.com (Y.H.); znytsmy@126.com (M.S.); jiali890411@163.com (L.J.); x15237127582@163.com (Y.X.); tan18853548898@163.com (Y.T.)
* Correspondence: 02862@zjhu.edu.cn (C.Q.); yjy@zjhu.edu.cn (J.Y.); Tel.: +86-18868215672 (C.Q.)

Abstract: In order to explore the effects of glutamate on the growth performance, antioxidant capacity and protein metabolism of juvenile Chinese mitten crab, 0%, 1% and 2% glutamate were supplemented to low protein (30%) and normal protein (35%) diets, respectively. There were 5 parallel tanks in each treatment, and the feeding duration was 8 weeks. The results showed that dietary glutamate did not significantly affect the weight gain of Chinese mitten crab. Diets supplemented with 2% glutamate significantly decreased the crude protein of crabs. The T-AOC of crabs fed the 30% protein diets was significantly lower than crabs fed the 35% protein diets. At 30% protein level, the superoxide dismutase (SOD) activity significantly increased with the increase in glutamate content. Dietary glutamate significantly down-regulated the relative expressions of *mTOR*, *PI3K*, *S6K1* and *4EBP* at 35% protein level. In conclusion, dietary glutamate cannot significantly increase the growth of Chinese mitten crab, but it can improve the antioxidant capacity in Chinese mitten crab under low protein conditions.

Keywords: glutamate; protein level; growth performance; antioxidant capacity; Chinese mitten crab

Key Contribution: Dietary glutamate can improve the antioxidant capacity in Chinese mitten crab fed with low protein diet.

Citation: Zheng, J.; He, Y.; Shi, M.; Jia, L.; Xu, Y.; Tan, Y.; Qi, C.; Ye, J. Effects of Dietary Glutamate on the Growth Performance and Antioxidant Capacity of Juvenile Chinese Mitten Crab (*Eriocheir sinensis*). *Fishes* **2024**, *9*, 306. https://doi.org/10.3390/fishes9080306

Academic Editor: Laura Susana López-Greco

Received: 16 May 2024
Revised: 19 July 2024
Accepted: 28 July 2024
Published: 3 August 2024

1. Introduction

Glutamate is a functional amino acid, which plays an important physiological role in the growth of animals [1–3]. Glutamate is the main source of energy for animals as it is an intermediate in the metabolism of amino acids [4,5]. Therefore, dietary supplementation with glutamate can improve the growth performance of animals. It has been reported that 1.5% dietary glutamate effectively improved the weight gain of Atlantic salmon [6]. Some similar studies were widely reported in tilapia and salmon, or other fish species [7,8]. In addition, glutamate can promote protein synthesis, thereby improving the efficiency of protein utilization in animals [9,10]. It has been reported that dietary glutamate improved the protein retention in godhead bream [11]. Glutamate occupies a central position in the metabolism of amino acids, although it is a non-essential amino acid [12,13]. Diets supplemented with an appropriate amount of non-essential amino acids can save the catabolism of some essential amino acids and improve feed utilization [14]. When there is insufficient protein in the diet or amino acid imbalance, alanine, glutamate, glutamine and aspartate in the gut of mice are preferentially used as energy substrates [15,16]. For example, juvenile herring preferentially use glutamate as an energy substrate, thereby saving the consumption of essential amino acids for protein synthesis [4]. Therefore, dietary supplementation with glutamate can improve growth performance and save dietary protein in fish and mammals.

Glutamate is also the metabolic precursor of glutathione, which is the biologically active molecule [3], so it plays an important role in antioxidant capacity [17]. Some studies have reported that dietary glutamate can provide materials for glutathione synthesis [18,19]. In addition, glutamine-derived glutamate can be converted to glutamate-γ-semialdehyde under the catalysis of dihydropyrrole-5-carboxylic acid synthase, which spontaneously generates dihydropyrrole-5-carboxylic acid and degrades to proline [20]. Proline has been reported to eliminate free radicals, which can improve the antioxidant capacity of animals [21]. Moreover, glutamate can increase the activities of several antioxidant enzymes in animals [22]. A study reported that dietary glutamate reduced the activity of plasma alanine aminotransferase (ALT) and up-regulated the expression of antioxidation-related genes in Atlantic salmon (*Salmo salar* L.), indicating that glutamate has a positive effect on the antioxidant capacity and liver health of Atlantic salmon [6]. Similarly, glutamate increased the activities of SOD, GPX, and GST in muscle of *Cyprinus carpio* var. Jian, thereby reducing muscle lipid peroxidation and improving muscle quality [23]. However, most studies have focused on mammals or fish, and this has not been reported in crustaceans.

In recent years, with the rapid development of aquaculture, feed protein resources such as fish meal and soybean meal have become increasingly scarce and their prices have been rising [24,25]. How to reduce the amount of feed protein and improve protein utilization is an important direction to maintain the sustainable development of aquaculture. Dietary protein deficiency can result in a large number of essential amino acids being used for the synthesis of non-essential amino acids, and thereby reducing utilization efficiency [26,27]. Therefore, this study aimed to investigate the effects of glutamate on the growth performance and antioxidant capacity of juvenile Chinese mitten crab *Eriocheir sinensis* fed with low protein diets or normal protein diets.

2. Materials and Methods

2.1. Experimental Diets

Six experimental diets were formulated by gradient supplementation, in which 0%, 1% and 2% glutamate were supplemented to low protein (30%) and normal protein (35%) diets, respectively. The formulation and proximate composition of the experimental diets are shown in Table 1.

The ingredients were finely ground using a pulverizer (2500Y, Anhui Hualing Xichu Equipment Co., Ltd., Anhui, China). Thereafter, the crushed ingredients were sieved using a 60-mesh strainer (Huafeng Hardware Instrument Co., Ltd., Zhejiang, China). The ingredients were weighed according to the formula and mixed adequately. Then, distilled water was added into the mixed ingredients. Subsequently, the dough was pelleted using a screw-press pelletizer (F-26, South China University of Technology, Guangzhou, China). Finally, the diets were oven-dried and stored at −20 °C.

Table 1. Formulation and proximate composition of the experimental diets (dry matter, %).

Ingredients	Experimental Diets					
	30% Protein 0% Glu	30% Protein 1% Glu	30% Protein 2% Glu	35% Protein 0% Glu	35% Protein 1% Glu	35% Protein 2% Glu
Ingredients						
Fish meal	21	21	21	24.5	24.5	24.5
Gelatin	3	3	3	3.5	3.5	3.5
Casein	12	12	12	14	14	14
Corn starch	26	26	26	26	26	26
Fish oil	2.5	2.5	2.5	2.5	2.5	2.5
Soybean oil	2.5	2.5	2.5	2.5	2.5	2.5
Arginine	2	2	2	2	2	2
Methionine	0.5	0.5	0.5	0.5	0.5	0.5
Lysine	0.5	0.5	0.5	0.5	0.5	0.5
Vitamin premix [a]	1.5	1.5	1.5	1.5	1.5	1.5
Mineral premix [b]	1.5	1.5	1.5	1.5	1.5	1.5
Soybean lecithin	2	2	2	2	2	2
Cholesterol	0.5	0.5	0.5	0.5	0.5	0.5

Table 1. *Cont.*

Ingredients	Experimental Diets					
	30% Protein 0% Glu	30% Protein 1% Glu	30% Protein 2% Glu	35% Protein 0% Glu	35% Protein 1% Glu	35% Protein 2% Glu
Choline chloride	0.5	0.5	0.5	0.5	0.5	0.5
Betaine	2	2	2	2	2	2
Butylated hydroxytoluene	0.1	0.1	0.1	0.1	0.1	0.1
Sodium carboxymethyl cellulose	2	2	2	2	2	2
Glutamate	0	1	2	0	1	2
Cellulose	19.9	18.9	17.9	13.9	12.9	11.9
Proximate analysis (%)						
Moisture	6.36	6.69	6.32	6.31	6.50	6.70
Crude protein	30.52	30.75	31.52	35.58	36.58	37.62
Crude lipid	8.68	8.59	8.63	9.56	9.47	9.62
Ash	4.81	4.92	5.00	5.53	5.61	5.29

[a] Vitamin premix (per 100 g premix): Ca pantothenate, 0.3 g; para-aminobenzoic acid, 0.1 g; cholecalciferol, 0.0075 g; riboflavin, 0.0625 g; menadione, 0.05 g; ascorbic acid, 0.5 g; biotin, 0.005 g; retinol acetate, 0.043 g; folic acid, 0.025 g; pyridoxine hydrochloride, 0.225 g; thiamin hydrochloride, 0.15 g; niacin, 0.3 g; α-tocopherol acetate, 0.5 g; The remaining part will be used α-cellulose to 100 g. [b] Mineral premix (per 100 g premix): KI, 0.023 g; $CuCl_2 \cdot 2H_2O$, 0.015 g; $Ca(H_2PO_4)_2$, 26.5 g; $MnSO_4 \cdot 6H_2O$, 0.143 g; $AlCl_3 \cdot 6H_2O$, 0.024 g; KH_2PO_4, 21.5 g; NaH_2PO_4, 10.0 g; $CoCl_2 \cdot 6H_2O$, 0.14 g; KCl, 2.8 g; $ZnSO_4 \cdot 7H_2O$, 0.476 g; Calcium lactate, 16.50 g; $CaCO_3$, 10.5 g; $MgSO_4 \cdot 7H_2O$, 10.0 g; Fe-citrate, 1 g; The remaining part will be used α-cellulose to 100 g.

2.2. Feeding Trial and Sampling

All experiments on animals were approved by the Committee on the Ethics of Animal Experiments in Huzhou University and the Care and Use of Laboratory Animals in China. Juvenile crabs were obtained from a farm in Huzhou. Crabs were acclimatized to the experimental conditions in 300 L tanks ($100 \times 80 \times 60$ cm) before the feeding trial. A total of 600 female crabs (0.4 ± 0.01 g, mean $\pm$ SE) were weighed and put into 30 tanks ($100 \times 80 \times 60$ cm). Each treatment used a set of four parallel tanks, each tank containing 20 crabs. To improve the survival of crabs, three folded plastic nets were placed in each tank as shelters. The experimental crabs were fed twice daily at a ratio of 4% body weight (6:00 and 18:00, respectively). After feeding, 30% tank water was exchanged daily. During the feeding trial, the dissolved oxygen concentration of the experimental water was >7 mg/L. The water temperature was maintained at 25 °C to 27 °C, and ammonia nitrogen was <0.05 mg/L.

After a 56-d feeding trial, crabs were euthanized, and five crabs were sampled randomly from each tank for proximate nutrient composition. Where after, six crabs were sampled randomly and the hepatopancreases were frozen in liquid nitrogen and kept in an ultra-low temperature freezer (MD-86L456, Midea Group, Guangdong, China) for enzyme activity, gene expression and nutrient composition analyses.

Weight gain, specific growth rate and hepatopancreas index of crabs were calculated according to the formulas below:

$$\text{Weight gain (WG, \%)} = 100 \times (\text{final crab weight} - \text{initial crab weight})/\text{initial crab weight}$$

$$\text{Specific growth rate (SGR, \%)} = 100 \times (\text{LN final crab weight} - \text{LN initial crab weight})/\text{days}$$

$$\text{Hepatopancreas index (\%)} = 100 \times \text{hepatopancreas weight of crab}/\text{whole crab weight}$$

2.3. Chemical Composition Analysis

The proximate nutrient compositions of diets and crabs were estimated using the methods described as AOAC [28]. Firstly, the moisture of diets and crabs was measured by drying 8 h at 105 °C in an oven. After drying, the crude protein contents of diets and crabs were measured using the kjeldah method (Kjeltec™ 8200, Foss, Hoganas, Sweden). The crude lipid content of diets and crabs were analyzed using the Soxhlet extraction method (1000 mL, Fujian minbo toughened glass Co., Ltd., Fujian, China). Finally, the ash contents of diets and crabs were analyzed by ashing the samples at 550 °C for 6 h in a muffle furnace (PCD-E3000 Serials, Peaks, Japan). Four duplicate samples were analyzed for each treatment (n = 4).

2.4. Analysis of Biochemical Parameters in the Hepatopancreas

The biochemical parameters were analyzed using commercial assay kits (Nanjing Jiancheng Bioengineering Institute, Nanjing, China). The information for each kit is listed as follows: total antioxidant capacity (T-AOC; Cat. No. A015-2), superoxide dismutase (SOD; Cat. No. A001-1), glutathione peroxidase (GPX; Cat. No. A005-1-2), glutathione S-transferase (GST; Cat. No. A004-1-1), pyruvate (Pyruvate; Cat. No. A081-1-1), glutaminase (GLS; Cat. No.A124-1-1), glutamine synthetase (GS; Cat. No. A047-1-1), glutamate dehydrogenase (GDH; Cat. No. A125-1-1). Four duplicate samples were analyzed for each treatment (n = 4).

2.5. Analysis of Gene Expression

Total RNA was extracted from the hepatopancreas using the (RNAiso Plus, Takara, Japan). The total RNA concentration of each sample was measured using a Nano Drop 2000 (Thermo, USA). The samples were reverse transcribed using a commercial kit (PrimeScript™ RT master, Takara, Japan). The primers were designed based on the transcriptome sequencing and NCBI data base (Table 2). The RT–PCR amplification reactions were performed using a CFX96 Real-Time PCR system (Bio-rad, Richmond, CA, USA). PCR conditions were set according to the instruction from commercial kits (SYBR Premix, Takara, Japan). The relative gene expressions were calculated by geometric averaging of multiple internal control genes (*β-actin* and glyceraldehyde-phosphate dehydrogenase (*GAPDH*)) [29]. Four duplicate samples were analyzed for each treatment (n = 4).

Table 2. Sequences of primers.

Primers Name	Sequences (5′-3′)	Product Size	References
mTOR F	AGGTCCTGTTATGCTGTGGC	158 bp	MT920347.1
mTOR R	ATCTCGGGGATGTCCTGTGA		
PI3K F	GCTGTCAGTCCAGTTCGACA	111 bp	c147204_g1
PI3K R	ACAGTATGCTTGGTCAGGGC		
AMPD F	CACAACGTCCACTCCGAGAA	116 bp	c143453_g1
AMPD R	CGGAACAGGTTGTCGAGGAA		
AKT F	ATAAGGACCCCAACAAGCGG	134 bp	KY709138.1
AKT R	CACTTGGGGTTTGAAAGGCG		
S6K1 F	TGACTACCCGGACCTGCTAA	154 bp	XM_050855088.1
S6K1 R	TGCCACACCAATGAACCCTT		
4EBP F	GCTGTCTGCTCCCTCACTTT	163 bp	XM_050856547.1
4EBP R	ACCCGTCAGCTTCTTAAGCC		
GLDH F	GGCAACGATGTAACGTGTGG	116 bp	XM_050832606.1
GLDH R	CGAAGCATCTTGCCACCAAC		

2.6. Statistical Analysis

Statistical analysis was performed using SPSS 26.0 for Windows (SPSS, Michigan Avenue, Chicago, IL, USA). Two-way ANOVA was used to determine if there was any interaction between dietary protein level and glutamate level. At the same glutamate level, independent-samples T test was used to determine significant differences between crabs cultivated at different protein levels. At the same protein condition, one-way ANOVA was used to analyze the significant differences among crabs fed the diets with different glutamate levels. Significance was set at $p < 0.05$.

3. Results

3.1. Growth Performance

As showed in the Table 3. At 30% protein level, diets supplemented with 1% and 2% glutamate slightly increased the weight gain (WG) and specific growth rate (SGR) of crabs

($p > 0.05$). However, at 35% protein level, the WG and SGR of crabs significantly decreased with the increasing dietary glutamate, and the WG and SGR of crabs fed the 2% glutamate diet was significantly lower than that of the control crabs ($p < 0.05$). At 30% protein level, the hepatopancreas index (HSI) of crabs fed the 1% glutamate was significantly higher than that of crabs fed the diets containing 0% and 2% glutamate ($p < 0.05$). At the 35% protein level, dietary glutamate did not significantly affect the HSI of crabs ($p > 0.05$). There were no significant interactions between dietary protein level and glutamate levels based on the WG, SGR and HSI ($p > 0.05$).

Table 3. Effects of glutamate on the growth performance of juvenile Chinese mitten crab.

	Parameters		
Diets	**Weight Gain (%)**	**Specific Growth Rate (% Day^{-1})**	**Hepatopancreas Index (%)**
30% Protein-0% Glu	283.59 ± 76.72	2.29 ± 0.37	8.62 ± 2.72 [b]
30% Protein-1% Glu	327.91 ± 73.45	2.48 ± 0.32	12.4 ± 0.95 [a]
30% Protein-2% Glu	328.53 ± 67.74	2.49 ± 0.29	8.46 ± 1.63 [b]
35% Protein-0% Glu	375.2 ± 59.48 [A]	2.68 ± 0.22 [A]	8.98 ± 1.6
35% Protein-1% Glu	315.42 ± 66.24 [AB]	2.44 ± 0.27 [AB]	8.28 ± 1.34
35% Protein-2% Glu	265.83 ± 106.38 [B]	2.17 ± 0.55 [B]	9.83 ± 0.75
Two-way ANOVA (*p* value)			
Protein	NS	NS	NS
Glu	NS	NS	NS
Protein × Glu	NS	NS	NS

The different lowercase letters in the table indicate that there are significant differences among crabs fed with different glutamate diets at 30% protein ($p < 0.05$). Different capital letters indicate that there are significant differences among crabs fed with different glutamate diets at 35% protein ($p < 0.05$). NS means no significant differences.

3.2. Nutrient Composition of Crabs

As shown in Table 4, dietary glutamate did not significantly affect the moisture, ash, and crude lipid of crabs fed with diets for both 30% and 35% protein levels ($p > 0.05$). There was a significant main effect of dietary glutamate on crude protein content ($p < 0.05$). At 30% protein level, the crude protein content of crabs fed the 2% glutamate diets was significantly higher than that of crabs fed the diets containing 0% and 1% glutamate ($p < 0.05$). At 35% protein level, the crude protein content of crabs fed the 0% glutamate diets was significantly higher than that of crabs fed diets containing 2% glutamate ($p < 0.05$).

Table 4. Effect of glutamate on the whole-body composition of juvenile Chinese mitten crab.

	Parameters			
Diets	**Moisture (%)**	**Ash (%)**	**Crude Protein (%)**	**Crude Lipid (%)**
30% Protein–0% Glu	66.18 ± 2.16	11.88 ± 0.21	14.19 ± 0.13 [a]	6.12 ± 0.23
30% Protein–1% Glu	66.54 ± 3.15	12.01 ± 0.42	14.14 ± 0.33 [a]	5.89 ± 0.16
30% Protein–2% Glu	66.28 ± 1.11	11.96 ± 0.25	13.34 ± 0.24 [b]	5.95 ± 0.32
35% Protein–0% Glu	66.82 ± 3.74	12 ± 0.33	13.94 ± 0.26 [A]	6.1 ± 0.48
35% Protein–1% Glu	66.13 ± 1.65	12.08 ± 0.49	13.76 ± 0.11 [AB]	6.03 ± 0.23
35% Protein–2% Glu	66.19 ± 3.69	11.96 ± 0.47	13.26 ± 0.43 [B]	5.81 ± 0.67
Two-way ANOVA (*p* value)				
Protein	NS	NS	NS	<0.05
Glu	NS	NS	<0.01	NS
Protein × Glu	NS	NS	NS	NS

The different lowercase letters in the table indicate that there are significant differences among crabs fed with different glutamate diets at 30% protein ($p < 0.05$). Different capital letters indicate that there are significant differences among crabs fed with different glutamate diets at 35% protein ($p < 0.05$). NS means no significant differences.

3.3. The Actiities of Enzymes Related to Antioxidant Capacity in the Hepatopancreas

Dietary glutamate did not significantly affect the total antioxidant capacity (T-AOC) of crabs fed with the diets under both 30% and 35% protein levels ($p > 0.05$; Figure 1A). However, the crude protein content of diets has a significant main effect on the T-AOC ($p < 0.05$; Figure 1A). The T-AOC of crabs fed the 30% protein diets was significantly lower than that of crabs fed the 35% protein diets ($p < 0.05$; Figure 1A). At 30% protein level, the superoxide dismutase (SOD) activity significantly increased with the increase in glutamate content ($p < 0.05$; Figure 1B) but, at 35% protein level, dietary glutamate did not significantly affect the SOD of crabs ($p > 0.05$). Under the 2% glutamate condition, the SOD activity in the hepatopancreas of crabs fed 30% protein was significantly higher than that of crabs fed 35% protein ($p < 0.05$; Figure 1B). At 30% protein level, the glutathione peroxidase (GPX) and glutathione S-transferase (GST) of crabs fed the 1% glutamate diets were significantly lower than those of crabs fed the 0% and 2% glutamate diets ($p < 0.05$; Figure 1C,D). At 35% protein level, dietary glutamate did not significantly affect the activities of GPX and GST ($p > 0.05$; Figure 1C).

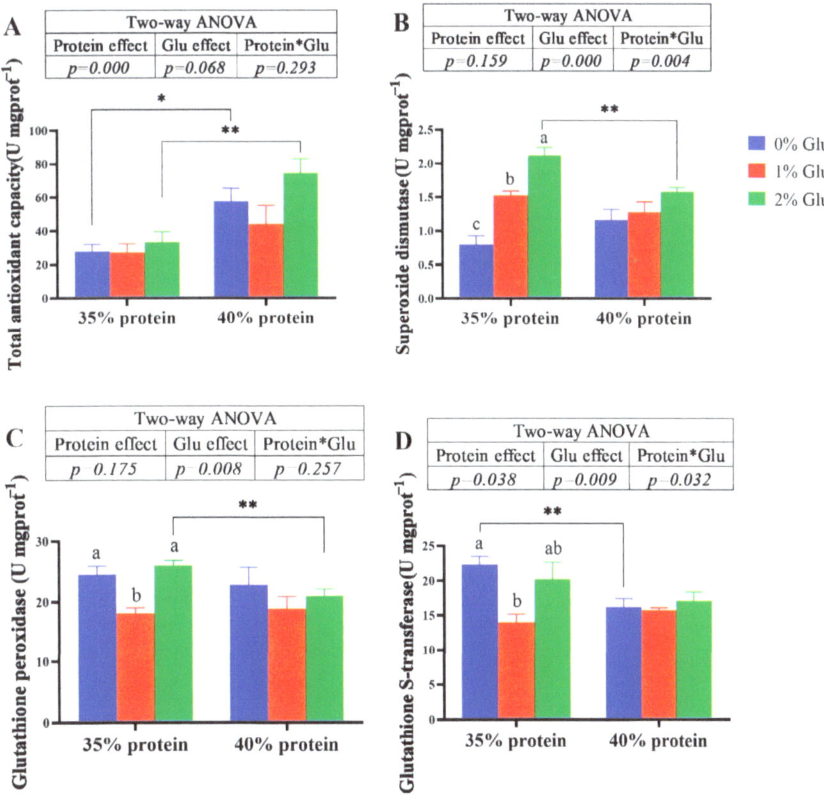

Figure 1. Effects of glutamate on antioxidant capacity in the hepatopancreas of juvenile Chinese mitten crab. The asterisk (*) indicates that there are significant differences among different protein level groups with the same amount of glutamate (* means $p < 0.05$, ** means $p < 0.01$). The different lowercase letters in the table indicate that there are significant differences among crabs fed with different glutamate diets at 30% protein ($p < 0.05$).

3 4. The Amino Acid Metabolism in the Hepatopancreas

Dietary glutamate did not significantly affect the pyruvate, glutamine synthetase and glutamate dehydrogenase of crabs fed with the diets under both 30% and 35% protein levels ($p > 0.05$; Figure 2A,C,D). However, the crude protein content of diets has a significant main effect on the pyruvate ($p < 0.05$, Figure 2A). At 30% protein level, the activity of glutaminase was significantly higher in crabs fed the 1% glutamate diets ($p < 0.05$; Figure 2B). At 35% protein level, dietary glutamate did not affect the activity of glutaminase ($p > 0.05$; Figure 2B).

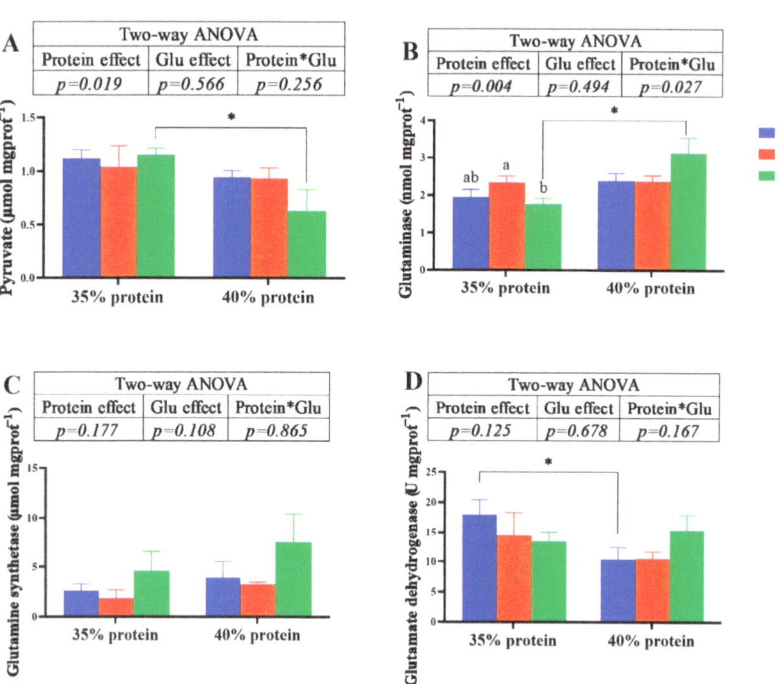

Figure 2. Effects of glutamate on the amino acid metabolism in the hepatopancreas of juvenile Chinese mitten crab. (**A**) The content of pyruvate; (**B**) The activity of glutaminase; (**C**) The activity of glutamine synthetase; (**D**) The activity of glutamate dehydrogenase. The asterisk (*) indicates that there are significant differences among different protein level groups with the same amount of glutamate (* means $p < 0.05$). The different lowercase letters in the table indicate that there are significant differences among crabs fed with different glutamate diets at 30% protein ($p < 0.05$).

3.5. Protein Metabolism in the Hepatopancreas

At 30% protein level, dietary glutamate did not affect the relative expressions of *mTOR*, *S6K1* and *4EBP* ($p > 0.05$; Figure 3A,C,D). However, at 35% protein level, dietary glutamate significantly down-regulated the relative expressions of *mTOR*, *PI3K*, *S6K1* and *4EBP* ($p < 0.05$; Figure 3A–D). At 30% protein level, dietary glutamate significantly down-regulated the relative expressions of *PI3K* ($p < 0.05$; Figure 3B). The crude protein content of diets has a significant main effect on the relative expressions of *4EBP* ($p < 0.05$; Figure 3D). The relative expressions of *4EBP* of crabs fed the 30% protein diets was significantly lower than that of crabs fed the 35% protein diets ($p < 0.05$; Figure 3D).

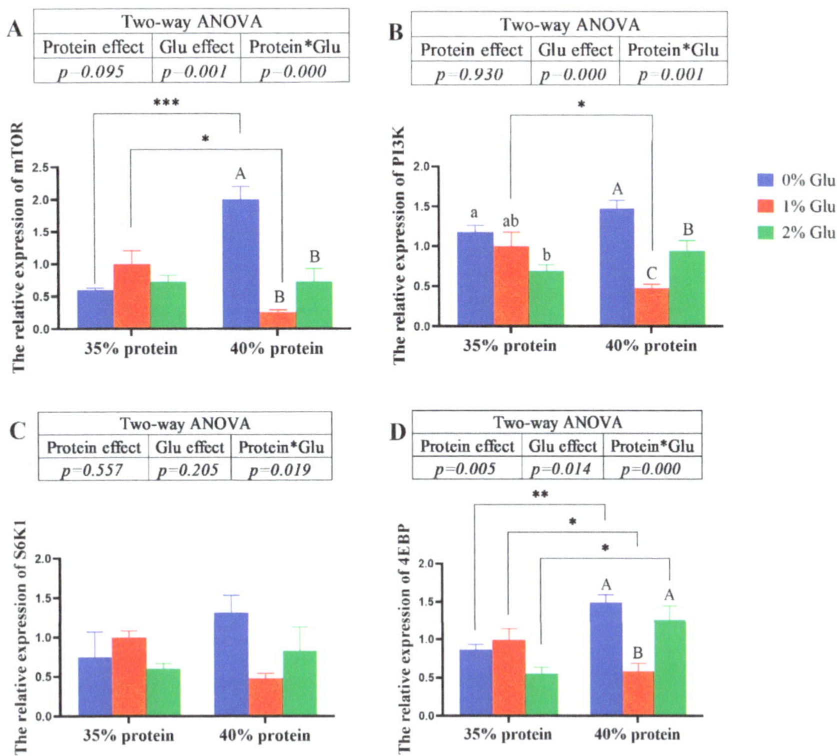

Figure 3. Effects of glutamate on the protein metabolism related genes in the hepatopancreas of juvenile Chinese mitten crab. (**A**) The relative expression of *mTOR* (Mammalian target of rapamycin); (**B**) The relative expression of *PI3K* (Phosphatidylinositide 3-kinases); (**C**) The relative expression of *S6K1* (Ribosomal protein S6 kinase 1); (**D**) The relative expression of *4EBP* (e IF4E-binding protein). The different lowercase letters in the table indicate that there are significant differences among crabs fed with different glutamate diets at 30% protein ($p < 0.05$). Different capital letters indicate that there are significant differences among crabs fed with different glutamate diets at 35% protein ($p < 0.05$). The asterisk (*) indicates that there are significant differences among different protein level groups with the same amount of glutamate (* means $p < 0.05$, ** means $p < 0.01$, *** means $p < 0.001$,).

4. Discussion

Glutamate plays an important physiological role in the growth of animals [1–3]. Some previous studies reported that dietary glutamate can improve the weight gain of fish [6–8]. However, in the present study, dietary glutamate did not significantly affect the weight gain and specific growth rate. Moreover, dietary glutamate significantly decreased the growth of Chinese mitten crab. This result was different from that for grass carp [30], golden head snapper [11] and Atlantic salmon [6]. Moreover, a previous study in weaned piglet reported that dietary supplementation with more than 3.2% glutamate reduced growth performance and presented a toxicity effect [30]. On the other hand, dietary excessive concentrations of sodium glutamate caused intestinal mucosa stress in fish [31]. The differences may be caused by species specificity. Unfortunately, it is difficult to find an article on glutamate toxicity effects in crustaceans. Therefore, it is necessary to further study the toxic effect of glutamate in shrimp and crab. Some previous studies reported that glutamate can significantly increase the hepatosomic index of aquatic animals [7]. Similar results were also found in the present study, where 1% glutamate supplemented to a diet with 30% protein level significantly increased the hepatopancreas index of juvenile Chinese mitten crab. Growth is closely related to the accumulation of nutrients in animals.

A study reported that dietary glutamate increased the crude lipid content in the gold head snapper (*Sparus aurata*) [11]. It has also been reported that dietary glutamate can improve protein content and lipid content in grass carp (*Ctenopharyngodon idella*) [32]. In the present study, dietary glutamate did not significantly affect the moisture, ash, and crude lipid of crabs fed with the diets under both 30% and 35% protein levels. Even worse, 2% glutamate significantly decreased the crude protein of Chinese mitten crab. The differences may be caused by species specificity. Unfortunately, glutamate has been poorly studied in crustaceans. Therefore, the expression of genes involved in protein synthesis in the hepatopancreas of the crab was further investigated in the present study. The results indicated that dietary glutamate significantly down-regulated the relative expressions of *mTOR*, *PI3K*, *S6K1* and *4EBP* of crabs fed the diets containing 35% protein, which indicated that dietary glutamate inhibits protein synthesis by inhibiting the *mTOR* pathway in Chinese mitten crab. These results were consistent with those of body composition. In summary, dietary glutamate decreased the growth and protein content of Chinese mitten crab by regulating the *mTOR* pathway.

Previous studies have demonstrated that a low protein diet can lead to oxidative stress in Chinese mitten crabs [33]. In the present study, the T-AOC of crabs fed the 30% protein diets was significantly lower than that of crabs fed the 35% protein diets, which is consistent with other studies. GSH–PX and GSH–ST are required by the endogenous antioxidant defense system to scavenge oxygen free radicals and maintain cellular redox balance [34]. In the present study, dietary glutamate did not significantly affect the GSH–PX and GSH–ST, which indicated that glutamate did not improve the GSH enzyme system. However, SOD activity significantly increased with the increase of glutamate content. This result indicated that dietary glutamate could improve the activity of SOD, thereby increasing the antioxidant capacity in Chinese mitten crab under low protein conditions.

5. Conclusions

Dietary glutamate cannot significantly increase the growth of Chinese mitten crab, but it can decrease the growth and protein content of Chinese by regulating the *mTOR* pathway. However, dietary glutamate can improve the activity of SOD, thereby increasing the antioxidant capacity in Chinese mitten crab under low protein conditions.

Author Contributions: J.Z., writing—original draft and editing, Y.H., writing—review and editing. M.S., writing—review and editing. L.J., writing—review and editing. Y.X., writing—review and editing. Y.T., writing—review and editing. C.Q., supervision. J.Y., supervision. All authors have read and agreed to the published version of the manuscript.

Funding: This research was supported by grants from the Zhejiang Province R&D Plan (2022C02058), Zhejiang Provincial Natural Science Foundation of China under Grant No. LTGN23C190003, and the Huzhou Natural Science Foundation (2021YZ14) and the Graduate Research Innovation Project of Huzhou University (2023KYCX68).

Institutional Review Board Statement: All experiments on animals were approved by the Committee on the Ethics of Animal Experiments in Huzhou University and the Care and Use of Laboratory Animals in China. approval code: 20220701; approval date: 20 July 2022.

Informed Consent Statement: Not applicable.

Data Availability Statement: Data are contained within the article.

Conflicts of Interest: The authors declare no conflicts of interest.

References

1. Brosnan, J.T.; Brosnan, M.E. Glutamate: A truly functional amino acid. *Amino Acids* **2013**, *45*, 413–418. [CrossRef] [PubMed]
2. Lallès, J.P.; Bosi, P.; Smidt, H.; Stokes, C.R. Weaning—A challenge to gut physiologists. *Livest. Sci.* **2007**, *108*, 82–93. [CrossRef]
3. Burrin, D.G.; Stoll, B. Metabolic fate and function of dietary glutamate in the gut. *Am. J. Clin. Nutr.* **2009**, *90*, 850–856. [CrossRef] [PubMed]

4. Conceio, L.E.C.; Rnnestad, I.; Tonheim, S.K. Metabolic budgets for lysine and glutamate in unfed herring (*Clupea harengus*) larvae. *Aquaculture* **2002**, *206*, 305–312. [CrossRef]
5. Van Waarde, A. Aerobic and anaerobic ammonia production by fish. *Comp. Biochem. Physiol. B* **1983**, *74*, 675–684. [CrossRef] [PubMed]
6. Larsson, T.; Koppang, E.O.; Espe, M.; Terjesen, B.F.; Krasnov, A.; Moreno, H.M.; Rørvik, K.A.; Thomassen, M.; Mørkøre, T. Fillet quality and health of Atlantic salmon (*Salmo salar* L.) fed a diet supplemented with glutamate. *Aquaculture* **2014**, *426*, 288–295. [CrossRef]
7. Oehme, M.; Grammes, F.; Takle, H.; Zambonino-Infante, J.L.; Refstie, S.; Thomassen, M.S.; Rørvik, K.-A.; Terjesen, B.F. Dietary supplementation of glutamate and arginine to Atlantic salmon (*Salmo salar* L.) increases growth during the first autumn in sea. *Aquaculture* **2010**, *310*, 156–163. [CrossRef]
8. Silva, L.C.R.D.; Furuya, W.M.; Natali, M.R.M.; Schamber, C.R.; Santos, L.D.D.; Vidal, L.V.O. Productive performance and intestinal morphology of Nile tilapia juvenile fed diets with L-glutamine and L-glutamate. *Rev. Bras. Zootec.* **2010**, *39*, 1175–1179. [CrossRef]
9. Maclennan, P.A.; Brown, R.; Rennie, M.J. A positive relationship between protein synthetic rate and intracellular glutamine concentration in perfused rat skeletal muscle. *FEBS Lett.* **1987**, *215*, 187–191. [CrossRef]
10. Coffier, M.S.; Claeyssens, S.; Hecketsweiler, B.; Lavoinne, A.; Ducrotté, P.; Déchelotte, P. Enteral glutamine stimulates protein synthesis and decreases *ubiquitin* mRNA level in human gut mucosa. *Am. J. Physiol.-Gastrointest. Liver Physiol.* **2003**, *285*, G266–G273. [CrossRef]
11. Caballero-Solares, A.; Viegas, I.; Salgado, M.C.; Siles, A.M.; Saez, A.; Metón, I.; Baanante, I.V.; Fernández, F. Diets supplemented with glutamate or glutamine improve protein retention and modulate gene expression of key enzymes of hepatic metabolism in gilthead seabream (*Sparus aurata*) juveniles. *Aquaculture* **2015**, *444*, 79–87. [CrossRef]
12. Brosnan, M.E.; Brosnan, J.T. Hepatic glutamate metabolism: A tale of 2 hepatocytes. *Am. J. Clin. Nutr.* **2009**, *90*, 857–861. [CrossRef] [PubMed]
13. Wu, G. Functional amino acids in growth, reproduction, and health. *Adv. Nutr.* **2010**, *1*, 31–37. [CrossRef] [PubMed]
14. Abboudi, T.; Mambrini, M.; Larondelle, Y.; Rollin, X. The effect of dispensable amino acids on nitrogen and amino acid losses in Atlantic salmon (*Salmo salar*) fry fed a protein-free diet. *Aquaculture* **2009**, *65*, 345–353. [CrossRef]
15. Tanaka, H.; Shibata, K.; Mori, M.; Ogura, M. Metabolism of essential amino acids in growing rats at graded levels of soybean protein isolate. *J. Nutr. Sci. Vitaminol.* **1995**, *41*, 433–443. [CrossRef] [PubMed]
16. Nakamura, H.; Kawamata, Y.; Kuwahara, T.; Torii, K.; Sakai, R. Nitrogen in dietary glutamate is utilized exclusively for the synthesis of amino acids in the rat intestine. *Am. J. Physiol.-Endocrinol. Metab.* **2013**, *304*, E100–E108. [CrossRef] [PubMed]
17. Espinosa, D.C.; Miguel, V.; Mennerich, D.; Kietzmann, T.; Sánchez-Pérez, P.; Cadenas, S.; Lamas, S. Antioxidant responses and cellular adjustments to oxidative stress. *Redox Biol.* **2015**, *6*, 183–197. [CrossRef] [PubMed]
18. He, L.; Wu, J.; Tang, W.; Zhou, X.; Lin, Q.; Luo, F.; Yin, Y.; Li, T. Prevention of oxidative stress by α-ketoglutarate via activation of *CAR* signaling and modulation of the expression of key antioxidant-associated targets in vivo and in vitro. *J. Agric. Food Chem.* **2018**, *66*, 11273–11283. [CrossRef] [PubMed]
19. Cai, G.Y. Effect of Glutamate on Antioxoidant Level and Subseouent Development of In Vitro Cultured Mouse Embryos in the Blocking Stage. Master's Thesis, Yanbian University, Yanji, China, 2017. (In Chinese).
20. Fujita, T.; Yanaga, K. Association between glutamine extraction and release of citrulline and glycine by the human small intestine. *Life Sci.* **2007**, *80*, 1846–1850. [CrossRef]
21. Kaul, S.; Sharma, S.S.; Mehta, I.K. Free radical scavenging potential of L-proline: Evidence from in vitro assays. *Amino Acids* **2008**, *34*, 315–320. [CrossRef]
22. Meister, A. Glutathione-ascorbic acid antioxidant system in animals. *J. Biol. Chem.* **1994**, *269*, 9397–9402. [CrossRef]
23. Zhao, Y.; Li, J.-Y.; Yin, L.; Feng, L.; Liu, Y.; Jiang, W.-D.; Wu, P.; Zhao, J.; Chen, D.-F.; Zhou, X.-Q.; et al. Effects of dietary glutamate supplementation on flesh quality, antioxidant defense and gene expression related to lipid metabolism and myogenic regulation in Jian carp (*Cyprinus carpio* var. Jian). *Aquaculture* **2019**, *502*, 212–222. [CrossRef]
24. Canton, H. *The Europa Directory of International Organizations 2021*, 23rd ed.; Routledge: London, UK, 2021; pp. 297–305.
25. Tao, S.; Zhang, Q.; Zhang, J. Feed market situation, prospect and countermeasures in 2021. *Chin. J. Anim. Sci.* **2022**, *86*, 45–49. (In Chinese)
26. Huang, J.; Deng, H. Effects of low protein diet on nutrient digestion and nitrogen emission of growing pigs. *Livest. Poult. Inventory* **2017**, *45*, 21–28. (In Chinese)
27. Gloaguen, M.; Floc'H, N.L.; Corrent, E.; Primot, Y.; Milgen, J.V. The use of free amino acids allows formulating very low crude protein diets for piglets. *J. Anim. Sci.* **2014**, *92*, 637–641. [CrossRef]
28. AOAC. *Official Methods of Analysis of AOAC International*; Association of Official Analytical Chemists: Washington, DC, USA, 2005.
29. Vandesompele, J.; Preter, K.D.; Pattyn, F.; Poppe, B.; Roy, N.V.; Paepe, A.D.; Speleman, F. Accurate normalization of real-time quantitative RT-PCR data by geometric averaging of multiple internal control genes. *Genome Biol.* **2002**, *3*, research0034.1. [CrossRef] [PubMed]
30. Li, Y.; Han, H.; Yin, J.; Zheng, J.; Zhu, X.; Li, T.; Yin, Y. Effects of glutamate and aspartate on growth performance, serum amino acids, and amino acid transporters in piglets. *Food Agric. Immunol.* **2018**, *54*, 675–687. [CrossRef]

31. Ladeira, A.; Rusth, R.; Carneiro, C.; Campelo, D.; Morante, V.; Luz, R.; Carneiro, A.; Salaro, A. Dietary monosodium glutamate supplementation during the feed training of pacamã (*Lophiosilurus alexandri*): Growth performance and intestinal histomorphometry. *Aquac. Res.* **2021**, *52*, 356–363. [CrossRef]
32. Zhao, Y.; Hu, Y.; Zhou, X.Q.; Zeng, X.Y.; Feng, L.; Liu, Y.; Jiang, W.D.; Li, S.H.; Li, D.B.; Wu, X.Q. Effects of dietary glutamate supplementation on growth performance, digestive enzyme activities and antioxidant capacity in intestine of grass carp (*Ctenopharyngodon idella*). *Aquac. Nutr.* **2015**, *21*, 935–941. [CrossRef]
33. Zhu, S.; Long, X.; Turchini, G.M.; Deng, D.; Cheng, Y.; Wu, X. Towards defining optimal dietary protein levels for male and female sub-adult Chinese mitten crab, *Eriocheir sinensis* reared in earthen ponds: Performances, nutrient composition and metabolism, antioxidant capacity and immunity. *Aquaculture* **2021**, *536*, 736442. [CrossRef]
34. Cabrini, L.; Bergami, R.; Fiorentini, D.; Marchetti, M.; Landi, L.; Tolomelli, B. Vitamin B6 deficiency affects antioxidant defences in rat liver and heart. *IUBMB Life* **1998**, *46*, 689–697. [CrossRef] [PubMed]

 *fishes*

Article

Effects of Dietary L-glutamic acid on the Growth Performance, Gene Expression Associated with Muscle Growth-Related Gene Expression, and Intestinal Health of Juvenile Largemouth Bass (*Micropterus salmoides*)

Feifan Jiang [1], Wenqing Huang [2], Meng Zhou [1], Hongyan Gao [1], Xiaozhou Lu [1], Zhoulin Yu [1], Miao Sun [1] and Yanhua Huang [1,*]

[1] Institute of Animal Health Breeding Innovation, College of Animal Science and Technology, Zhongkai College of Agricultural Engineering, Guangzhou 510225, China; feifan1999813@outlook.com (F.J.); zhoumeng@zhku.edu.cn (M.Z.); gaohongyan@zhku.edu.cn (H.G.); l18216164395@outlook.com (X.L.); yuzhoulin99@outlook.com (Z.Y.); sunmiao545@gmail.com (M.S.)

[2] Guangzhou Fishteach Biotechnology Co., Ltd., Guangzhou 510640, China; wenqingfishteach@outlook.com

* Correspondence: huangyanhua@zhku.edu.cn

Abstract: The present research examined the impact of L-glutamic acid (Glu) supplementation on the growth performance, muscle composition, gene expression correlated with muscle growth, and intestinal health of largemouth bass. There were 525 fish in total, which were distributed randomly into five groups. Each group had three replicates, and each replicate consisted of 35 fish. Groups with control and experimental diets were assigned glutamic acid amounts of 0.2%, 0.4%, 0.6%, and 0.8%. The findings demonstrated that glutamic acid supplementation enhanced growth performance, feed intake (FI), and condition factor (CF), with the best value being attained at 0.4% Glu. The mean muscle fiber area was increased and the muscle fiber density was decreased in the 0.6% Glu group. The levels of total amino acids and specific amino acids, such as glutamic acid, aspartic acid, leucine, valine, alanine, and glycine, were shown to be higher in the 0.6% Glu group. In the 0.6% Glu group, the mRNA expression levels of *atrogin-1*, *murf-1*, *foxo3a*, and *4e-bp1* were decreased compared to the control group. Conversely, the mRNA expression levels of *myf5*, *myog*, *myod*, *s6k1*, *tor*, *akt*, and *pi3k* were increased in the 0.6% Glu group compared to the control group. The 0.4% Glu group had higher intestinal amylase, lipase, and protease activities and greater villus height, villus width, and muscle thickness. In summary, Glu can support largemouth bass growth, muscular development, intestinal digestion, and absorption.

Keywords: glutamic acid; largemouth bass; muscle; intestinal health

Key Contribution: This study was conducted to investigate the effects of dietary HMB on growth performance, muscle development and intestinal health of largemouth bass and to explore the possible mechanisms.

Citation: Jiang, F.; Huang, W.; Zhou, M.; Gao, H.; Lu, X.; Yu, Z.; Sun, M.; Huang, Y. Effects of Dietary L-glutamic acid on the Growth Performance, Gene Expression Associated with Muscle Growth-Related Gene Expression, and Intestinal Health of Juvenile Largemouth Bass (*Micropterus salmoides*). Fishes **2024**, *9*, 312. https://doi.org/10.3390/fishes9080312

Academic Editor: Sung Hwoan Cho

Received: 10 July 2024
Revised: 30 July 2024
Accepted: 31 July 2024
Published: 6 August 2024

1. Introduction

Recently, due to the progress of the economic landscape, there has been a growing demand for aquatic items among individuals [1]. The aquaculture sector has risen rapidly, with worldwide fish production rising from 2.6 million tons in 1970 to 87.5 million tons in 2020 [2]. As people's demand for aquatic products continues to expand, various high-density, high-yield factory farming methods continue to innovate [3]. The increase in the total output of the aquaculture industry has made the competition in the aquaculture market more intense. People are eager to find ways to increase the output of aquatic products, enhance the health of aquatic products, and improve the meat quality of aquatic products [4].

In a previous study, amino acids, as the main component of proteins, undertake a variety of biological functions, including maintaining cell growth [5] and promoting the synthesis of physiological factors [6] and oxidation energy supply [7]. In the study of carnivorous fish, some amino acids can improve the growth performance of the fish, promote muscle development, improve muscle quality [8], improve intestinal digestion [9], and promote intestinal health [10]. L-glutamic acid (Glu) is a functional feed additive [11]. It can provide energy for animal tissues and participate in many physiological regulation processes [12]. Research has demonstrated that adding glutamate to the diet can enhance the growth performance of *Ctenopharyngodon idella*, enhance the antioxidant capacity of the intestinal tract, and facilitate digestion and absorption [13]. And studies have shown that glutamate can affect the growth, muscle development, and muscle quality of *Ctenopharyngodon idellus* by different regulation of protein metabolism, muscle development, and antioxidant-related genes [14]. In summary, glutamic acid as a functional additive has achieved good research results, but it has not been reported in the study of largemouth bass.

As a carnivorous freshwater fish, largemouth bass has high economic benefits [15], which can be well adapted to a variety of breeding modes [16]. It also has the advantages of rapid growth, delicious flavor, and is boneless between the muscles, which makes it popular in the Chinese market [17]. According to the statistics of the Fisheries Administration of the Ministry of Agriculture and Rural Affairs and the Chinese Aquatic Association, the output of largemouth bass reached 802,500 tons in 2022 [18]. Nevertheless, research is scarce about the impact of glutamic acid as a dietary supplement on the growth performance, intestinal health, and muscular growth and development of largemouth bass. Therefore, the present research was performed to examine the impacts of dietary glutamic acid on the developmental performance, intestinal health, and muscular development and growth of largemouth bass.

2. Generally Used Wording ID: Materials and Methods

2.1. Preparation of Experimental Diet

Five experimental diets were established, each containing equal amounts of nitrogen and lipids. The diets included fish meal, chicken meal, cottonseed meal, soybean meal, and soybean protein concentrate as the primary sources of protein. Fish oil and soybean oil were used as the primary sources of fat, while high-gluten flour served as the main source of sugar. Glutamic acid was added at 0.2%, 0.4%, 0.6%, and 0.8%. From McLean Biochemical Technology Co., Ltd Shanghai China., glutamic acid was acquired. After being crushed, the feed components were run through an 80-mesh sieve. The low-dose raw components were gradually pre-mixed before being combined in a V-type mixer for a thorough mixing. The oil-kneading machine was then filled with fish and soybean oils to begin the kneading process. After the oil was mixed, 30% water was added to the mixer to stir and mix, and then the SLX-80 twin-screw extruder (South China University of Technology, Guangzhou, China) was used to make a pellet feed with a particle size of 2.0 mm. The pellet feed was allowed to naturally cool to 55 °C before being wrapped in a bag and refrigerated at −20 °C for subsequent use (Table 1).

2.2. Experimental Design and Feeding Management

The feeding experiment was conducted in an indoor circulating water system at the Guangdong Academy of Agricultural Sciences' Baiyun Experimental Base. Largemouth bass juveniles were purchased from Renzhi (Guangzhou China) Technology Co., Ltd. A total of 525 healthy fish, each weighing 8.00 ± 0.00 g, were selected and categorized into 5 groups. Each group had 3 repetitions, resulting in a total of 15 replicates. For 56 days, the fish were fed satiating meals twice per day (7:30 and 17:30). The ideal environmental conditions are natural light in summer, water temperature 27~31 °C, no heating equipment, ammonia nitrogen concentration below 0.20 mg/L, nitrite concentration below 0.01 mg/L, dissolved oxygen concentration above 5.0 mg/L, and pH levels between 7.8 and 8.2.

Table 1. Formulation and composition of experimental diets (air-dry basis).

Item	CON	0.2% Glu	0.4% Glu	0.6% Glu	0.8% Glu
Fish meal	40	40	40	40	40
Chicken meal	10	10	10	10	10
Cotton meal	7.5	7.5	7.5	7.5	7.5
Soybean meal	10.5	10.5	10.5	10.5	10.5
Soy protein	8	8	8	8	8
Flour	16	16	16	16	16
Fish oil	3.5	3.5	3.5	3.5	3.5
Microcrystalline cellulose	1.3	1.1	0.9	0.7	0.5
Mineral mixture	1	1	1	1	1
Vitamin mix	0.2	0.2	0.2	0.2	0.2
Calciumbiphosphate	1.5	1.5	1.5	1.5	1.5
Choline chloride	0.5	0.5	0.5	0.5	0.5
Glutamic acid	0	0.2	0.4	0.6	0.8
Total	100	100	100	100	100
Analyzed chemical composition					
Crude protein	48.26	48.31	49.01	48.22	49.23
Crude lipid	6.14	6.07	6.19	6.03	6.12
Moisture	8.05	8.04	8.03	8.02	8.01
Crude ash	12.57	12.59	12.27	12.42	12.34

Each kilogram of premix contained: vitamin A 4,000,000 IU, vitamin D3 2,000,000 IU, vitamin E 30 g, vitamin K3 10 g, vitamin B_1 5 g, vitamin B_2 15 g, vitamin B6 8 g, calcium pantothenate 25 g, folic acid 2.5 g, biotin 0.08 g, nicotinic acid 40 g. Vitamin B_{12} 0.02 g, inositol 150 g, $MgSO_4 \cdot H_2O$ 12 g, KCl 90 g, Met-Cu 3 g, $FeSO_4 \cdot H_2O$ 1 g, $ZnSO_4 \cdot H_2O$ 10 g, Ca (IO_3) 2 0.06 g, Met-Co 0.16 g, and $NaSeO_3$ 0.003 6 g. The fish oil was provided by China Oil and Foodstuffs Corporation.

2.3. Sample Analysis

2.3.1. Evaluation of the Growth Performance and Physical Composition

Following the feeding experiment, the fish were deprived of food for 24 h. They were then weighed and numbered in each barrel, and the final average weight and survival rate were determined. From each replicate, a total of 10 fish were selected at random, and 3 of them were utilized to determine the overall body composition. Six fish were used for measuring body weight and body length, dissecting visceral mass, separating liver and weighing, and calculating body index. Additionally, 3 fish were picked from each cage. The foregut and liver were then separated and preserved in 4% paraformaldehyde. These samples were then maintained to section the intestinal and hepatic tissues.

2.3.2. Nutrient Composition of Diets and Muscle

The moisture content was evaluated by subjecting the sample to oven drying at a temperature of 105 °C until a consistent weight was achieved. The Kjeldahl method was employed to determine the protein content in both whole fish and muscle [19]. The ether Soxhlet extraction method was used to determine the crude fat content. The ash content was measured by subjecting the sample to combustion in a muffle furnace until a steady weight was achieved at a temperature of 550 °C [20].

2.3.3. Histological Analysis

One fish was randomly selected from each replicate, and the middle part of the dorsal muscle and the foregut were sliced on ice. The muscle and intestine sections were subjected to histological processing using the paraffin procedure and stained with the hematoxylin and eosin (H&E) staining method. The tissue was washed with a 0.01 mol/L PBS solution, treated with 4% paraformaldehyde for 10 min to fix it, slowly dehydrated in 75% ethanol, and finally embedded in paraffin and sliced into sections that were 5 μm thick. Finally, stained with eosin and hematoxylin, the intestinal and muscle tissues were observed under a microscope and photographed. The intestinal villus height, muscle thickness, villus width, muscle fiber density, muscle fiber number, and average muscle fiber area were measured [21].

2.3.4 Analysis of Muscle Amino Acid Content

The analysis revealed the presence of fifteen different amino acids, namely arginine (Arg), lysine (Lys), histidine (His), phenylalanine (Phe), tyrosine (Tyr), leucine (Leu), isoleucine (Ile), methionine (Met), valine (Val), alanine (Ala), glycine (Gly), glutamic acid (Glu), serine (Ser), threonine (Thr), and asparagine (Asp). The mobile phase A was 40 mmol/L sodium dihydrogen phosphate (pH = 7.8); the detection signal was UV 339 nm, fluorescence (EX = 266 nm, EM = 305 nm). Acetonitrile, methanol, and water comprised the mobile phase B; the ratios were 45, 45, and 10, respectively. Every experimental technique was carried out strictly in line with the standard instructions; the measured findings deviate by less than 10% of the arithmetic mean value, and the correlation coefficient of every amino acid's linear regression equation is >0.99.

2.3.5. Determination of Intestinal Enzyme Activity

The foregut was immediately frozen in a refrigerator at −80 °C. After weighing, the samples were homogenized with a high-speed tissue homogenizer and centrifuged with a refrigerated centrifuge (at 3000 r/min for 15 min at 4 °C). Trypsin activity was determined using r-toluenesul-phonyl-l-arginine methyl esther as a substrate in 0.05-mol/LTriseHCl buffer, (pH = 9.0). Amylase was assayed using 1% solublestarch as a substrate in 0.02 mol/L phosphate buffer, (pH = 8.0). Lipase activity was measured at 405 nm by the rate of methyl-resorufin formation.

2.3.6. Real-Time Quantitative PCR Analysis

The protein–nucleic acid approach was utilized to ascertain the total RNA concentration after largemouth bass muscle was treated with Trizol reagent to extract RNA. The process of reverse transcription was employed to convert the total RNA into complementary DNA (cDNA) using the TaKaRa reverse transcription kit (Takara, Tokyo, Japan) [22]. Primer Premier 5.0 software was used for primer design, cDNA sequences from Gen Bank or published papers, and primer sequences from Wang et al. (2021) [23] (Table 2). Using β-actin as an internal reference, the $2^{-\Delta\Delta Ct}$ technique was used to calculate the relative expression of the gene based on the control group.

Table 2. Primer sequences of target genes used in real-time quantitative PCR.

Target Gene	Primer Sequence (5–3′)	Number/Source
pi3k	F-CGCAAGACCAGAGATCAGTATC R-CGTCGTCCACCATAGAGTATTG	XM_038733022.1
akt	F-CACCGTAGAACCGAGCCCGCT R-CGCCATGAAGATCCTAAAGAA	[23]
tor	F-TCAGGACCTCTTCTCATTGGC R-CCTCTCCCACCATGTTTCTCT	[23]
S6k1	F-GCCAATCTCAGCGTTCTCAAC R-CTGCCTAACATCATCCTCCTT	XM_038708508.1
4ebp1	F-ACGAGGTCTGCCCAACATTC R-CAGCGTTGCTGCTATCAGGT	XM_038703879.1
Foxo3a	F-AAGAAGAAAGCCTCGCTACAG R-GTGGGACTTCCTGTCCATTT	XM_038728762.1
myod	F-CCTGCCGCTGATGATTTCTAT R-AGTCGTCCGGCTTCAGTA	EU367961.1
myog	F-GTGACAGGAACAGAGGACAAA R-ACGATCCATGGTAACAGTCTTC	XM_038697403.1
Myf5	F-GGCTGAAGAAGGTCAACCA R-GTCCTGCAGACTCTCAATGTAA	EU555403.1

Table 2. *Cont.*

Target Gene	Primer Sequence (5–3′)	Number/Source
murf-1	F-ACGCCAAAGAGCTGAAGTGT R-TGTCCGAACACCTTGCACAT	XM_038728309.1
atrogin-1	F-CCAAATCAACAGGCCCACAT R-GACAGACGCTGCATGATGTT	XM_038711973.1
β-actin	F-ACTGCTGCTTCCTCTTCATC R-GGATACCGCAAGACTCCATAC	MH018565.1

pi3k, phosphatidylinositol 3-kinase; AKT, protein kinase B; tor, target of rapamycin; S6K1, 70 kDa ribosomal protein S6 kinase 1; 4E-BP1, eukaryotic translation initiation factor 4E-binding protein 1; FoxO3a, forkhead box O3a; MyoD, myoblast determination protein; MyoG, myogenin; Myf5, myogenic factor 5; MuRF-1, muscle-specific RING finger protein 1; β-Actin, muscle-specific F-box protein-regulated; F, forward; R, reverse.

2.3.7. Statistical Analysis

Using SPSS 26.0, a one-way ANOVA was run on each result, and then the Tukey multiple comparisons procedure was applied. The statistical significance threshold was set at $p < 0.05$, and the data were presented as mean ± SD.

3. Results

3.1. Growth Performance

Table 3 displays growth performance and body indexes. Compared to the control group, the FBW, SGR, and WGR of the 0.2% and 0.4% Glu groups were considerably greater ($p < 0.05$). Significant differences were seen between the FCR of the 0.2% and 0.4% Glu groups and that of the control group ($p < 0.05$). Additionally, there was no statistically significant difference in the VSI between the groups, and the HSIs of the 0.4%, 0.6%, and 0.8% Glu groups were all considerably greater than those of the control group ($p > 0.05$).

Table 3. Effects of dietary Glu supplementation on growth performance of largemouth bass.

Item	CON	0.2% Glu	0.4% Glu	0.6% Glu	0.8% Glu
IBW (g/fish)	8.00 ± 0.00	8.00 ± 0.00	8.00 ± 0.00	8.00 ± 0.00	8.00 ± 0.00
FBW (g/fish)	47.81 ± 1.63 [c]	56.13 ± 2.64 [ab]	58.94 ± 5.61 [a]	49.17 ± 2.48 [c]	52.70 ± 1.70 [bc]
FCR	1.31 ± 0.05 [a]	1.12 ± 0.06 [c]	1.20 ± 0.06 [abc]	1.27 ± 0.08 [ab]	1.18 ± 0.07 [bc]
FI (g/fish)	52.18 ± 1.65 [a]	53.72 ± 4.86 [a]	60.79 ± 4.43 [b]	52.33 ± 0.40 [a]	52.51 ± 1.35 [a]
WGR (%)	497.32 ± 20.37 [c]	601.10 ± 33.08 [ab]	636.16 ± 69.94 [a]	514.22 ± 30.70 [c]	558.38 ± 21.37 [bc]
SGR (%/day)	3.19 ± 0.06 [c]	3.48 ± 0.08 [ab]	3.56 ± 0.17 [a]	3.24 ± 0.09 [c]	3.36 ± 0.06 [bc]
CF (g/cm³)	2.13 ± 0.36 [ab]	2.17 ± 0.16 [ab]	2.39 ± 0.21 [a]	2.11 ± 0.33 [b]	2.13 ± 0.56 [ab]
VSI (%)	7.22 ± 0.72	6.98 ± 0.64	6.89 ± 0.6	7.17 ± 0.64	6.7 ± 1.32

IBW, initial body weight; FBW, final body weight; FCR, feed conversion ratio; FI, feed intake; SGR, specific growth rate; CF, condition factor; VSI, viscerosomatic index; HSI, hepatosomatic index; Means in the same row with different superscripts are significantly different (mean ± SD; ANOVA, $p < 0.05$; $n = 3$).

3.2. Body Composition

Table 4 displays the body composition of the largemouth bass. The levels of protein, fat, moisture, and ash content were not significantly different across the groups ($p > 0.05$).

Table 4. Effects of dietary Glu on body composition of largemouth bass.

Item	CON	0.2% Glu	0.4% Glu	0.6% Glu	0.8% Glu
Crude protein	57.29 ± 0.96	58.96 ± 1.35	57.60 ± 0.55	58.34 ± 2.38	58.55 ± 1.51
Crude lipid	21.29 ± 2.57	21.22 ± 1.90	21.95 ± 0.77	22.83 ± 2.46	21.88 ± 1.06
Moisture	69.91 ± 1.09	69.7 ± 0.15	69.47 ± 0.83	70.09 ± 1.17	68.97 ± 1.18
Crude ash	14.75 ± 0.96	13.63 ± 0.54	13.52 ± 0.17	13.21 ± 0.33	13.75 ± 0.56

3.3. Muscle Amino Acid Composition and Inosine Monophosphate Content

Table 5 displays the inosine monophosphate concentration and amino acid composition of largemouth bass muscle. There was a significant difference between the crude protein levels, total amino acid content, and inosine monophosphate amounts of the 0.2%, 0.4%, 0.6%, and 0.8% Glu groups and the control group ($p < 0.05$). The crude protein level of the 0.6% Glu group was the highest, while the total amino acid and inosine monophosphate contents were the highest at 0.4% Glu. The levels of glutamic acid, aspartic acid, and tyrosine in the 0.4% Glu and 0.6% Glu groups were considerably greater than those in the control group ($p < 0.05$). After adding glutamic acid, the contents of phenylalanine, alanine, and glycine were also increased. The levels of serine, threonine, valine, methionine, isoleucine, leucine, and lysine in Glu solutions with concentrations of 0.2%, 0.4%, 0.6%, and 0.8% were considerably greater than those in the control group ($p < 0.05$).

Table 5. Effects of dietary Glu on muscle amino acid composition of largemouth bass.

Item	CON	0.2% Glu	0.4% Glu	0.6% Glu	0.8% Glu
Crude protein	84.91 ± 0.73 [a]	86.89 ± 0.03 [b]	87.02 ± 0.04 [bc]	87.56 ± 0.03 [c]	86.76 ± 0.05 [b]
Inosine monophosphate	2.12 ± 0.06 [a]	2.80 ± 0.07 [bc]	3.10 ± 0.11 [bc]	2.69 ± 0.01 [bc]	2.58 ± 0.06 [abc]
Aspartic acid	9.13 ± 0.11 [a]	9.41 ± 0.13 [bc]	9.47 ± 0.06 [c]	9.40 ± 0.07 [bc]	9.38 ± 0.06 [bc]
Glutamic acid	14.43 ± 0.26 [a]	14.75 ± 0.12 [ab]	14.89 ± 0.03 [b]	14.92 ± 0.31 [b]	14.87 ± 0.09 [b]
Phenylalanine	3.83 ± 0.12	3.91 ± 0.07	3.96 ± 0.04	3.91 ± 0.58	3.91 ± 0.02
Alanine	6.04 ± 0.10	6.07 ± 0.18	6.15 ± 0.13	6.21 ± 0.09	6.20 ± 0.02
Glycine	4.15 ± 0.05	4.23 ± 0.15	4.29 ± 0.08	4.24 ± 0.13	4.31 ± 0.02
Tyrosine	4.15 ± 0.05 [a]	4.23 ± 0.15 [c]	4.29 ± 0.08 [c]	4.24 ± 0.13 [c]	4.31 ± 0.02 [bc]
Serine	3.39 ± 0.12 [a]	3.49 ± 0.05 [b]	3.51 ± 0.02 [b]	3.49 ± 0.04 [b]	3.50 ± 0.00 [b]
Glycine	4.15 ± 0.05 [a]	4.23 ± 0.15 [b]	4.29 ± 0.08 [b]	4.24 ± 0.13 [b]	4.31 ± 0.02 [b]
Threonine	3.89 ± 0.02 [a]	4.02 ± 0.04 [b]	4.05 ± 0.04 [b]	4.02 ± 0.05 [b]	4.03 ± 0.01 [b]
Histidine	2.14 ± 0.08	2.21 ± 0.04	2.18 ± 0.06	2.20 ± 0.06	2.18 ± 0.01
Arginine	5.56 ± 0.45	5.70 ± 0.09	5.73 ± 0.07	5.73 ± 0.02	5.73 ± 0.07
Valine	4.10 ± 0.08 [a]	4.23 ± 0.06 [b]	4.25 ± 0.05 [b]	4.23 ± 0.05 [b]	4.17 ± 0.02 [ab]
Methionine	2.69 ± 0.03 [a]	2.79 ± 0.03 [b]	2.81 ± 0.03 [b]	2.79 ± 0.01 [b]	2.78 ± 0.11 [b]
Isoleucine	4.27 ± 0.05 [a]	4.43 ± 0.06 [b]	4.45 ± 0.04 [b]	4.41 ± 0.05 [b]	4.37 ± 0.01 [b]
Leucine	7.07 ± 0.14 [a]	7.35 ± 0.08 [b]	7.38 ± 0.12 [b]	7.35 ± 0.08 [b]	7.35 ± 0.10 [b]
Lysine	8.18 ± 0.08 [a]	8.48 ± 0.08 [b]	8.58 ± 0.11 [b]	8.53 ± 0.22 [b]	8.66 ± 0.02 [b]
Total amino acids	81.80 ± 0.71 [a]	84.12 ± 0.64 [b]	84.78 ± 0.21 [b]	84.45 ± 0.33 [b]	84.46 ± 0.02 [b]
IMP (mg/kg)	2.12 ± 0.06 [a]	2.80 ± 0.07 [bc]	3.10 ± 0.11 [bc]	2.69 ± 0.01 [bc]	2.58 ± 0.06 [abc]

Means in the same row with different superscripts are significantly different (mean ± SD; ANOVA, $p < 0.05$; $n = 3$).

3.4. Proximate Composition and Histological Traits of Muscle Fibers

Muscle sections are shown in Figure 1. In comparison to the control group, the muscle fiber size of the 0.4% and 0.6% Glu groups was substantially larger ($p < 0.05$). In comparison to the control group, the muscle fiber density and number of the 0.6% and 0.7% Glu groups were much lower ($p < 0.05$).

3.5. Muscle Growth and Development-Related Gene Expression

pi3k, *akt*, and *tor* mRNA expression levels in the 0.4% Glu, 0.6% Glu, and 0.8% Glu groups were considerably greater than those in the control group (Figures 2–4). In comparison to the control group, the mRNA expression levels of *s6k1*, *myod*, *myog*, and *myf5* in the 0.4% and 0.6% Glu groups were considerably higher ($p < 0.05$). 4e-bp1, foxo3a, murf-1, and atrogin-1 had significantly reduced mRNA expression levels in the 0.6% Glu group compared to the control group ($p < 0.05$).

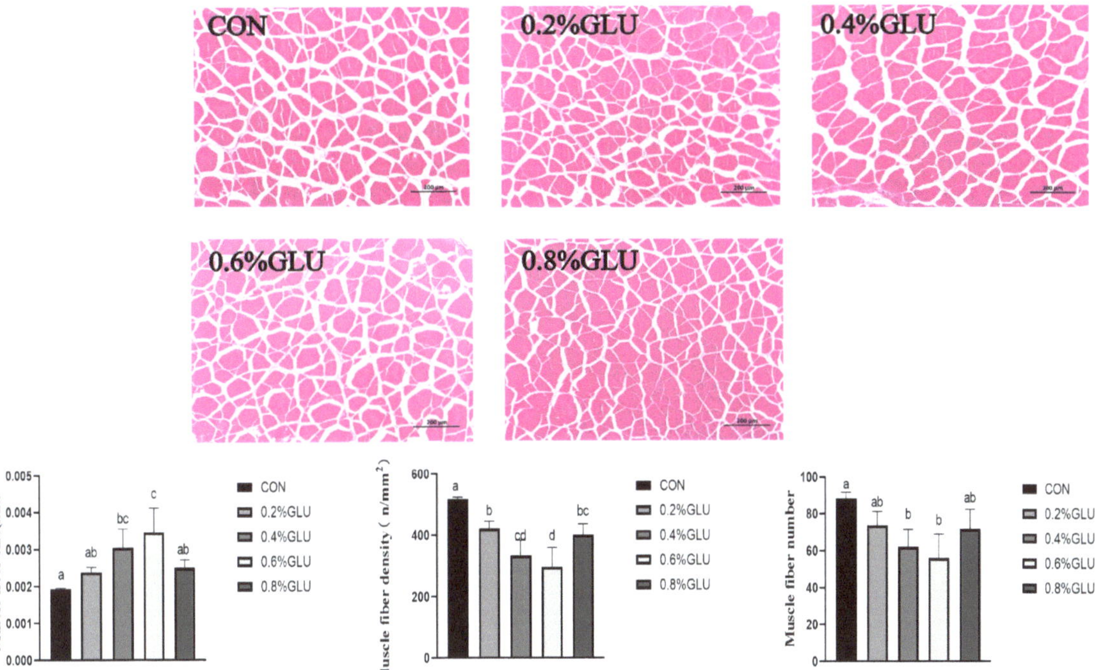

Figure 1. Effects of dietary Glu on muscle histology of largemouth bass. Means in the same row with different superscripts are significantly different (mean ± SD; ANOVA, $p < 0.05$; $n = 3$).

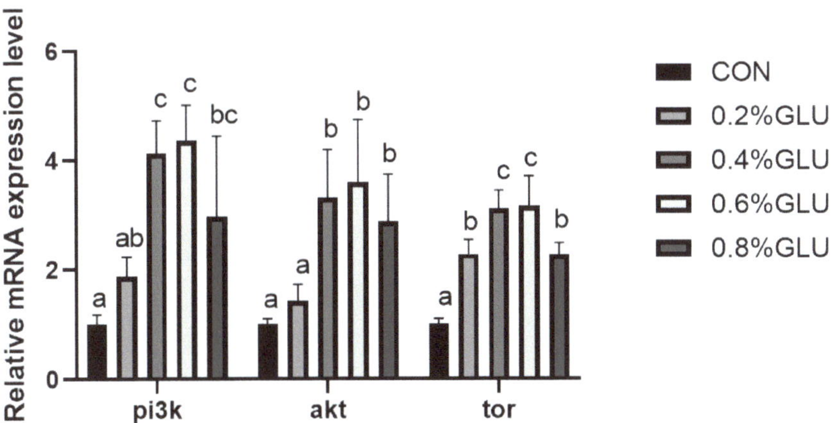

Figure 2. Effects of dietary Glu supplementation on the relative mRNA expression levels of phosphatidyl inositol 3 kinase (*pi3k*), protein kinase B (*akt*), and target of rapamycin (*tor*) in the muscle of largemouth bass. Means in the same row with different superscripts are significantly different (mean ± SD; ANOVA, $p < 0.05$; $n = 3$).

3.6. Intestinal Enzyme Activity

Protease, lipase, and amylase activity in the 0.2% and 0.4% Glu groups were significantly higher than those in the control group ($p < 0.05$). The highest concentrations of lipase, protease, and amylase were found in the 0.6%, 0.4%, and 0.6% Glu groups, respectively (Table 6).

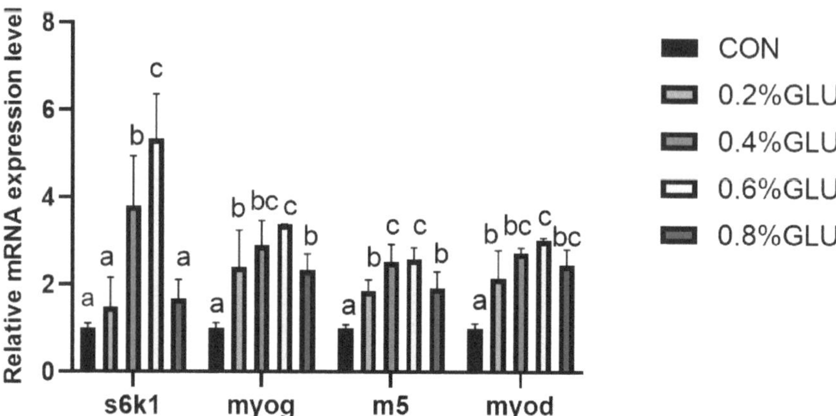

Figure 3. Relative mRNA expression level of 70 kDa ribosomal protein S6 kinase 1 (*S6K1*), myoblast determination protein (*myod*), myogenin (*myog*), myogenic factor 5 (*myf5*) in the muscle of largemouth bass fed Glu diets. Means in the same row with different superscripts are significantly different (mean ± SD; ANOVA, $p < 0.05$; $n = 3$).

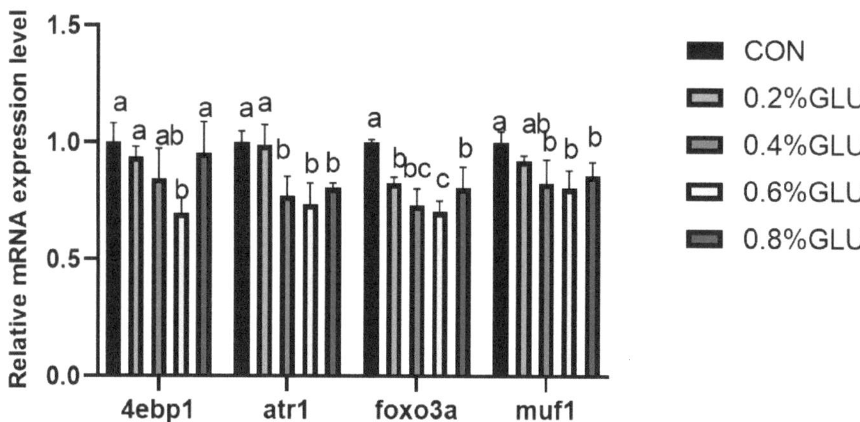

Figure 4. Relative mRNA expression level of eukaryotic translation initiation factor 4E-binding protein 1 (*4e-bp1*), forkhead boxO3a (*foxo3a*), muscle-specific RING finger protein 1 (*murfF-1*), and atrogin-1 in the muscle of largemouth bass fed Glu diets. Means in the same row with different superscripts are significantly different (mean ± SD; ANOVA, $p < 0.05$; $n = 3$).

Table 6. Effects of dietary glutamate on intestinal villus height, villus width, and muscle thickness of largemouth bass.

Item	CON	0.2%Glu	0.4%Glu	0.6%Glu	0.8%Glu
Amylase	369.57 ± 12.46 [b]	440.88 ± 6.81 [d]	410.48 ± 15.85 [cd]	378.68 ± 0.85 [bc]	305.37 ± 42.09 [a]
Lipase	372.90 ± 37.93 [a]	444.75 ± 10.25 [b]	495.73 ± 15.04 [c]	499.83 ± 26.72 [c]	464.88 ± 5.11 [bc]
Protease	731.68 ± 25.56 [a]	1041.18 ± 57.12 [b]	1398.53 ± 12.05 [c]	1184.85 ± 156.94 [b]	841.77 ± 60.19 [a]

Means in the same row with different superscripts are significantly different (mean ± SD; ANOVA, $p < 0.05$; $n = 3$).

3.7. Intestinal Tissue Structure

Compared to the control group, the villus height of the 0.4% Glu group was substantially higher ($p < 0.05$). The width of the 0.6% Glu group was substantially greater than that of the control group ($p < 0.05$). The thickness of the intestinal walls did not significantly differ across the groups (Table 7 and Figure 5).

Table 7. Effects of dietary glutamate on intestinal villus height, villus width, and muscle layer thickness of largemouth bass after 8 weeks of feeding.

Item	CON	0.2% Glu	0.4% Glu	0.6% Glu	0.8% Glu
Villus height (μm)	225.38 ± 31.76 [a]	261.52 ± 32.16 [ab]	308.3 ± 29.11 [b]	257.29 ± 43.62 [ab]	266.28 ± 16.09 [ab]
Villus width (μm)	44.64 ± 6.17 [ab]	45.27 ± 3.60 [ab]	41.26 ± 7.40 [a]	51.43 ± 3.61 [b]	46.82 ± 1.63 [ab]
Intestinal wall thickness (μm)	64.38 ± 4.80	71.93 ± 2.66	76.66 ± 14.19	65.05 ± 10.26	64.14 ± 6.37

Means in the same row with different superscripts are significantly different (mean ± SD; ANOVA, $p < 0.05$; $n = 3$).

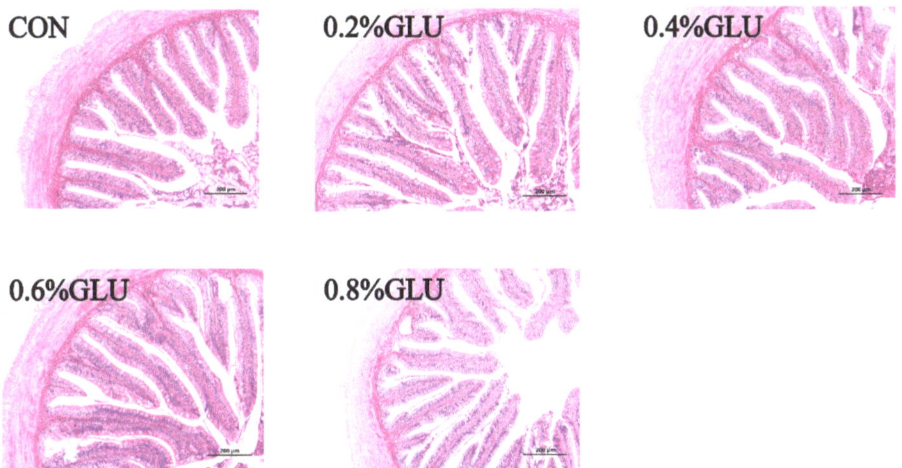

Figure 5. Effects of dietary Glu supplementation on intestinal morphology of largemouth bass.

4. Discussion

This research shows that adding glutamic acid to largemouth bass can improve the growth performance of largemouth bass. Additionally, it boosted their feed intake and condition factor. This was similar to the results of Jian carp (*Cyprinus carpio* var. Jian) with dietary glutamate levels of 68.4 and 83.4 g/kg [24]. Administration of 0.25% of glutamic acid to *Ctenopharyngodon idellus* also had the same effect [14]. In recent years, more and more studies have shown that non-essential amino acids such as glutamic acid, glycine [25], and proline [26] can also promote the growth of fish, thereby enhancing the growth performance of fish. The potential cause for the enhancement in growth performance in largemouth bass due to glutamic acid supplementation may be attributed to the observed increase in meal intake. This glutamic acid experiment was carried out in summer and achieved results such as improving fish growth performance.

In this study, the addition of glutamic acid to the feed can increase the crude protein content of largemouth bass. In a prior study, it was found that adding 4% glutamic acid boosted the protein content of *Sparus aurata* [27]. Similarly, adding 3.2% glutamic acid to the diet of triploid crucian carp resulted in the same finding. The protein level in aquatic animals is strongly linked to protein turnover, which is a key element influencing muscle growth and the quality of meat [28]. The important gene-level pathway is the pi3k/akt signaling pathway [29]. mTOR phosphorylation can be enhanced by the control of protein phosphorylation in the pi3k/akt pathway [30]. In addition, s6k1 and 4e-bp1 are downstream targets of tor signal transduction [31]. Two downstream effectors can be directly phosphorylated by tor, which stimulates the start and translation of protein transcription and controls cell activity to control protein synthesis. Conversely, foxo3a functions as a downstream effector of the pi3k signaling pathway and is essential for the degradation of proteins [32]. Dephosphorylation of foxo3a is induced by inhibition of akt phosphorylation, and dephosphorylated foxo3a enters the nucleus upon activation by stimulating the production of muscle ring-finger gene 1 (murf1) and muscle atrophy Fbox (mafbx/atrogin-1), causing the myofibrillar protein to degrade [33]. The mRNA expression of s6k1, tor, akt, pi3k, myod, myog, and myf5 was elevated in this study by dietary glutamate. Atrogin-1, murf-1, foxo3a, and 4e-bp1 all had downregulated mRNA expression levels. The findings demonstrated that by stimulating the pi3k/akt pathway, Glu might increase the synthesis of muscle protein and decrease its decomposition.

Precursors of flavor compounds include glutamic acid, aspartic acid, leucine, and valine. These compounds are strongly linked to the development of muscle flavor [34]. The sweet amino acids alanine and glycine may help to increase the sweetness of fish during chewing [35]. In this study, the inclusion of glutamic acid in the meal raised the amount of umami amino acids, including aspartic acid, leucine, valine, alanine, and glycine, as well as the overall amino acid content of the largemouth bass muscle. This is consistent with the result of adding 3.2% Glu to crucian carp. Inosine monophosphate is an important flavor substance in fish muscle [36]. The inclusion of glutamic acid in the food resulted in an elevation of inosine monophosphate levels in the muscle tissue of largemouth bass observed in this study.

Muscle growth is related to the number and area of muscle fibers [37]. Diet plays an important role in the balance between muscle fiber hyperplasia and hypertrophy [38]. The study found that incorporating glutamic acid into the diet increased the average size of muscle fibers in largemouth bass and decreased muscle fiber density. This suggests that glutamic acid effectively stimulates muscle growth in largemouth bass. In short, the inclusion of glutamic acid in the meal resulted in an augmentation of the protein composition in the muscle of largemouth bass. This led to the stimulation of muscle growth and development, as well as an increase in the concentration of umami amino acids.

The gastrointestinal system is the primary site for the breakdown and assimilation of nutrients in fish [39]. The morphological structure of the digestive tract is an important basis for maintaining the growth status of largemouth bass [40]. The intestine is the main digestive organ of fish [41]. The integrity of the intestinal structure has a direct impact on the process of digesting and absorbing nutrients [42], as well as on the growth and development of fish [43]. Muscle layer thickness (MT) [44], villus height (VH), and villus width (VW) [45] are usually used to evaluate the intestinal digestion ability of animals [46]. The increase in villus height [47] and muscle layer thickness [48] can increase the surface area of the fish intestine and improve its ability to digest and absorb feed nutrients [49]. The study found that adding glutamic acid to the diet of largemouth bass elevated the thickness of the intestinal muscle, as well as the height and width of the villi. The activity of digestive enzymes is also crucial to the digestive capacity of largemouth bass [50]. After entering the body of largemouth bass, macromolecular nutrients such as protein and fatty acids need to be decomposed into small molecules by digestive enzymes in the digestive tract before they can be absorbed by the digestive tract [51]. As the first step to obtain nutrients, digestive enzyme activity is closely related to nutrient utilization efficiency [52]. The presence of lipase [53], protease [54], and amylase [55] in the fish gut can impact the process of absorbing and utilizing nutrients [56]. As a result, this has an impact on the proliferation and development of the fish [57]. These findings demonstrated that the inclusion of glutamic acid in the meal enhanced the enzymatic activity of amylase, protease, and lipase in the intestine of largemouth bass. The aforementioned findings suggest that including Glu in the meal enhances the digestive capacity and fosters intestinal well-being in largemouth bass. This is comparable to the effects of glutamic acid on Jian carp (Cyprinus carpio var. Jian) [24].

5. Conclusions

In summary, our findings showed that largemouth bass can benefit from Glu in terms of growth performance, muscular development, and intestinal digestion. Taking growth performance as the index, the optimum addition amount of Glu was 0.4%, and taking muscle development as the index, the optimum addition amount of Glu was 0.6%.

Author Contributions: Conceptualization, F.J., H.G. and X.L.; Methodology, F.J., X.L. and Z.Y.; Software, W.H., H.G., Z.Y. and M.S.; Validation, M.S.; Formal analysis, F.J., H.G., Z.Y., M.S. and Y.H.; Investigation, M.Z. and X.L.; Resources, W.H.; Data curation, M.S.; Writing—original draft, F.J.; Writing—review and editing, F.J.; Visualization, F.J.; Supervision, W.H. and Y.H.; Project administration, M.Z. and Y.H.; Funding acquisition, M.Z. and Y.H. All authors have read and agreed to the published version of the manuscript.

Funding: The research was supported by the innovation team project of Guangdong Provincial Department of Education, the innovation team of functional feed and animal immune regulation (2020KCXTD019), and the research and development of special feed for California bass (D122222L6).

Institutional Review Board Statement: The animal management procedures followed the guidelines of the Animal Care and Use Committee of Zhongkai Agricultural Engineering University (ethical approval number: ZHKUMO-2022-055).

Data Availability Statement: The datasets of the current study are available from the corresponding author upon reasonable request.

Conflicts of Interest: Author Wenqing Huang was employed by the company Guangzhou Fishteach Biotechnology. The remaining authors declare that the research was conducted in the absence of any commercial or financial relationships that could be construed as a potential conflict of interest.

References

1. Pincinato, R.B.M. Market aspects and external economic effects of aquaculture. *Aquac. Econ. Manag.* **2021**, *25*, 127–134. [CrossRef]
2. Afewerki, S.; Asche, F.; Misund, B.; Thorvaldsen, T.; Tveteras, R. Innovation in the Norwegian aquaculture industry. *Rev. Aquac.* **2023**, *15*, 759–771. [CrossRef]
3. Xiao, J.; Zhang, Y. Marine factory farming techniques and equipment. *IOP Conf. Ser. Earth Environ. Sci.* **2020**, *615*, 012013. [CrossRef]
4. He, J.; Feng, P.; Lv, C.; Lv, M.; Ruan, Z.; Yang, H.; Ma, H.; Wang, R. Effect of a fish–rice co-culture system on the growth performance and muscle quality of tilapia (*Oreochromis niloticus*). *Aquac. Rep.* **2020**, *17*, 100367. [CrossRef]
5. Yu, J.; Yang, H.; Wang, Z.; Dai, H.; Xu, L.; Ling, C. Effects of arginine on the growth performance, hormones, digestive organ development and intestinal morphology in the early growth stage of layer chickens. *Ital. J. Anim. Sci.* **2018**, *17*, 1077–1082. [CrossRef]
6. Garcia, I.S.; Teixeira, S.A.; Costa, K.A.; Marques, D.B.D.; Rodrigues, G.d.A.; Costa, T.C.; Guimarães, J.D.; Otto, P.I.; Saraiva, A.; Ibelli, A.M.G.; et al. l-Arginine supplementation of gilts during early gestation modulates energy sensitive pathways in pig conceptuses. *Mol. Reprod. Dev.* **2020**, *87*, 819–834. [CrossRef] [PubMed]
7. Cheng, C.; Liu, Z.; Zhou, Y.; Wei, H.; Zhang, X.; Xia, M.; Deng, Z.; Zou, Y.; Jiang, S.; Peng, J. Effect of oregano essential oil supplementation to a reduced-protein, amino acid-supplemented diet on meat quality, fatty acid composition, and oxidative stability of *Longissimus thoracis* muscle in growing-finishing pigs. *Meat Sci.* **2017**, *133*, 103–109. [CrossRef]
8. Wu, Y.-Y.; Dai, Y.-J.; Xiao, K.; Wang, X.; Wang, M.-M.; Huang, Y.-Y.; Guo, H.-X.; Li, X.-F.; Jiang, G.-Z.; Liu, W.-B. Effects of different dietary ratio lysine and arginine on growth, muscle fiber development and meat quality of *Megalobrama amblycephala*. *Aquac. Rep.* **2022**, *26*, 101322. [CrossRef]
9. Ding, L.; Chen, J.; He, F.; Chen, Q.; Li, Y.; Chen, W. Effects of dietary arginine supplementation on growth performance, antioxidant capacity, intestinal digestive enzyme activity, muscle transcriptome, and gut health of *Siniperca chuatsi*. *Front. Mar. Sci.* **2024**, *10*, 1305192. [CrossRef]
10. Yu, Y.; Huang, D.; Zhang, L.; Chen, X.; Wang, Y.; Zhang, L.; Ren, M.; Liang, H. Dietary arginine levels affect growth performance, intestinal antioxidant capacity and immune responses in largemouth bass (*Micropterus salmoides*). *Aquac. Rep.* **2023**, *32*, 101703. [CrossRef]
11. Hou, Y.; Wu, G. l-Glutamate nutrition and metabolism in swine. *Amino Acids* **2018**, *50*, 1497–1510. [CrossRef]
12. Hu, C.J.; Jiang, Q.Y.; Zhang, T.; Yin, Y.L.; Li, F.N.; Deng, J.P.; Wu, G.Y.; Kong, X.F. Dietary supplementation with arginine and glutamic acid modifies growth performance, carcass traits, and meat quality in growing-finishing pigs1. *J. Anim. Sci.* **2017**, *95*, 2680–2689. [CrossRef] [PubMed]

13. Zhao, Y.; Hu, Y.; Zhou, X. Q.; Zeng, X.-Y.; Feng, L.; Liu, Y.; Jiang, W.-D.; Li, S.-H.; Li, D.-B.; Wu, X.-Q.; et al. Effects of dietary glutamate supplementation on growth performance, digestive enzyme activities and antioxidant capacity in intestine of grass carp (*Ctenopharyngodon idella*). *Aquac. Nutr.* **2015**, *21*, 935–941. [CrossRef]
14. Cai, Y.; He, L.; Cao, S.; Zeng, P.; Xu, L.; Luo, Y.; Tang, X.; Wang, Q.; Liu, Z.; He, Z.; et al. Insights into Dietary Different Co-Forms of Lysine and Glutamate on Growth Performance, Muscle Development, Antioxidation and Related Gene Expressions in Juvenile Grass Carp (*Ctenopharyngodon idellus*). *Mar. Biotechnol.* **2024**, *26*, 74–91. [CrossRef]
15. Song, R.; Yao, X.; Jing, F.; Yang, W.; Wu, J.; Zhang, H.; Zhang, P.; Xie, Y.; Pan, X.; Zhao, L.; et al. Effects of Five Lipid Sources on Growth, Hematological Parameters, Immunity and Muscle Quality in Juvenile Largemouth Bass (*Micropterus salmoides*). *Animals* **2024**, *14*, 781. [CrossRef]
16. Li, Y.; Qin, J.; Zheng, X.; Wang, Y. Production performance of largemouth bass *Micropterus salmoides* and water quality variation in monoculture, polyculture and integrated culture. *Aquac. Res.* **2019**, *50*, 423–430. [CrossRef]
17. Zheng, Z.; Nie, Z.; Zheng, Y.; Tang, X.; Sun, Y.; Zhu, H.; Gao, J.; Xu, P.; Xu, G. Effects of Submerged Macrophytes on the Growth, Morphology, Nutritional Value, and Flavor of Cultured Largemouth Bass (*Micropterus salmoides*). *Molecules* **2022**, *27*, 927. [CrossRef]
18. Xu, J.-M.; Gao, W.-R.; Liang, P.; Cai, G.-H.; Yang, H.-L.; Lin, J.-B.; Sun, Y.-Z. Pleurotus eryngii root waste and soybean meal co-fermented protein improved the growth, immunity, liver and intestinal health of largemouth bass (*Micropterus salmoides*). *Fish. Shellfish. Immunol.* **2024**, *149*, 109551. [CrossRef] [PubMed]
19. Ido, A.; Ali, M.-F.-Z.; Takahashi, T.; Miura, C.; Miura, T. Growth of Yellowtail (*Seriola quinqueradiata*) Fed on a Diet Including Partially or Completely Defatted Black Soldier Fly (*Hermetia illucens*) Larvae Meal. *Insects* **2021**, *12*, 722. [CrossRef]
20. Jastaniah, S.D.; Mansour, A.A.; Al-Tarawni, A.H.; El-Haroun, E.; Munir, M.B.; Saghir, S.A.M.; Abdul Kari, Z.; Téllez-Isaías, G.; Bottje, W.G.; Al-Farga, A.; et al. The effects of nano-curcumin on growth performance, feed utilization, blood biochemistry, disease resistance, and gene expression in European seabass (*Dicentrarchus labrax*) fingerlings. *Aquac. Rep.* **2024**, *36*, 102034. [CrossRef]
21. Chen, M.; Li, Q.; Yang, L.; Lin, W.; Qin, Z.; Liang, S.; Lin, L.; Xie, X. Effects of diet containing germinated faba bean (*Vicia faba* L.) on the intestinal health and gut microbial communities of *Nile tilapia* (*Oreochromis niloticus*). *Aquac. Rep.* **2024**, *36*, 102053. [CrossRef]
22. Che, M.; Lu, Z.; Liu, L.; Li, N.; Ren, L.; Chi, S. Dietary lysophospholipids improves growth performance and hepatic lipid metabolism of largemouth bass (*Micropterus salmoides*). *Anim. Nutr.* **2023**, *13*, 426–434. [CrossRef]
23. Wang, W.; Yang, P.; He, C.; Chi, S.; Li, S.; Mai, K.; Song, F. Effects of dietary methionine on growth performance and metabolism through modulating nutrient-related pathways in largemouth bass (*Micropterus salmoides*). *Aquac. Rep.* **2021**, *20*, 100642. [CrossRef]
24. Zhao, Y.; Zhang, T.-R.; Li, Q.; Feng, L.; Liu, Y.; Jiang, W.-D.; Wu, P.; Zhao, J.; Zhou, X.-Q.; Jiang, J. Effect of dietary L-glutamate levels on growth, digestive and absorptive capability, and intestinal physical barrier function in *Jian carp* (*Cyprinus carpio* var. *Jian*). *Anim. Nutr.* **2020**, *6*, 198–209. [CrossRef]
25. Xie, S.; Tian, L.; Niu, J.; Liang, G.; Liu, Y. Effect of N-acetyl cysteine and glycine supplementation on growth performance, glutathione synthesis, and antioxidative ability of grass carp, *Ctenopharyngodon idella*. *Fish. Physiol. Biochem.* **2017**, *43*, 1011–1020. [CrossRef] [PubMed]
26. Wei, Z.; Deng, K.; Zhang, W.; Mai, K. Interactions of dietary vitamin C and proline on growth performance, anti-oxidative capacity and muscle quality of large yellow croaker *Larimichthys crocea*. *Aquaculture* **2020**, *528*, 735558. [CrossRef]
27. Caballero-Solares, A.; Viegas, I.; Salgado, M.C.; Siles, A.M.; Sáez, A.; Metón, I.; Baanante, I.V.; Fernández, F. Diets supplemented with glutamate or glutamine improve protein retention and modulate gene expression of key enzymes of hepatic metabolism in gilthead seabream (*Sparus aurata*) juveniles. *Aquaculture* **2015**, *444*, 79–87. [CrossRef]
28. Wei, Z.; Zhuang, Y.; Liu, X.; Zou, D.; Mai, K.; Sun, Z.; Ye, C. Leucine promotes protein synthesis of juvenile white shrimp *Litopenaeus vannamei* through TOR signaling pathway. *Aquaculture* **2023**, *564*, 739060. [CrossRef]
29. Zhu, X.; Ren, L.; Liu, J.; Chen, L.; Cheng, J.; Chu, W.; Zhang, J. Transcriptome analysis provides novel insights into the function of PI3K/AKT pathway in maintaining metabolic homeostasis of Chinese perch muscle. *Aquac. Rep.* **2021**, *21*, 100838. [CrossRef]
30. Luo, C.; Zhao, S.; Dai, W.; Zheng, N.; Wang, J. Proteomic analyses reveal GNG12 regulates cell growth and casein synthesis by activating the Leu-mediated mTORC1 signaling pathway. *Biochim. Et Biophys. Acta (BBA)—Proteins Proteom.* **2018**, *1866*, 1092–1101. [CrossRef]
31. Gao, Q.; Hou, B.; Yang, H.; Jiang, X. Distinct role of 4E-BP1 and S6K1 in regulating autophagy and hepatitis B virus (HBV) replication. *Life Sci.* **2019**, *220*, 1–7. [CrossRef]
32. Li, J.; Long, H.; Cong, Y.; Gao, H.; Lyu, Q.; Yu, S.; Kuang, Y. Quercetin prevents primordial follicle loss via suppression of PI3K/Akt/Foxo3a pathway activation in cyclophosphamide-treated mice. *Reprod. Biol. Endocrinol.* **2021**, *19*, 63. [CrossRef]
33. Liu, H.-W.; Chen, Y.-J.; Chang, Y.-C.; Chang, S.-J. Oligonol, a Low-Molecular Weight Polyphenol Derived from Lychee, Alleviates Muscle Loss in Diabetes by Suppressing Atrogin-1 and MuRF1. *Nutrients* **2017**, *9*, 1040. [CrossRef] [PubMed]
34. Roobab, U.; Zeng, X.-A.; Ahmed, W.; Madni, G.M.; Manzoor, M.F.; Aadil, R.M. Effect of Pulsed Electric Field on the Chicken Meat Quality and Taste-Related Amino Acid Stability: Flavor Simulation. *Foods* **2023**, *12*, 710. [CrossRef]
35. Dong, M.; Zhang, Y.-Y.; Huang, X.-H.; Xin, R.; Dong, X.-P.; Konno, K.; Zhu, B.-W.; Fisk, I.; Qin, L. Dynamic sensations of fresh and roasted salmon (*Salmo salar*) during chewing. *Food Chem.* **2022**, *368*, 130844. [CrossRef]

36. Ackroff, K.; Sclafani, A. Flavor Preferences Conditioned by Dietary Glutamate. *Adv. Nutr.* **2016**, *7*, 845S–852S. [CrossRef] [PubMed]
37. Lin, F.; Lin, J.; Liu, X.; Yuan, Y.; Liu, G.; Ye, X. Effects of temperature on muscle growth and collagen deposition in zebrafish (*Danio rerio*). *Aquac. Rep.* **2022**, *22*, 100952. [CrossRef]
38. Bao, S.-T.; Liu, X.-C.; Huang, X.-P.; Guan, J.-F.; Xie, D.-Z.; Li, S.-A.; Xu, C. Magnesium supplementation in high carbohydrate diets: Implications on growth, muscle fiber development and flesh quality of *Megalobrama amblycephala*. *Aquac. Rep.* **2022**, *23*, 101039. [CrossRef]
39. Zhou, Y.; Jiang, W.-D.; Zhang, J.-X.; Feng, L.; Wu, P.; Liu, Y.; Jiang, J.; Kuang, S.-Y.; Tang, L.; Peng, Y.; et al. Cinnamaldehyde improves the growth performance and digestion and absorption capacity in grass carp (*Ctenopharyngodon idella*). *Fish. Physiol. Biochem.* **2020**, *46*, 1589–1601. [CrossRef]
40. Yang, W.; Wu, J.; Song, R.; Li, Z.; Jia, X.; Qian, P.; Zhang, H.; Zhang, P.; Xue, X.; Li, S.; et al. Effects of dietary soybean lecithin on growth performances, body composition, serum biochemical parameters, digestive and metabolic abilities in largemouth bass *Micropterus salmoides*. *Aquac. Rep.* **2023**, *29*, 101528. [CrossRef]
41. Zhang, H.; Ding, Q.; Wang, A.; Liu, Y.; Teame, T.; Ran, C.; Yang, Y.; He, S.; Zhou, W.; Olsen, R.E.; et al. Effects of dietary sodium acetate on food intake, weight gain, intestinal digestive enzyme activities, energy metabolism and gut microbiota in cultured fish: Zebrafish as a model. *Aquaculture* **2020**, *523*, 735188. [CrossRef]
42. Volatiana, J.A.; Wang, L.; Gray, N.; Tong, S.; Zhang, G.; Shao, Q. Tributyrin-supplemented high-soya bean meal diets of juvenile black sea bream, Acanthopagrus schlegelii: Study on growth performance and intestinal morphology and structure. *Aquac. Res.* **2020**, *51*, 135–146. [CrossRef]
43. Lei, W.; Li, J.; Fang, P.; Wu, S.; Deng, Y.; Luo, A.; He, Z.; Peng, M. Effects of Dietary Bile Acids on Growth Performance, Lipid Deposition, and Intestinal Health of Rice Field Eel (*Monopterus albus*) Fed with High-Lipid Diets. *Aquac. Nutr.* **2023**, *2023*, 3321734. [CrossRef] [PubMed]
44. Amin, A.; El Asely, A.; Abd El-Naby, A.S.; Samir, F.; El-Ashram, A.; Sudhakaran, R.; Dawood, M.A.O. Growth performance, intestinal histomorphology and growth-related gene expression in response to dietary *Ziziphus mauritiana* in Nile tilapia (*Oreochromis niloticus*). *Aquaculture* **2019**, *512*, 734301. [CrossRef]
45. Barca, A.; Abramo, F.; Nazerian, S.; Coppola, F.; Sangiacomo, C.; Bibbiani, C.; Licitra, R.; Susini, F.; Verri, T.; Fronte, B. *Hermetia illucens* for Replacing Fishmeal in Aquafeeds: Effects on Fish Growth Performance, Intestinal Morphology, and Gene Expression in the Zebrafish (*Danio rerio*) Model. *Fishes* **2023**, *8*, 127. [CrossRef]
46. Wang, Y.; Wu, J.; Li, L.; Yao, Y.; Chen, C.; Hong, Y.; Chai, Y.; Liu, W. Effects of Tannic Acid Supplementation of a High-Carbohydrate Diet on the Growth, Serum Biochemical Parameters, Antioxidant Capacity, Digestive Enzyme Activity, and Liver and Intestinal Health of Largemouth Bass, *Micropterus salmoides*. *Aquac. Nutr.* **2024**, *2024*, 6682798. [CrossRef] [PubMed]
47. Xavier, M.J.; Navarro-Guillén, C.; Lopes, A.; Colen, R.; Teodosio, R.; Mendes, R.; Oliveira, B.; Valente, L.M.P.; Conceição, L.E.C.; Engrola, S. Effects of dietary curcumin in growth performance, oxidative status and gut morphometry and function of gilthead seabream postlarvae. *Aquac. Rep.* **2022**, *24*, 101128. [CrossRef]
48. Chen, X.-C.; Huang, X.-Q.; Tang, Y.-W.; Zhang, L.; Lin, F. Effects of dietary nucleotides on growth performance, immune response, intestinal morphology and disease resistance of juvenile largemouth bass, *Micropterus salmoides*. *J. Fish. Biol.* **2022**, *101*, 204–212. [CrossRef]
49. Sun, C.; Zhang, M.; Feng, D.; Wang, S.; Li, M. Effects of dietary D-mannose supplementation on growth performance, intestinal digestive capacity, gut microbiota, and ammonia tolerance of largemouth bass *Micropterus salmoides*. *Aquac. Rep.* **2024**, *36*, 102054. [CrossRef]
50. Wang, S.; Han, Z.; Turchini, G.M.; Wang, X.; Fang, Z.; Chen, N.; Xie, R.; Zhang, H.; Li, S. Effects of Dietary Phospholipids on Growth Performance, Digestive Enzymes Activity and Intestinal Health of Largemouth Bass (*Micropterus salmoides*) Larvae. *Front. Immunol.* **2022**, *12*, 827946. [CrossRef]
51. Vogt, G. Synthesis of digestive enzymes, food processing, and nutrient absorption in decapod crustaceans: A comparison to the mammalian model of digestion. *Zoology* **2021**, *147*, 125945. [CrossRef] [PubMed]
52. Dai, B.; Hou, Y.; Hou, Y.; Qian, L. Effects of multienzyme complex and probiotic supplementation on the growth performance, digestive enzyme activity and gut microorganisms composition of snakehead (*Channa argus*). *Aquac. Nutr.* **2019**, *25*, 15–25. [CrossRef]
53. Fang, H.; Xie, J.; Liao, S.; Guo, T.; Xie, S.; Liu, Y.; Tian, L.; Niu, J. Effects of Dietary Inclusion of Shrimp Paste on Growth Performance, Digestive Enzymes Activities, Antioxidant and Immunological Status and Intestinal Morphology of Hybrid Snakehead (*Channa maculata* ♀ × *Channa argus* ♂). *Front. Physiol.* **2019**, *10*, 1027. [CrossRef] [PubMed]
54. Lakwani, M.A.S.; Kenanoğlu, O.N.; Taştan, Y.; Bilen, S. Effects of black mustard (*Brassica nigra*) seed oil on growth performance, digestive enzyme activities and immune responses in rainbow trout (*Oncorhynchus mykiss*). *Aquac. Res.* **2022**, *53*, 300–313. [CrossRef]
55. Huang, B.; Zhang, S.; Dong, X.; Chi, S.; Yang, Q.; Liu, H.; Tan, B.; Xie, S. Effects of fishmeal replacement by black soldier fly on growth performance, digestive enzyme activity, intestine morphology, intestinal flora and immune response of pearl gentian grouper (*Epinephelus fuscoguttatus* ♀ × *Epinephelus lanceolatus* ♂). *Fish. Shellfish. Immunol.* **2022**, *120*, 497–506. [CrossRef] [PubMed]

56. Ruenkoed, S.; Nontasan, S.; Phudkliang, J.; Phudinsai, P.; Pongtanalert, P.; Panprommin, D.; Mongkolwit, K.; Wangkahart, E. Effect of dietary gamma aminobutyric acid (GABA) modulated the growth performance, immune and antioxidant capacity, digestive enzymes, intestinal histology and gene expression of Nile tilapia (*Oreochromis niloticus*). *Fish. Shellfish. Immunol.* **2023**, *141*, 109056. [CrossRef]
57. Sokooti, R.; Chelemal Dezfoulnejad, M.; Javaheri Baboli, M.; Askary Sary, A.; Mabudi, H. The effects of probiotics-supplemented diets on Asian sea bass (*Lates calcarifer*): Growth performance, microbial flora, digestive enzymes activity, serum biochemical and non-specific immune indices. *Aquac. Res.* **2022**, *53*, 5500–5509. [CrossRef]

Article

Dietary β-1,3-Glucan Promotes Growth Performance and Enhances Non-Specific Immunity by Modulating Pattern Recognition Receptors in Juvenile Oriental River Prawn (*Macrobrachium nipponense*)

Tailei Xu, Junbao Wang, Hao Xu, Zifan Wang, Yujie Liu, Hongfeng Bai, Yixiang Zhang, Youqin Kong, Yan Liu and Zhili Ding *

College of Life Science, Huzhou University, Huzhou 313000, China; xtl1920147742@outlook.com (T.X.); wfurlong@163.com (J.W.); waibixu@163.com (H.X.); 158913777@163.com (Z.W.); 03235@zjhu.edu.cn (Y.L.); bhf2003@zjhu.edu.cn (H.B.); yxzhang@zjhu.edu.cn (Y.Z.); susankuq@zjhu.edu.cn (Y.K.); 02870@zjhu.edu.cn (Y.L.)
* Correspondence: dingzhili@zjhu.edu.cn; Tel.: +86-138-6724-3500

Abstract: As a typical pathogen-associated molecular pattern (PAMP), β-1,3-glucan can engage with pattern recognition receptors (PRRs) to initiate an immune response. In this study, we investigated the effects of dietary β-1,3-glucan on growth performance, antioxidant capacity, immune response, intestinal health, and bacterial resistance in juvenile *Macrobrachium nipponense*. Prawns were fed with five experimental diets containing 0%, 0.05%, 0.1%, 0.2%, and 0.4% β-1,3-glucan for eight weeks. The findings demonstrated that the inclusion of β-1,3-glucan improved weight gain and survival rate in prawns. Prawns fed with β-1,3-glucan exhibited elevated activities of hepatopancreatic ACP (acid phosphatase), AKP (alkaline phosphatase), and SOD (superoxide dismutase), while MDA (malondialdehyde) content was reduced. Expression levels of PRRs related genes including *LGBP* (lipopolysaccharide and β-1,3-glucan binding protein), *lectin*, and *LBP* (lipopolysaccharide-binding protein) were significantly increased in prawns fed with β-1,3-glucan. Intestinal flora analysis revealed suppression of Cyanobacteria abundance at the Phylum level and enhancement in Rhodobacter abundance at the genus level in prawns fed with a 0.2% β-1,3-glucan diet. Furthermore, prawns fed with 0.1%, 0.2%, and 0.4% β-1,3-glucan demonstrated significantly higher survival rates following *Aeromonas hydrophila* infection. In conclusion, β-1,3-glucan can activate PRRs to improve immune responses in *M. nipponese*. Within the range of β-1,3-glucan concentrations set in this experiment, it is recommended to add 0.18% of β-1,3-glucan to the diet, taking into account the positive effect of β-1,3-glucan on the survival rate of *M. nipponensecu*.

Keywords: *Macrobrachium nipponense*; β-1,3-glucan; PRRs; non-specific immunity; intestinal flora; *Aeromonas hydrophila*

Key Contribution: The potential mechanism of β-1,3-glucan in enhancing the immune capacity of *M. nipponense* was investigated with a focus on pattern recognition protein receptors and intestinal health. The optimal levels of β-1,3-glucan supplementation were determined based on survival rates.

Citation: Xu, T.; Wang, J.; Xu, H.; Wang, Z.; Liu, Y.; Bai, H.; Zhang, Y.; Kong, Y.; Liu, Y.; Ding, Z. Dietary β-1,3-Glucan Promotes Growth Performance and Enhances Non-Specific Immunity by Modulating Pattern Recognition Receptors in Juvenile Oriental River Prawn (*Macrobrachium nipponense*). *Fishes* **2024**, *9*, 379. https://doi.org/10.3390/fishes9100379

Academic Editor: David Sánchez Peñaranda

Received: 29 August 2024
Revised: 22 September 2024
Accepted: 23 September 2024
Published: 26 September 2024

1. Introduction

The aquaculture industry is experiencing rapid global growth and it is recognized as the fastest-growing sector in worldwide food production, accompanied by a surge in demand for fish products in the consumer market [1]. Ensuring the sustainable health of aquatic organisms is an important endeavor for the human food supply chain. Disease outbreaks are widely acknowledged as a major impediment to the growth of the aquaculture industry, resulting in substantial global losses annually [2]. Antibiotics are now widely used in the aquaculture industry to minimize losses of aquatic products caused by bacterial pathogens [3]. Yet, antibiotic-resistant (AMR) bacteria are also born out of the feed industry's practice of indiscriminate and constant abuse of production [4], its genes and residues come with it [5]. When AMR bacteria evolve and spread in natural water bodies, they not only cause disease and death in aquatic animals but also have a profoundly negative impact on economic development [6]. Leading some countries to impose restrictions or bans on the widespread antimicrobial agent usage in aquaculture [7]. Consequently, there is an imperative need for non-toxic, environmentally friendly, and highly efficient biological agents that can serve as alternatives to antibiotics within the aquaculture industry. Immunostimulants represent one such option that can effectively mitigate the outbreak of aquatic animal diseases [8]. Understanding natural immune responses in crustaceans plays an important role in studies applying immunotherapy to enhance disease resistance among cultured animals.

Crustaceans are known to lack an acquired immune system, relying mainly on innate immunity [9]. PRRs (pattern recognition receptors) enable the immune system to distinguish between "self" and not-self", playing a crucial role in initiating the immune response [10]. Various PRRs have been established in crustaceans, including *LBPs* (lipopolysaccharide-binding proteins) [11], *LGBPs* (lipopolysaccharide and β-1,3-glucan binding proteins) [12], *GNBPs* (gram-negative binding proteins) [13], *PGRPs* (peptidoglycan recognition proteins binding proteins) [14], *bGBPs* (β-glucan binding proteins) [15], *CTLs* (C-type lectins) [16], *TEPs* (sulfate-containing proteins) [17], *SCRs* (scavenger receptors) [18], *FBGLs* (fibrinogen-like structural domain immunoselectors) [19], and *TLRs* (Toll-like receptors) [20]. Upon recognition of corresponding PAMPs, PRR-PAMP interactions trigger a cascade of immune reactions. Phagocytosis mediated primarily by hyalinocytes eliminates foreign invasive substances [21]; primarily encapsulation and cytotoxicity predominantly carried out by semigranular cells counteract parasites and fungi invading the host [22]; activation of Toll and IMD (immune deficiency) signaling pathways regulates antimicrobial peptides production [23,24]; *lectin* secretion and proPO-AS (Prophenoloxidase activating system) activation occur as well [25,26]. Furthermore, we observed a correlation between immunization and growth performance in crustaceans [27–29]. Therefore, it is worthwhile to explore the research directions that stimulate the interaction between PRRs and PAMPs using immunostimulants to enhance environmental resistance and growth performance in crustaceans.

Immunostimulants are effective in boosting non-specific immune responses and improving overall immune system function [30]. Numerous studies have demonstrated that incorporating immunostimulants into feeds leads to improved immunity in crustaceans [31–33]. One prominent immunostimulant in crustaceans is β-1,3-glucans, a carbohydrate group synthesized by the cell walls of plants, bacteria, fungi, and some freshwater algae. These glucans are well-sourced and inexpensive [34]. They can be either branched or linear polymers (polysaccharides) composed of repeating glucose units linked by β-glycosidic bonds. Remarkably, previous studies have shown that β-1,3-glucan acts as a typical PAMP interacting with PRRs present in immune cells to trigger an innate immune response [35,36]. As a dietary additive, β-1,3-glucan is widely used in aquatic animals to bolster immunity, enhance disease resistance, and stimulate growth [37]. For instance, feeding juvenile tiger shrimp (*Penaeus esculentus*) with diets containing β-glucan and mannan oligosaccharide improves immune response and survival when exposed to WSSV (*white Spot Syndrome Virus*) infection [38], while feeding *M. nipponense* diets enriched with β-1,3-glucan not only improves its innate immunity and growth performance but also modulates the diversity of gut microbiota for maintaining gut health [39,40]. In another study involving *Eriocheir sinensis* fed β-1,3-glucan followed by *Vibrio parahaemolyticus* infestation after knocking out *LGBP* at the entry point of PRRs for β-1,3-glucan resulted in increased mortality rates [41]. Similarly, higher expression levels of LGBP were detected in the hepatopancreas of *Litopenaeus vannamei* fed a diet supplemented with β-glucan [42]. These findings highlight the prominent research interest surrounding the use of β-1,3-glucans as trace additives in aquaculture. Therefore, β-1,3-glucan represents a promising micro-additive for immune-enhancing feeds with the potential to reduce and replace the dependence on antibiotics in the aquaculture industry.

In China and other Asian countries, the oriental river prawn, *M. nipponense*, is widely distributed in freshwater basins. Specifically in China, the annual production of this prawn is 240,739 tons, making it one of the most commercially viable species [43–45]. However, intensive prawn farming in *M. nipponense* has led to deteriorating water quality and environmental stress, making prawns susceptible to bacterial diseases, mass death events caused by *Aeromonas veronii* [46], and "red gill disease" from *M. nipponense* [47]. Due to multiple adverse effects associated with antibiotic usage [48], many countries have prohibited their use in aquatic organisms intended for human consumption like fish and shrimp due to concerns over negative impacts on human health, development of antibiotic-resistant strains, reduced efficacy from repeated applications, and environmental contamination issues [49–51]. Consequently, finding viable alternatives that are effective yet somewhat economical and environmentally safe has become an urgent requirement for restraining antibiotic usage in *M. nipponense* farming practices. This study sought to evaluate the impacts of β-1,3-glucan diets on growth capability, antioxidant power, immune response, digestive well-being, and immunity against *M. nipponense* bacterial infections. Practical insight into alternative treatment options for ensuring sustainability within *M. nipponense* aquaculture is provided by the results obtained.

2. Materials and Methods

2.1. The Origin of Prawns and β-1,3-Glucan

The prawns selected in this experiment were healthy, with similar growth age, and no obvious scar. The prawns were all supplied by Wuyue Agricultural Co., Ltd. (Huzhou, China). The β-1,3-glucan utilized in the study was obtained from Nanjing Taixin Biotechnology Co., Ltd. (Nanjing, China) with a purity of 95.21%.

2.2. Experimental Design and Dietary Composition

The basal diet was made as an isonitrogenous and isocaloric diet using fishmeal, soybean meal, and rapeseed meal as protein sources, and fish oil and soybean oil as lipid sources. The experimental diets were prepared by supplementing the basic diet with 0% (control), 0.05%, 0.1%, 0.2%, and 0.4% β-1,3-glucan, referred to as B0, B1, B2, B3, and B4, respectively (Table 1).

In order to create experimental diets, all ingredients need to be processed individually. The initial step involved grinding the larger particles of the ingredients. Subsequently, a more homogeneous mixture of particles and powders was obtained using a 212 μm sieve screen. Following this, it was ensured that the individually bagged powders exhibited uniform and consistent quality. The individual ingredients were accurately weighed according to the list of diet ingredients (Table 1) using a 0.1 mg electronic scale. The finished weighed solid ingredients were added to the mixing vessel with accurately weighed semi-solid and liquid ingredients (fish oil, soybean oil, and distilled water). The mix was produced in preliminary strip form using a twin-screw extruder (School of Chemical Engineering, South China University of Technology, Guangzhou, China) and finally dried in a forced-air oven at 40 °C. The dried feeds were stored in sealed bags at −20 °C.

The determination of moisture, crude lipid, and crude protein contents in each experimental diet was conducted by using standard procedures [52]. To determine moisture, the sample was dried at 105 °C for 24 h to a constant weight. The Dumas nitrogen determination apparatus combustion method was utilized to determine the crude protein content. The ash content was determined by setting the parameter 550 °C in a Muffle furnace for 6 h and Soxhlet extraction method was employed to measure total lipid (Table 1).

correlated and analyzed using the RDP Classifier, which sets a confidence threshold of 70%. Finally, community diversity and species richness estimates were created in Mothur in version v.1.30.1.

2.8. mRNA Expression Analysis

The trizol method was used to extract the total RNA from the hepatopancreas samples of 5 prawns from each tank. The Thermo NanoDrop 2000 nucleic acid and protein analyzer (Thermo Fisher Scientific, Waltham, MA, USA) was used to determine the concentration and purity of RNA. Total RNA was reverse transcribed to cDNA under the instructions of the Reverse Transcription Kit (Takara, Japan) and stored in a refrigerator at -20 °C for future use. Quantitative real-time polymerase chain reaction (qRT-PCR) was used to examine the mRNA expression of immune-related genes in the hepatopancreas. The analyzed genes include lipopolysaccharide and *LGBP*, *lectin*, *LBP*, tumor *p53*, and cysteine-aspartic acid protease 3 (*caspase 3*). The mixture reaction system was 20 L, including 7.6 μL H_2O, primers (10 μM) of 0.2 L each, 10 μL 2× SYBR Green Premix Ex Taq, and 2 μL cDNA. The qRT-PCR instruments were as follows: 10 min at 95 °C, followed by 40 cycles of 10 s at 94 °C, 30 s at 58 °C, and 32 s at 72 °C. The mixture reaction system was used to generate the appropriate melting curves after the qRT-PCR reaction to determine the correctness of the results. Three replicate assays were completed each time including a negative control lacking cDNA, and amplification efficiencies were guaranteed to be between 95% and 103%, with correlation coefficients greater than 0.98 for each gene. 18S ribosomal RNA (18S rRNA) was used as an internal reference gene The primer sequences of the genes used can be seen in Table 2. Gene mRNA expression was calculated using the $2^{-\Delta\Delta CT}$ [53] quantification method. The cDNA sequences of the oriental river prawn from Genbank were used to design the primers and then synthesized by Sangon Biotech Co., Ltd. (Shanghai, China).

Table 2. Primers used in this study.

Primer	Sequence (5′→3′)	GenBank	Product
(Mn) *18S* rRNA-S (Mn) *18S* rRNA-A	TACTGCTGAGCCGAAGAT CCACGGACTATTACTACCTAC	EU118285.1	159
(Mn) *lectin*-S (Mn) *lectin*-A	AAGGGCAAGGTGTCTCTTCG CCTCCCATGGTGTCCATGTC	PP516428	160
(Mn) *LBP*-S (Mn) *LBP*-A	GTCTGTCTAGCAAGGGCGTT AGTGTTGATGCGATGAGCGA	PP516429	159
(Mn) *LGBP*-S (Mn) *LGBP*-A	CTGCTGATATCGTCGACCCC GGCATAGCTGATGCTACGGT	AGF86400.1	167
(Mn) *P53*-S (Mn) *P53*-A	TGCTTGCTCACAGCGATAAACTT AGTCGCCGAGTGTCAAGTCAATAT	KT963043.1	112
(Mn) *caspase 3*-S (Mn) *caspase 3*-A	TTGTCATGCAGTACTTGACTGAAGC CCTCATGGGTTGTGCATCATTATA	KX651496.1	171

2.9. Data Analysis and Processing

The Kolmogorov–Smirnov test was used to determine data normality, while Levene's test was used to test for variance homogeneity. After confirming the parametric assumptions, the effect of different diets that contain β-1,3-glucans on the measured parameters of *M. nipponense* was evaluated using an analysis of variance (ANOVA).

After the prerequisite ANOVA showed significant differences, Tukey's multiple range test was used to compare means between groups. Third orthogonal polynomials were used to analyze the survival rate of bioinfestation experiments to correctly estimate the optimal amount of β-1,3-glucans to be added. The cubic equation in this experiment had the highest R^2 value and the lowest *p*-value when analyzed as a mapping exercise, so polynomial regression analyses were used to determine the optimal ratio of β-1,3-glucan required for optimal survival. The Mantle–Cox test was employed to evaluate survival rates for prawns infected with *A. hydrophila*. Statistical analyses were performed using SPSS 25.0 (Chicago, IL, USA). The mean ± standard deviation (Mean ± SD) is the way results are expressed.

Alpha diversity indexes including Chao1 reflected community richness while Shannon and Simpson indices represented community diversity levels. Coverage analysis determined whether the identified 16S rRNA gene sequences accounted for a majority presence of bacteria in samples.

3. Results

3.1. Growth Performance of Prawns

Different levels of dietary β-1,3-glucan significantly influenced the growth performance of prawns (Table 3; $p < 0.05$). The results indicate that there were no significant differences in FBW, WG, and SGR among prawns fed B0, B3, and B4 diets ($p > 0.05$), while both B1 and B2 showed higher values compared to B4, with B2 being notably superior to the other groups ($p < 0.05$). Prawns fed on the B2 and B3 diets exhibited an increase in survival rate compared to those fed on the B0, B1, and B4 diets ($p < 0.05$). Overall, prawn FBW, WG, SGR, and SR showed an increasing trend followed by a decline with the increasing β-1,3-glucan supplementation in their diets. Utilizing the SR data for linear regression analysis revealed the optimal amount of dietary β-1,3-glucan for juvenile *M. nipponense* was 0.18% (Figure 1).

Table 3. Growth performance of juvenile *M. nipponense* fed with experimental diets containing different concentrations of β-1,3-glucan for 8 weeks.

Diet	IBW (g)	FBW (g)	WG (%)	SGR (%/day)	SR (%)
B0	0.10 ± 0.01	0.51 ± 0.05 [ab]	391.00 ± 47.33 [ab]	2.91 ± 0.17 [abc]	58.10 ± 0.83 [b]
B1	0.10 ± 0.01	0.57 ± 0.01 [bc]	446.00 ± 5.20 [bc]	3.10 ± 0.02 [bc]	60.48 ± 2.18 [b]
B2	0.10 ± 0.01	0.60 ± 0.01 [c]	480.67 ± 14.47 [c]	3.20 ± 0.04 [c]	69.52 ± 2.97 [c]
B3	0.10 ± 0.01	0.50 ± 0.07 [ab]	375.52 ± 78.21 [ab]	2.84 ± 0.28 [ab]	69.05 ± 0.82 [c]
B4	0.10 ± 0.01	0.46 ± 0.04 [a]	341.62 ± 42.74 [a]	2.72 ± 0.17 [a]	48.09 ± 6.75 [a]

IBW: initial body weight, FBW: final body weight, WG: weight gain, SGR: specific growth rate, SR: survival rate. The value is the average of the three replicates for each treatment ± SD. The presence of different superscript letters in the same column indicates significant differences ($p < 0.05$).

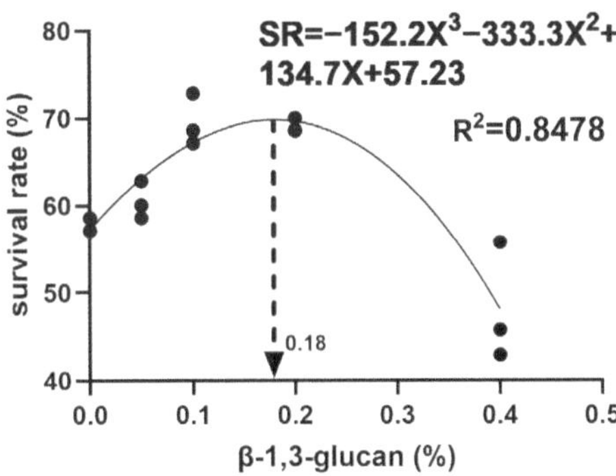

Figure 1. Polynomial regression analysis on survival rate to estimate the optimal β-1,3-glucan supplementation levels in the diet of oriental river prawns (*M. nipponense*).

3.2. Survival Rate of the Prawns after A. hydrophila Challenge Test

The cumulative survival rates of *M. nipponense* over a 72 h period are depicted in Figure 2 prawn mortality was primarily observed within the first 12 h, with only isolated individual deaths occurring after 24 h, specifically in groups B2 and B3. The descending order of survival rates for each group was as follows: B2 (79%), B3 (76%), B4 (69%), B1 (63%), and B0 (49%). Notably, groups B2, B3, and B4 exhibited higher survival rates compared to group B0 ($p < 0.05$).

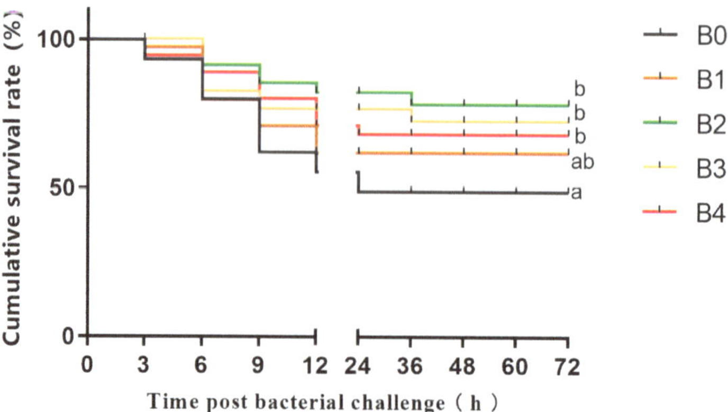

Figure 2. Survival rate of juvenile *M. nipponense* infected with *A. hydrophila* for 72 h. Different colors represent varying concentrations of β-1,3-glucan. The presence of different letters in the figure indicates significant differences ($p < 0.05$).

3.3. Innate Immune Response and Antioxidant Enzyme Activity in the Hepatopancreas of Prawns

The dietary intake of β-1,3-glucan significantly impacts the activities of immune and antioxidant-related enzymes in prawn hepatopancreas ($p < 0.05$; Figure 3). Feeding prawns with diets B1, B2, B3, and B4 resulted in a decrease in MDA content in hepatopancreas compared to those fed on diet B0 ($p < 0.05$; Figure 3A). Prawns fed on diets B1, B2, and B4 exhibited higher SOD activities than those fed on diet B0 ($p < 0.05$; Figure 3B). Additionally, prawns fed on diet B1 showed a higher SOD activity compared to other groups ($p < 0.05$), while prawns fed on diet B3 did not show a significant difference in SOD activity compared to the control group.

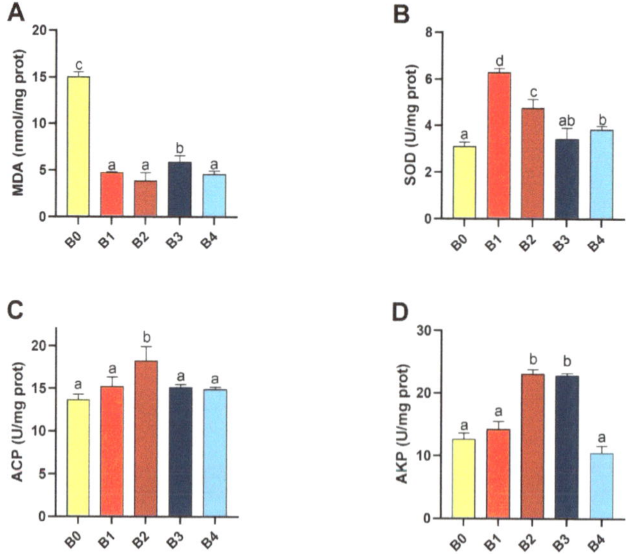

Figure 3. Antioxidant enzyme activities and innate immune response in the hepatopancreas of juvenile *M. nipponense* fed different concentrations of β-1,3-glucan. (A) Malondialdehyde (MDA) content, (B) superoxide dismutase (SOD) activity, (C) Acid phosphatase (ACP) activity, (D) Alkaline phosphatase (AKP) activity. Values are means for three replicates for each treatment ± SD. Bars with different superscript letters are statistically significant ($p < 0.05$).

Prawns fed on diet B2 demonstrated higher ACP activity in the hepatopancreas than all other experimental diets ($p < 0.05$; Figure 3C). Furthermore, there was no significant difference in ACP activity among prawns fed on other diets. Prawns fed on both diet types B2 and B3 displayed an increase in AKP activity compared to all other experimental diets ($p < 0.05$; Figure 3D). Moreover, there was no statistically significant difference observed in the AKP activity of prawns that were administered either B2 or B3 diets.

3.4. The mRNA Expression of Genes Related to PRRs and Apoptosis

The mRNA expression of genes related to PRRs and apoptosis in the hepatopancreas was significantly affected by feeding prawns with different levels of dietary β-1,3-glucan ($p < 0.05$; Figure 4). Compared to the control diet, prawns fed on the β-1,3-glucan diet showed an upregulation in mRNA relative expression of the *LGBP* gene in the hepatopancreas ($p < 0.05$; Figure 4A). Furthermore, compared to those fed on all other diets, prawns fed on diet B2 exhibited an increase in mRNA expression of the *LGBP* gene ($p < 0.05$). The *lectin* gene expression in prawns fed on diets B1 and B2 was elevated compared to those fed on diets B0, B3, and B4 ($p < 0.05$; Figure 4B). Additionally, the *lectin* gene expression was increased in prawns fed on diet B2 compared to those fed on diet B1 ($p < 0.05$; Figure 4B).

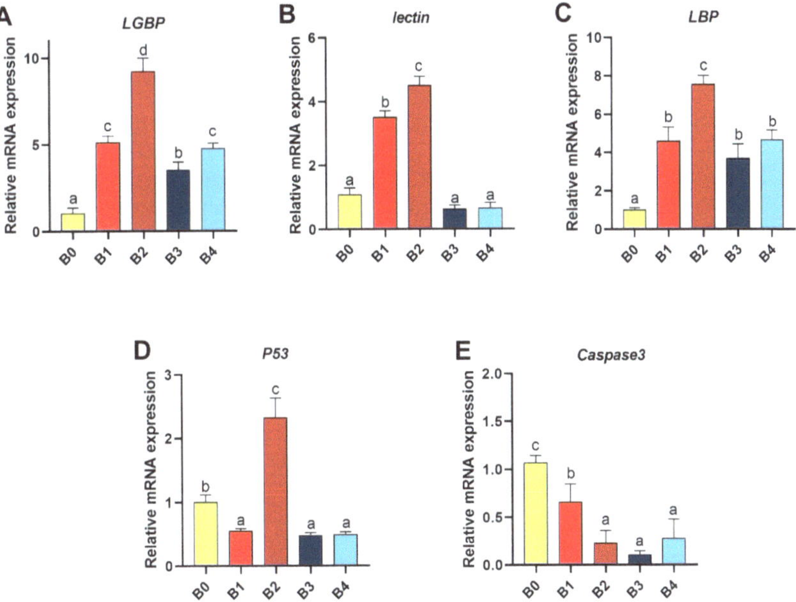

Figure 4. Effect of β-1,3-glucan on mRNA expression of immune-related genes in the hepatopancreas of juvenile oriental river prawns (*M. nipponense*) fed diets supplemented with different concentrations of β-1,3-glucan for eight weeks. (**A**) lipopolysaccharide and β-1,3-glucan binding protein (*LGBP*), (**B**) lectin, (**C**) lipopolysaccharide-binding protein (*LBP*), (**D**) tumor suppressor protein p53 (*p53*), (**E**) cysteine-aspartic acid protease 3 (*caspase 3*). Values are means ± SD for three replicates for each treatment. Bars with different superscript letters are statistically significant ($p < 0.05$).

The prawns fed on the β-1,3-glucan diet upregulated the mRNA relative expression of the *LBP* gene in the hepatopancreas compared to the control diet ($p < 0.05$; Figure 4C). Moreover, prawns fed on diet B2 exhibited an enhancement in mRNA expression of the *LBP* gene compared to those fed on all other diets ($p < 0.05$).

Interestingly, *P53* gene mRNA expression was down-regulated in both diets B1, B3, and B4 compared to diet B0 ($p < 0.05$; Figure 4D). However, prawns fed on diet B2 showed an increase in mRNA expression of the *P53* gene compared to those fed on all other diets

($p < 0.05$). Prawns fed with the β-1,3-glucan diet exhibited an increase in *Caspase3* mRNA expression compared to those in the control group ($p < 0.05$; Figure 4E).

3.5. Gut Microbiota

The groups B0, B2, and B4 were selected for high-throughput sequencing analysis based on the increasing dietary β-1,3-glucan additions. The results presented in Table 4 revealed a total of 274,131 sequences across the three groups (B0, B2, and B4), with an average of 91,377 sequences per sample. The coverage within each group exceeded 99.8%, indicating that the obtained 16S rRNA gene sequences represented the majority of bacterial populations present in the samples. No significant difference was observed in sequence numbers among the B0, B2, and B4 groups. Similarly, there were no significant variations in Chao indices or Shannon and Simpson diversity measures between the three groups.

Table 4. Diversity of intestinal flora of juvenile oriental river prawn (*M. nipponense*) in groups B0, B2, and B4.

Diet	Sequences	Chao	Shannon	Simpson	Coverage (%)
B0	89,538 ± 16,475	925.66 ± 97.81	4.53 ± 0.12	0.045 ± 0.02	99.87 ± 0.06
B2	94,935 ± 27,033	763.73 ± 58.44	4.47 ± 0.21	0.033 ± 0.01	99.93 ± 0.02
B4	89,658 ± 56,480	829.08 ± 77.77	4.40 ± 0.12	0.044 ± 0.01	99.90 ± 0.03

The value is the average of the three replicates for each treatment ± SD.

Using a relative abundance threshold of 1%, we classified the dominant bacterial phyla and genera in the gut of prawns fed on experimental diets (Tables 5 and 6) based on phylum (Figure 5A) and genus (Figure 5B). The dominant phyla observed were Proteobacteria, followed by Cyanobacteria, Actinobacteria, and Patescibacteria (Table 5; Figure 5A). Although not statistically significant, a slight decrease in Actinobacteria and Patescibacteria abundance was observed in prawns fed diets B2 and B4 compared to those fed diet B0 (Table 5).

Table 5. The relative abundance of major phylum in intestinal flora of juvenile *M. nipponense* in groups B0, B2, and B4 fed experimental diets.

Phylum	Experimental Diet		
	B0	**B2**	**B4**
Proteobacteria	39.31 ± 14.65	56.99 ± 12.11	54.01 ± 13.98
Cyanobacteria	23.27 ± 7.37	15.55 ± 17.82	24.44 ± 13.09
Actinobacteria	10.24 ± 4.10	5.90 ± 2.56	5.98 ± 0.51
Patescibacteria	8.40 ± 4.05	5.67 ± 2.34	4.39 ± 1.06

The value is the average of the three replicates for each treatment ± SD.

Table 6. The relative abundance of major genera in intestinal flora of juvenile *M. nipponense* in groups B0, B2, and B4 fed experimental diets.

Genus	Experimental Diet		
	B0	**B2**	**B4**
Chloroplast_norank	12.03 ± 8.31	10.48 ± 12.76	10.60 ± 2.89
Candidatus_Hepatincola_norank	10.37 ± 10.87	7.29 ± 3.28	16.91 ± 2.87
Rhodobacteraceae_unclassified	2.24 ± 0.90 [a]	8.54 ± 3.78 [b]	4.20 ± 3.21 [ab]
Saccharimonadales_norank	7.44 ± 3.66	5.14 ± 2.07	4.00 ± 1.23
Rhodobacter	2.48 ± 0.67 [a]	9.67 ± 3.18 [b]	6.36 ± 4.47 [ab]

The value is the average of the three replicates for each treatment ± SD. The presence of different superscript letters in the same column indicates significant differences ($p < 0.05$).

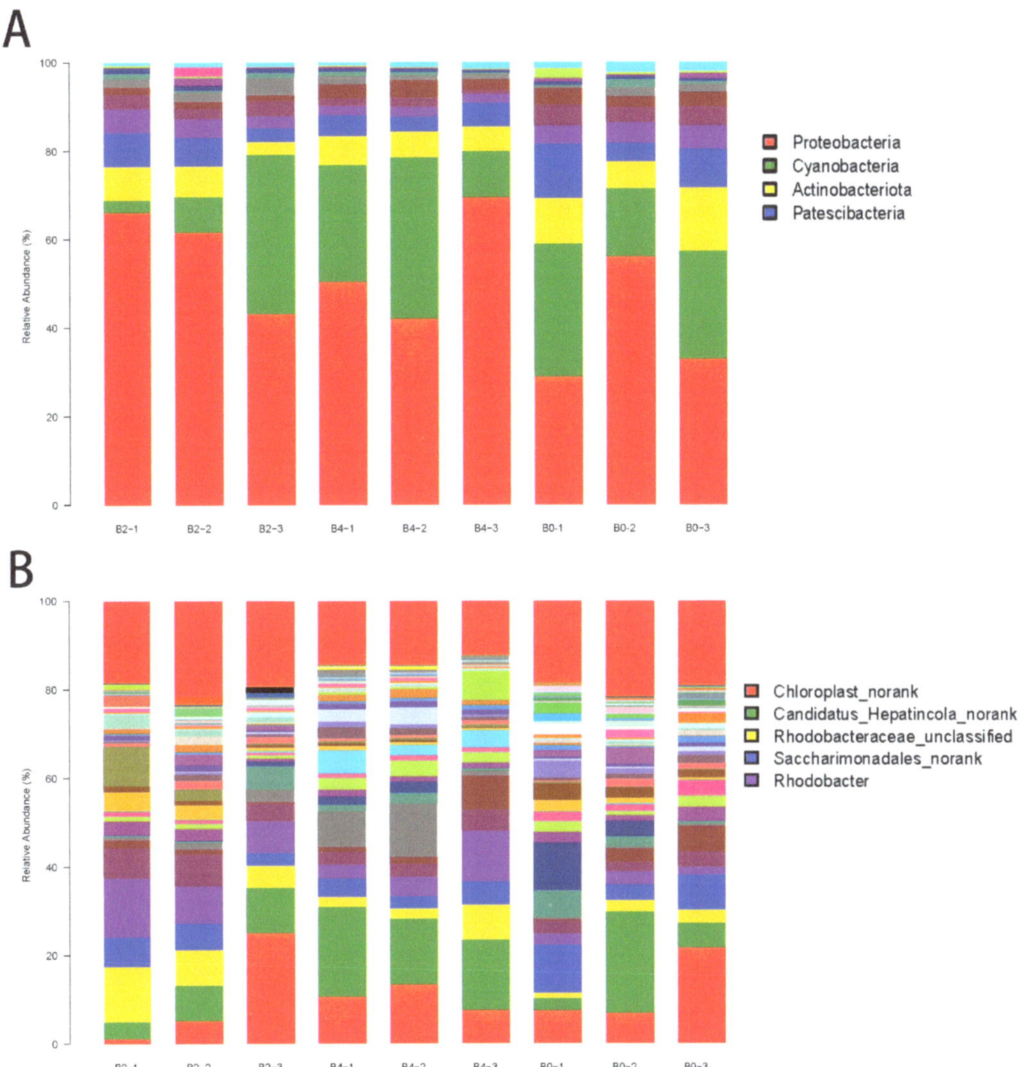

Figure 5. Histogram of intestinal microorganism distribution in juvenile oriental river prawns (*M. nipponense*) supplemented with different dietary concentrations of β-1,3-glucan. (**A**) phylum taxonomic level; (**B**) genus taxonomic level.

The dominant genera in the gut of the prawns fed on the experimental diets were primarily *Chloroplast_norank*, followed by *Candidatus_Hepatincola_norank*, *Rhodobacteraceae_ unclassified*, *Saccharimonadales_norank,* and *Rhodobacter* (Table 6; Figure 5B). There were no significant differences observed in the abundance of Chloroplast, Candidatus, and *Saccharimonadales_norank* among prawns fed on all diets. Interestingly, the prawns fed on diet B2 showed a higher abundance of *Rhodobacteraceae_unclassified* and *Rhodobacter* compared to those fed on diets B0 and B4 ($p < 0.05$).

4. Discussion

In the present study, the inclusion of dietary β-1,3-glucan enhanced the weight gain and survival rate of *M. nipponense.* The improved growth performance and survival rate observed in *M. nipponense* may be attributed to enhanced absorption and accumulation of nutrients as well as improved innate immune response facilitated by β-1,3-glucan supplementation. Despite being underutilized in practical agriculture due to a lack of scientific strategies for its application, our study demonstrates that incorporating β-1,3-glucan into feed rations can effectively enhance growth performance and survival rates, highlighting its potential for superior production applications. Similar findings have been reported with *L. vannamei* where the addition of 0.02% or 0.04% β-glucan resulted in increased weight gain [35,54]. Furthermore, β-glucan has also shown significant improvements in weight gain and survival rate of *Penaeus monodon* [55]. Notably, the present study revealed a trend wherein the survival rate of *M. nipponense* initially increased but then decreased with increasing levels of added β-1,3-glucan. Therefore, we recommend 0.18% β-1,3-glucan in *M. nipponense* diets to ensure optimal survival rate.

Oxygen is known to produce ROS, which are harmful to living organisms. [56]. Lipid peroxidation is a common injury phenomenon induced by ROS, resulting from the oxidation of polyunsaturated fatty acids that are susceptible to attack by free radicals [57]. Therefore, organisms possess enzymatic antioxidant defenses to minimize the detrimental effects of ROS [58]. SOD plays a crucial role as an important component of the antioxidant enzyme system in biological systems and acts as a key enzyme in the elimination of ROS [59]. MDA serves as an end product of lipid peroxidation and indirectly reflects the tissue damage caused by peroxidation [60]. Hence, MDA content and SOD activity in the hepatopancreas can serve as an indicator for assessing the antioxidant capacity of prawns. In this study, feeding prawns with a β-1,3-glucan diet resulted in increased SOD activity but decreased MDA levels in *M. nipponense*. These findings demonstrate that β-1,3-glucan enhances the antioxidant capacity of *M. nipponense*. Similar results regarding enhanced antioxidant capacity after β-1,3-glucan supplementation have also been reported in swimming crabs (*Portunus trituberculatus*) [61] and *Apostichopus japonicus* [62]. Additionally, our experiment results indicate that β-1,3-glucan can increase AKP and ACP activities in the hepatopancreas of *M. nipponense*. The activity level initially increases with increasing amounts of added β-1,3-glucan but eventually decreases thereafter; with 0.2% being associated with the highest observed activity level among all tested concentrations. This observation can be attributed to crustaceans possessing a large and complex innate immune system [63], wherein AKP and ACP play vital roles as phosphatases involved in non-specific immunity responses [64]. Previous studies have shown increased ACP and AKP activities when red swamp crayfish (*Procambarus clarkii*) [36] and Ussuri catfish (*Pseudobagrus ussuriensis*) [65] were fed diets supplemented with β-glucan. These findings demonstrate that the inclusion of appropriate β-1,3-glucan enhances both antioxidant capacity and nonspecific immunity in *M. nipponense*. Notably, the improved non-specific immunity holds promise for reducing antibiotic usage during infectious disease outbreaks in prawn farming, thereby promoting sustainable development of prawn aquaculture.

Cumulative survival following bacterial infection serves as a straightforward indicator to assess disease resistance in aquatic animals and evaluate the efficacy of cultivating biological health and enhancing immunity [12]. Importantly, *M. nipponense* fed with β-1,3-glucan diets exhibited a distinct increase in cumulative survival after *A. hydrophila* infection; moreover, cumulative survival displayed an increasing trend followed by a decrease upon the addition of β-1,3-glucan. These observations suggest that appropriate amounts of addition of β-1,3-glucan can enhance the disease resistance of *M. nipponense*. Similarly, *Micropterus salmoides* showed higher survival after *Aeromonas schubertii* infection when fed diets supplemented with β- glucan [66], while juvenile tiger shrimp demonstrated increased survivability after WSSV infection when fed diets enriched with β-glucan [67]. Collectively, these results indicated that to some extent β-1,3-glucan has the potential to mitigate aquatic disease outbreaks caused by pathogenic microorganisms. The fundamental role of the immune system is to distinguish between "self" and "non-self". Crustaceans, known for their innate immunity, heavily rely on the recognition of "non-self" [25]. Recent findings indicate that this recognition is facilitated by certain PRRs, which can be soluble or membrane-bound [68]. A critical step in initiating the innate immune response involves the mutual recognition and interaction between PRRs and PAMPs present on the surface of pathogens and absent in the host [69]. β-1,3-glucan found on fungal cell walls represents an important class of PAMPs [70], capable of binding to common crustaceans PRRs such as *LBP* [11], *bGBP* [14], *LGBP* [12], and *lectin* [71]; this interaction subsequently triggers proPO (Prophenoloxidase), a widely distributed enzyme in blood and inner tissues of carapace, leading to activation of proPO-AS [41]. Therefore, quantitatively analyzing mRNA expression levels of *LBP*, *LGBP* and *lectin* in prawn hepatopancreas can serve as an effective tool for assessing prawn immune response status. Our results further demonstrate that feeding the prawns with β-1,3-glucan upregulates mRNA expression levels of *LBP*, *LGBP,* and *lectin* genes. Interestingly, numerous studies have investigated the relationship between β-glucan and aquatic animals, with PO activity often being associated with certain PRRs-related genes. Therefore, we hypothesized that upregulating the expression of *LBP*, *LGBP*, and *lectin* genes could enhance the immunity of *M. nipponense* by increasing PO activity. For instance, *L. vannamei* fed a diet supplemented with β-glucan exhibited simultaneous enhancements in both PO activity and *LGBP* gene expression [12], while *Macrobrachium rosenbergii* fed a similar diet showed concurrent increases in both PO activity and *lectin* gene expression [72]. Furthermore, our study found that the β-1,3-glucan group had a significantly higher cumulative survival rate 72 h after *A. hydrophila* infection compared to the control group. Hence, we postulated that *M. nipponense* could activate proPO-AS by enhancing the expression of *LBP*, *LGBP*, and *lectin* genes through β-1,3-glucan supplementation to resist *A. hydrophila* infestation. Previous research has also demonstrated significant induction of crustacean *LGBP* transcripts following challenges with bacteria such as *Vibrio harveyi* [73], *Vibrio alginolyticus* [74], and *Vibrio anguillarum* [75]. According to the results of the present study, the addition of moderate amounts of β-1,3-glucan to the diet improved the non-specific immunity of *M. nipponense*.

Caspases are closely associated with eukaryotic apoptosis and play a role in regulating cell growth, differentiation, and apoptosis [76]. Apoptosis is considered to be a form of programmed cell death that primarily functions to eliminate damaged, dangerous, and non-functional cells in order to maintain organismal health [77,78]. The caspase protease family consists of numerous members, among which downstream *caspase3* acts as the executors of apoptosis by enzymatically cleaving specific proteins [79]. In this study, the gene expression of *M. nipponense caspsae3* was significantly reduced in the group fed β-1,3-glucan compared to the control group. Therefore, it can be inferred that the addition of β-1,3-glucan to feed can attenuate apoptosis. *P53* is a protein involved in inhibiting cell cycle progression in DNA-damaged cells [80], which is regulated by activating cellular repair mechanisms for DNA damage repair or inducing apoptosis if repair is not feasible. It plays a pivotal role in gene expression related to cell cycle, genetic stability maintenance, and apoptosis [81]. In our experiment, the inclusion of 0.05%, 0.2%, and 0.4% β-1,3-glucan significantly downregulated *P53* gene expression in prawns consistently with *caspase3* trend. We speculate that the addition of β-1,3-glucan to the diet could attenuate apoptosis in *M. nipponense* by enhancing the nonspecific immunity characteristic of crustaceans, which is consistent with the findings shown above for ACP and AKP. Furthermore, a correlation between the enhancement of specific immunity and the reduction in apoptosis, as indicated by ACP and AKP levels, has been observed in several other studies [82,83]. Interestingly, the inclusion of 0.1% β-1,3-glucan significantly up-regulated *P53* gene expression possibly due to its upstream position relative to *caspase3*, as well as its involvement in antioxidant defense regulation, DNA repair processes, and apoptotic pathways [84].

The intestine is an important organ for the digestion and absorption of nutrients in shrimp [85,86]. Gut microorganisms are essential for nutritional digestion and absorption, as well as immune function and disease resistance [87]. Feeding probiotics to *Oreochromis niloticus* has been shown to enhance intestinal villi development, growth performance, and mucosal immunity [88]. Similarly, improving gut flora has been found to enhance immune responses and disease resistance in various finfish species [89]. Identifying up- and down-regulated microorganisms is an important step in understanding how β-glucan affects the gut microbiota and how this change in the microbiota affects the host by altering nutrients, the immune system, and compounds absorbed through the epithelial barrier [90]. In this study, the addition of β-1,3-glucan did not affect the gut flora abundance at phylum level, we speculate that the addition of β-1,3-glucan at the phylum level maintains the stability and diversity of the intestinal flora of *M. nipponense*. At the genus level, this experiment demonstrated that the addition of β-1,3-glucan significantly increased the abundance of *Rhodobacteraceae_unclassified* in the intestinal tract of *M. nipponense*. Consistent with previous studies on β-glucan, it has been consistently observed that supplementation with β-glucan enhances the abundance of *Rhodobacteraceae*, which may represent a key characteristic associated with competitive advantage induced by β-glucan [91,92]. Moreover, a prior investigation on *N. californicus* revealed that *Rhodobacteraceae* can serve as a probiotic in the diet of *N. californicus* and improve resistance to acute low-salinity challenge in shrimp *Penaeus vannamei* through higher survival rates and elevated levels of T-AOC activity as well as SOD, HSP70, and Relish gene transcripts [93]. Therefore, dietary supplementation with β-1,3-glucan could enhance *Rhodobacteraceae* abundance for gut health improvement in *M. nipponense*. However, further exploration is required to fully understand this potential mutualistic relationship between *M. nipponense* and specific taxa within its intestinal flora.

5. Conclusions

The present study demonstrated that the addition of appropriate amount of β-1,3-glucan to the diet improved the growth performance, survival, antioxidant capacity, and intestinal health of *M. nipponense*, within the concentration range of β-1,3-glucan set in this experiment. Moreover, β-1,3-glucan activation of pattern recognition receptors initiated diverse immune responses to effectively enhance the survival and disease resistance in *M. nipponense*.

Author Contributions: T.X. and J.W. writing—original draft and editing, H.X., Z.W., Y.L. (Yujie Liu), H.B., Y.Z. and Y.K. writing—review and editing, Y.L. (Yan Liu), supervision. Z.D., supervision. All authors have read and agreed to the published version of the manuscript.

Funding: The research was supported by "Pioneer" and "Leading Goose" R&D Program of Zhejiang (No: 2024C02012).

Institutional Review Board Statement: All experiments on animals were approved by the Committee on the Ethics of Animal Experiments at Huzhou University and the Care and Use of Laboratory Animals in China. Approval code: 20220701; approval date: 20 July 2022.

Data Availability Statement: The data referenced in this article is included within the text.

Conflicts of Interest: The authors declare no conflicts of interest.

References

1. Bostock, J.; Mcandrew, B.; Richards, R.; Jauncey, K.; Telfer, T.; Lorenzen, K.; Little, D.; Ross, L.; Handisyde, N.; Gatward, I.; et al. Aquaculture: Global status and trends. *Philos. Trans. R. Soc. Lond. B Biol.* **2010**, *365*, 2897–2912. [CrossRef] [PubMed]
2. Lafferty, K.D.; Harvell, C.D.; Conrad, J.M.; Friedman, C.S.; Kent, M.L.; Curis, A.M.; Powell, E.N.; Rondeau, D.; Saksida, S.M. Infectious diseases affect marine fisheries and aquaculture economics. *Ann. Rev. Mar. Sci.* **2015**, *7*, 471–496. [CrossRef] [PubMed]
3. Defoirdt, T.; Sorgeloos, P.; Bossier, P. Alternatives to antibiotics for the control of bacterial disease in aquaculture. *Curr. Opin. Microbiol.* **2011**, *14*, 251–258. [CrossRef] [PubMed]
4. Chuah, L.O.; Effarizah, M.E.; Goni, A.M.; Rusul, G. Antibiotic application and emergence of multiple antibiotic resistance (MAR) in global catfish aquaculture. *Curr. Environ. Health Rep.* **2016**, *3*, 118–127. [CrossRef] [PubMed]
5. Limbu, S.M.; Chen, L.Q.; Zhang, M.L.; Du, Z.Y. A global analysis on the systemic effects of antibiotics in cultured fish and their potential human health risk: A review. *Rev. Aquac.* **2021**, *13*, 1015–1059. [CrossRef]
6. Cassini, A.; Högberg, L.D.; Plachouras, D.; Quattrocchi, A.; Hoxha, A.; Simonsen, G.S.; Colomb-Cotinat, M.; Kretzschmar, M.E.; Devleesschauwer, B.; Cecchini, M.; et al. Attributable deaths and disability-adjusted life-years caused by infections with antibiotic-resistant bacteria in the EU and the European Economic Area in 2015: A population-level modelling analysis. *Lancet Infect. Dis.* **2019**, *19*, 56–66. [CrossRef]

7.   Khan, M.; Lively, J.A. Determination of sulfite and antimicrobial residue in imported shrimp to the USA. *Aquacult. Rep.* **2020**, *18*, 100529. [CrossRef]

8.   Wang, W.; Sun, J.; Liu, C.; Xue, Z. Application of immunostimulants in aquaculture: Current knowledge and future perspectives. *Aquac. Res.* **2017**, *48*, 1–23. [CrossRef]

9.   Bouallegui, Y. A comprehensive review on crustaceans' immune system with a focus on freshwater crayfish in relation to crayfish plague disease. *Front. Immunol.* **2021**, *12*, 667787. [CrossRef] [PubMed]

10.   Tran, N.T.; Kong, T.; Zhang, M.; Li, S. Pattern recognition receptors and their roles on the innate immune system of mud crab (*Scylla paramamosain*). *Dev. Comp. Immunol.* **2020**, *102*, 103469. [CrossRef]

11.   Hu, B.Q.; Wen, C.G.; Liu, Y. Identification and expression analysis on two forms bactericidal permeability-increasing protein (*bpi*)/lipopolysaccharide-binding protein (*lbp*) of triangle mussel, *hyriopsis cumingii*. *Fish Shellfish Immunol.* **2016**, *53*, 111–112. [CrossRef]

12.   Gao, X.; Ke, C.; Zhang, M.; Li, X.; Wu, F.; Liu, Y. Effects of the probiotic Bacillus *amyloliquefaciens* on the growth, immunity, and disease resistance of *Haliotis discus hannai*. *Fish Shellfish Immunol.* **2019**, *94*, 617–627. [CrossRef]

13.   Jin, P.; Zhou, L.; Song, X.; Qian, J.; Chen, L.; Ma, F. Particularity and universality of a putative Gram-negative bacteria-binding protein (*GNBP*) gene from amphioxus (*Branchiostoma belcheri*): Insights into the function and evolution of *GNBP*. *Fish Shellfish Immunol.* **2012**, *33*, 835–845. [CrossRef] [PubMed]

14.   Lai, A.G.; Aboobaker, A.A. Comparative genomic analysis of innate immunity reveals novel and conserved components in crustacean food crop species. *BMC Genom.* **2017**, *18*, 389. [CrossRef] [PubMed]

15.   Hoeger, U.; Schenk, S. Crustacean hemolymph lipoproteins. In *Vertebrate and Invertebrate Respiratory Proteins, Lipoproteins and Other Body Fluid Proteins*; Subcellular Biochemistry; Springer: Berlin/Heidelberg, Germany, 2020; pp. 35–62. [CrossRef]

16.   Hang, Y.; Ni, M.; Zhang, P.; Bai, Y.; Zhou, B.; Zheng, J.; Cui, Z. Identification and functional characterization of C-type lectins and crustins provide new insights into the immune response of *Portunus trituberculatus*. *Fish Shellfish Immunol.* **2022**, *129*, 170–181. [CrossRef]

17.   Wu, C.; Noonin, C.; Jiravanichpaisal, P.; Söderhäll, I.; Söderhäll, K. An insect *TEP* in a crustacean is specific for cuticular tissues and involved in intestinal defense. *Insect Biochem. Mol. Biol.* **2012**, *42*, 71–80. [CrossRef]

18.   Mateu, E.; Díaz, I. The challenge of PRRS immunology. *Vet. J.* **2008**, *177*, 345–351. [CrossRef]

19.   Chen, Q.; Bai, S.; Dong, C. A fibrinogen-related protein identified from hepatopancreas of crayfish is a potential pattern recognition receptor. *Fish Shellfish Immunol.* **2016**, *56*, 349–357. [CrossRef]

20.   Habib, Y.J.; Zhang, Z. The involvement of crustacean toll-like receptors in pathogen recognition. *Fish Shellfish Immunol.* **2020**, *102*, 169–176. [CrossRef]

21.   Liu, S.; Zheng, S.; Li, Y.; Li, J.; Liu, H. Hemocyte-mediated phagocytosis in crustaceans. *Front. Immunol.* **2020**, *11*, 268. [CrossRef]

22.   Gallo, C.; Schiavon, F.; Ballarin, L. Insight on cellular and humoral components of innate immunity in Squilla mantis (Crustacea, Stomatopoda). *Fish Shellfish Immunol.* **2011**, *31*, 423–431. [CrossRef] [PubMed]

23.   Sánchez-Paz, A.; Muhlia-Almazán, A. Uncovering and defragmenting the role of the Toll pathway in the innate immune responses of cultured crustaceans against viral pathogens. *Rev. Aquac.* **2020**, *12*, 1818–1835. [CrossRef]

24.   Bai, L.; Zhou, K.; Li, H.; Qin, Y.; Wang, Q.; Li, W. Bacteria-induced IMD-Relish-AMPs pathway activation in Chinese mitten crab. *Fish Shellfish Immunol.* **2020**, *106*, 866–875. [CrossRef] [PubMed]

25.   Wang, X.; Wang, J. Pattern recognition receptors acting in innate immune system of shrimp against pathogen infections. *Fish Shellfish Immunol.* **2013**, *34*, 981–989. [CrossRef]

26.   Huang, Y.; Ren, Q. Research progress in innate immunity of freshwater crustaceans. *Dev. Comp. Immunol.* **2020**, *104*, 103569. [CrossRef] [PubMed]

27.   Lu, J.; Bu, X.; Xiao, S.; Lin, Z.; Wang, X.; Jia, Y.; Wang, X.; Qin, J.; Chen, L. Effect of single and combined immunostimulants on growth, anti-oxidation activity, non-specific immunity and resistance to *Aeromonas hydrophila* in Chinese mitten crab (*Eriocheir sinensis*). *Fish Shellfish Immunol.* **2019**, *93*, 732–742. [CrossRef] [PubMed]

28.   Ojerio, V.; Corre, V.; Toledo, N.; Andrino-Felarca, K.; Nievales, L.; Traifalgar, F. Alginic acid as immunostimulant: Effects of dose and frequency on growth performance, immune responses, and white spot syndrome virus resistance in tiger shrimp *Penaeus monodon* (Fabricius, 1798). *Aquac. Int.* **2018**, *26*, 267–278. [CrossRef]

29.   Montero-Rocha, A.; McIntosh, D.; Sanchez-Merino, R.; Flores, I. Immunostimulation of white shrimp (*Litopenaeus vannamei*) following dietary administration of Ergosan. *J. Invertebr. Pathol.* **2006**, *91*, 188–194. [CrossRef]

30.   Khanjani, M.H.; Sharifinia, M.; Ghaedi, G. β-glucan as a promising food additive and immunostimulant in aquaculture industry. *Ann. N. Y. Acad. Sci.* **2022**, *22*, 817–827. [CrossRef]

31.   Ibrahim, L.A.; ElSayed, E.E. The influence of water quality on fish tissues and blood profile in Arab al-Ulayqat Lakes, Egypt. *Egypt. J. Aquat. Res.* **2023**, *49*, 235–243. [CrossRef]

32.   Ding, Z.; Xiong, Y.; Zheng, J.; Zhou, D.; Kong, Y.; Qi, C.; Liu, Y.; Limbu, S.M. Modulation of growth, antioxidant status, hepatopancreas morphology, and carbohydrate metabolism mediated by alpha-lipoic acid in juvenile freshwater prawns *Macrobrachium nipponense* under two dietary carbohydrate levels. *Aquaculture* **2022**, *546*, 737314. [CrossRef]

33.   Shahbazi, S.; Bolhassani, A. Immunostimulants: Types and functions. *Can. J. Infect. Dis. Med. Microbiol.* **2016**, *4*, 45–51.

34.   Stier, H.; Ebbeskotte, V.; Gruenwald, J. Immune-modulatory effects of dietary Yeast Beta-1,3/1,6-D-glucan. *Nutr. J.* **2014**, *13*, 38. [CrossRef] [PubMed]

35. Qiao, Y.; Zhou, L.; Qu, Y.; Lu, K.; Han, F.; Li, E. Effects of different dietary β-glucan levels on antioxidant capacity and immunity, gut microbiota and transcriptome responses of white shrimp (*Litopenaeus vannamei*) under low salinity. *Antioxidants* **2022**, *11*, 2282. [CrossRef] [PubMed]

36. Huang, Q.; Zhu, Y.; Yu, J.; Fang, L.; Li, Y.; Wang, M.; Liu, J.; Yan, P.; Xia, J.; Liu, G.; et al. Effects of sulfated β-glucan from *Saccharomyces cerevisiae* on growth performance, antioxidant ability, nonspecific immunity, and intestinal flora of the red swamp crayfish (*Procambarus clarkii*). *Fish Shellfish Immunol.* **2022**, *127*, 891–900. [CrossRef]

37. Dawood, M.A.; Eweedah, N.M.; Moustafa, E.M.; Shahin, M.G. Synbiotic effects of *Aspergillus oryzae* and β-glucan on growth and oxidative and immune responses of Nile Tilapia, *Oreochromis niloticus*. *Probiotics Antimicrob. Proteins* **2020**, *12*, 172–183. [CrossRef] [PubMed]

38. Andrino, K.G.S.; Apines-Amar, M.J.S.; Janeo, R.L.; Corre, V.L., Jr. Effects of dietary mannan oligosaccharide (MOS) and β-glucan on growth, immune response and survival against white spot syndrome virus (WSSV) infection of juvenile tiger shrimp Penaeus monodon. *Aquac. Aquar. Conserv. Legis.* **2014**, *7*, 321–332.

39. Tian, J.; Yang, Y.; Du, X.; Xu, W.; Zhu, B.; Huang, Y.; Ye, Y.; Zhao, Y.; Li, Y. Effects of dietary soluble β-1,3-glucan on the growth performance, antioxidant status, and immune response of the river prawn (*Macrobrachium nipponense*). *Fish Shellfish Immunol.* **2023**, *138*, 108848. [CrossRef]

40. Tian, J.; Yang, Y.; Xu, W.; Du, X.; Ye, Y.; Zhu, B.; Huang, Y.; Zhao, Y.; Li, Y. Effects of β-1,3-glucan on growth, immune responses, and intestinal microflora of the river prawn (*Macrobrachium nipponense*) and its resistance against *Vibrio parahaemolyticus*. *Fish Shellfish Immunol.* **2023**, *142*, 109142. [CrossRef]

41. Zhang, X.; Zhu, Y.T.; Li, X.J.; Wang, S.C.; Li, D.; Li, W.W.; Wang, Q. Lipopolysaccharide and beta-1,3-glucan binding protein (*LGBP*) stimulates prophenoloxidase activating system in Chinese mitten crab (*Eriocheir sinensis*). *Dev. Comp. Immunol.* **2016**, *61*, 70–79. [CrossRef]

42. Wongsasak, U.; Chaijamrus, S.; Kumkhong, S.; Boonanuntanasarn, S. Effects of dietary supplementation with β-glucan and synbiotics on immune gene expression and immune parameters under ammonia stress in Pacific white shrimp. *Aquaculture* **2015**, *436*, 179–187. [CrossRef]

43. Fu, H.; Jiang, S.; Xiong, Y. Current status and prospects of farming the giant river prawn (*Macrobrachium rosenbergii*) and the oriental river prawn (*Macrobrachium nipponense*) in China. *Aquac. Res.* **2012**, *43*, 993–998. [CrossRef]

44. Yuan, J.; Wang, X.; Gu, Z.; Zhang, Y.; Wang, Z. Activity and transcriptional responses of hepatopancreatic biotransformation and antioxidant enzymes in the oriental river prawn *Macrobrachium nipponense* exposed to microcystin-LR. *Toxins* **2015**, *7*, 4006–4022. [CrossRef] [PubMed]

45. Gui, J.; Tang, Q.; Li, Z.; Liu, J.; De Silva, S.S. Culture of the oriental river prawn (*Macrobrachium nipponense*). In *Aquaculture in China: Success Stories and Modern Trends*; Wiley: Hoboken, NJ, USA, 2018; pp. 218–225. [CrossRef]

46. Pan, X.; Shen, J.; Li, J.; He, B.; Hao, G.; Xu, Y.; Yao, J. Identification and biological characteristics of the pathogen causing *Macrobrachium nipponense* soft-shell syndrome. *Microbiology* **2009**, *36*, 1571–1576. [CrossRef]

47. Gao, X.; Tong, S.; Zhang, S.; Chen, Q.; Jiang, Z.; Jiang, Z.; Wei, W.; Zhu, J.; Zhang, X. *Aeromonas veronii* associated with red gill disease and its induced immune response in *Macrobrachium nipponense*. *Rev. Aquac.* **2020**, *51*, 5163–5174. [CrossRef]

48. Lulijwa, R.; Rupia, E.J.; Alfaro, A.C. Antibiotic use in aquaculture, policies and regulation, health and environmental risks: A review of the top 15 major producers. *Rev. Aquac.* **2020**, *12*, 640–663. [CrossRef]

49. Alderman, D.J.; Hastings, T.S. Antibiotic use in aquaculture: Development of antibiotic resistance–potential for consumer health risks. *Int. J. Food Sci. Technol.* **1998**, *33*, 139–155. [CrossRef]

50. Cabello, F.C. Heavy use of prophylactic antibiotics in aquaculture: A growing problem for human and animal health and for the environment. *Environ. Microbiol.* **2006**, *8*, 1137–1144. [CrossRef]

51. Limbu, S.M.; Zhou, L.; Sun, S.X.; Zhang, M.L.; Du, Z.Y. Chronic exposure to low environmental concentrations and legal aquaculture doses of antibiotics cause systemic adverse effects in *Nile tilapia* and provoke differential human health risk. *Environ. Int.* **2018**, *115*, 205–219. [CrossRef]

52. Andersen, W.C.; Casey, C.R.; Nickel, T.J.; Young, S.L.; Turnipseed, S.B. Dye residue analysis in raw and processed aquaculture products: Matrix extension of AOAC international official method 2012.25. *J. AOAC Int.* **2018**, *101*, 1927–1939. [CrossRef]

53. Livak, K.J.; Schmittgen, T.D. Analysis of relative gene expression data using real-time quantitative PCR and the $2^{-\Delta\Delta CT}$ method. *Methods* **2001**, *25*, 402–408. [CrossRef] [PubMed]

54. Li, H.; Xu, C.; Zhou, L.I.; Dong, Y.; Su, Y.; Wang, X.; Qin, J.; Chen, L.; Li, E. Beneficial effects of dietary β-glucan on growth and health status of Pacific white shrimp *Litopenaeus vannamei* at low salinity. *Fish Shellfish Immunol.* **2019**, *91*, 315–324. [CrossRef] [PubMed]

55. Suphantharika, M.; Khunrae, P.; Thanardkit, P.; Verduyn, C. Preparation of spent brewer's yeast β-glucans with a potential application as an immunostimulant for black tiger shrimp, Penaeus monodon. *Bioresour. Technol.* **2003**, *88*, 55–60. [CrossRef] [PubMed]

56. Chowdhury, S.; Saikia, S.K. Oxidative Stress in Fish: A Review. *J. Sci. Res.* **2020**, *12*, 145–160. [CrossRef]

57. Das, U.N.; Das, U.N. Introduction to Free Radicals, Antioxidants, Lipid Peroxidation, and Their Effects on Cell Proliferation. In *Molecular Biochemical Aspects of Cancer*; Springer: Berlin/Heidelberg, Germany, 2020; pp. 41–65. [CrossRef]

58. Staerck, C.; Gastebois, A.; Vandeputte, P.; Calenda, A.; Larcher, G.; Gillmann, L.; Papon, N.; Bouchara, J.; Fleury, M. Microbial antioxidant defense enzymes. *Microb. Pathog.* **2017**, *110*, 56–65. [CrossRef]

59. El-Missiry, M.A. *Antioxidant Enzyme*; BoD–Books on Demand: Norderstedt, Germany, 2012. [CrossRef]
60. Lushchak, V.I. Environmentally induced oxidative stress in aquatic animals. *Aquat. Toxicol.* **2012**, *101*, 13–30. [CrossRef]
61. Perveen, S.; Yang, L.; Zhou, S.; Feng, B.; Xie, X.; Zhou, Q.; Qian, D.; Wang, C.; Yin, F. β-1,3-Glucan from *Euglena gracilis* as an immunostimulant mediates the antiparasitic effect against *Mesanophrys sp.* on hemocytes in marine swimming crab (*Portunus trituberculatus*). *Fish Shellfish Immunol.* **2021**, *114*, 28–35. [CrossRef]
62. Gu, M.; Ma, H.; Mai, K.; Zhang, W.; Bai, N.; Wang, X. Effects of dietary β-glucan, mannan oligosaccharide and their combinations on growth performance, immunity and resistance against *Vibrio splendidus* of sea cucumber, *Apostichopus japonicus*. *Fish Shellfish Immunol.* **2011**, *31*, 303–309. [CrossRef] [PubMed]
63. Buchmann, K. Evolution of Innate Immunity: Clues from Invertebrates via Fish to Mammals. *Front. Immunol.* **2014**, *5*, 459. [CrossRef]
64. Du, J.; Zhu, H.; Liu, P.; Chen, J.; Xiu, Y.; Yao, W.; Wu, T.; Ren, Q.; Meng, Q.; Gu, W.; et al. Immune responses and gene expression in hepatopancreas from *Macrobrachium rosenbergii* challenged by a novel pathogen *spiroplasma* MR-1008. *Fish Shellfish Immunol.* **2013**, *34*, 315–323. [CrossRef]
65. Bu, X.; Lian, X.; Wang, Y.; Luo, C.; Tao, S.; Liao, Y.; Yang, J.; Chen, A.; Yang, Y. Dietary yeast culture modulates immune response related to TLR2-MyD88-NF-kβ signaling pathway, antioxidant capability and disease resistance against *Aeromonas hydrophila* for Ussuri catfish (*Pseudobagrus ussuriensis*). *Fish Shellfish Immunol.* **2019**, *84*, 711–718. [CrossRef] [PubMed]
66. Zhang, Y.; Guo, M.; Li, N.; Dong, Z.; Cai, L.; Wu, B.; Xie, J.; Liu, L.; Ren, L.; Shi, B. New insights into β-glucan-enhanced immunity in largemouth bass *Micropterus salmoides* by transcriptome and intestinal microbial composition. *Front. Immunol.* **2022**, *13*, 1086103. [CrossRef] [PubMed]
67. Vu, N.T.; Nghia, N.T.; Thao, N.H.P. Synergic degradation of yeast β-glucan with a potential of immunostimulant and growth promotor for tiger shrimp. *Aquacult. Rep.* **2021**, *21*, 100858. [CrossRef]
68. Li, D.; Wu, M. Pattern recognition receptors in health and diseases. *Signal Transduct. Target. Ther.* **2021**, *6*, 291. [CrossRef]
69. Panigrahi, A.; Sivakumar, M.R.; Sundaram, M.; Saravanan, A.; Das, R.R.; Katneni, V.K.; Ambasankar, K.; Dayal, J.; Gopikrishna, G. Comparative study on phenoloxidase activity of biofloc-reared pacific white shrimp *Penaeus vannamei* and Indian white shrimp *Penaeus indicus* on graded protein diet. *Aquaculture* **2020**, *518*, 734654. [CrossRef]
70. Lee, D.H.; Kim, H.W. Innate immunity induced by fungal β-glucans via dectin-1 signaling pathway. *Int. J. Med. Mushrooms* **2014**, *16*, 1–16. [CrossRef] [PubMed]
71. Zhang, X.W.; Man, X.; Huang, X.; Wang, Y.; Song, Q.S.; Hui, K.M.; Zhang, H.W. Identification of a *C-type lectin* possessing both antibacterial and antiviral activities from red swamp crayfish. *Fish Shellfish Immunol.* **2018**, *77*, 22–30. [CrossRef]
72. Han, B.; Zhang, X.; Chang, E.; Wan, W.; Xu, J.; Zhao, C.; Miao, S. Effects of dietary *Saccharomyces cerevisiae* and β-glucan on the growth performance, antioxidant capacity and immunity response in *Macrobrachium rosenbergii*. *Aquacult. Nutr.* **2021**, *27*, 20–28. [CrossRef]
73. Amparyup, P.; Sutthangkul, J.; Charoensapsri, W.; Tassanakajon, A. Pattern recognition protein binds to lipopolysaccharide and β-1,3-glucan and activates shrimp prophenoloxidase system. *J. Biol. Chem.* **2016**, *291*, 10949. [CrossRef]
74. Cheng, W.; Liu, C.H.; Tsai, C.H.; Chen, J.C. Molecular cloning and characterisation of a pattern recognition molecule, lipopolysaccharide-and β-1,3-glucan binding protein (*LGBP*) from the white shrimp *Litopenaeus vannamei*. *Fish Shellfish Immunol.* **2005**, *18*, 297–310. [CrossRef]
75. Liu, F.; Li, F.; Dong, B.; Wang, X.; Xiang, J. Molecular cloning and characterisation of a pattern recognition protein, lipopolysaccharide and β-1,3-glucan binding protein (LGBP) from Chinese shrimp *Fenneropenaeus chinensis*. *Mol. Biol. Rep.* **2009**, *36*, 471–477. [CrossRef] [PubMed]
76. Fernando, P.; Megeney, L.A. Is caspase-dependent apoptosis only cell differentiation taken to the extreme? *FASEB J.* **2007**, *21*, 8–17. [CrossRef]
77. Yang, D.; Chai, L.; Wang, J.; Zhao, X. Molecular cloning and characterization of Hearm caspase-1 from *Helicoverpa armigera*. *Mol. Biol. Rep.* **2008**, *35*, 405–412. [CrossRef] [PubMed]
78. Zhang, X.; Wang, X.; Huang, Y.; Hui, K.; Shi, Y.; Wang, W.; Ren, Q. Cloning and characterization of two different ficolins from the giant freshwater prawn *Macrobrachium rosenbergii*. *Dev. Comp. Immunol.* **2014**, *44*, 359–369. [CrossRef] [PubMed]
79. Sun, S.; Xuan, F.; Fu, H.; Zhu, J.; Ge, X.; Wu, X. Molecular cloning, characterization and expression analysis of *caspase-3* from the oriental river prawn, *Macrobrachium nipponense* when exposed to acute hypoxia and reoxygenation. *Fish Shellfish Immunol.* **2017**, *62*, 291–302. [CrossRef] [PubMed]
80. Dhuppar, S.; Mazumder, A. Measuring cell cycle-dependent DNA damage responses and *p53* regulation on a cell-by-cell basis from image analysis. *Cell Cycle* **2018**, *17*, 1358–1371. [CrossRef] [PubMed]
81. Mai, W.; Yan, J.; Wang, L.; Zheng, Y.; Xin, Y.; Wang, W. Acute acidic exposure induces *p53*-mediated oxidative stress and DNA damage in tilapia (*Oreochromis niloticus*) blood cells. *Aquat. Toxicol.* **2010**, *100*, 271–281. [CrossRef]
82. Huang, P.; Du, J.; Cao, L.; Gao, J.; Li, Q.; Sun, Y.; Shao, N.; Zhang, Y.; Xu, G. Effects of prometryn on oxidative stress, immune response and apoptosis in the hepatopancreas of *Eriocheir sinensis* (Crustacea: Decapoda). *Ecotoxicol. Environ. Saf.* **2023**, *262*, 115159. [CrossRef]
83. Wang, C.; Pan, J.; Wang, X.; Cai, X.; Lin, Z.; Shi, Q.; Li, E.; Qin, J.G.; Chen, L. N-acetylcysteine provides protection against the toxicity of dietary T-2 toxin in juvenile Chinese mitten crab (*Eriocheir sinensis*). *Aquaculture* **2021**, *538*, 736531. [CrossRef]

84. Cheng, C.; Ma, H.; Ma, H.; Liu, C.; Deng, Y.; Feng, J.; Wang, L.; Cheng, Y.; Gou, Z. The role of tumor suppressor protein *p53* in the mud crab (*Scylla paramamosain*) after *Vibrio parahaemolyticus* infection. *Comp. Biochem. Physiol. C Toxicol. Pharmacol.* **2021**, *246*, 108976. [CrossRef] [PubMed]

85. Huang, Z.; Li, X.; Wang, L.; Shao, Z. Changes in the intestinal bacterial community during the growth of white shrimp, *Litopenaeus vannamei*. *Aquac. Res.* **2016**, *47*, 1737–1746. [CrossRef]

86. Hou, D.; Huang, Z.; Zeng, S.; Liu, J.; Weng, S.; He, J. Comparative analysis of the bacterial community compositions of the shrimp intestine, surrounding water and sediment. *J. Appl. Microbiol.* **2018**, *125*, 792–799. [CrossRef] [PubMed]

87. Dawood, M.A. Nutritional immunity of fish intestines: Important insights for sustainable aquaculture. *Rev. Aquac.* **2021**, *13*, 642–663. [CrossRef]

88. Pirarat, N.; Pinpimai, K.; Endo, M.; Katagiri, T.; Ponpornpisit, A.; Chansue, N.; Maita, M. Modulation of intestinal morphology and immunity in nile tilapia (*Oreochromis niloticus*) by *Lactobacillus rhamnosus* GG. *Res. Vet. Sci.* **2011**, *91*, e92–e97. [CrossRef] [PubMed]

89. Hoseinifar, S.H.; Van Doan, H.; Dadar, M.; Ringø, E.; Harikrishnan, R. Feed additives, gut microbiota, and health in finfish aquaculture. In *Microbial Communities in Aquaculture Ecosystems: Improving Productivity and Sustainability*; Springer: Berlin/Heidelberg, Germany, 2019; pp. 121–142. [CrossRef]

90. Eichmiller, J.; Hamilton, M.; Staley, C.; Sadowsky, M.; Sorensen, P.W. Environment shapes the fecal microbiome of invasive carp species. *Microbiome* **2016**, *4*, 44. [CrossRef]

91. Yang, G.; Xu, Z.; Tian, X.; Dong, S.; Peng, M. Intestinal microbiota and immune related genes in sea cucumber (*Apostichopus japonicus*) response to dietary β-glucan supplementation. *Biochem. Biophys. Res. Commun.* **2015**, *458*, 98–103. [CrossRef]

92. Yamazaki, Y.; Meirelles, P.M.; Mino, S.; Suda, W.; Oshima, K.; Hattori, M.; Thompson, F.; Sakai, Y.; Sawabe, T.; Sawabe, T. Individual *Apostichopus japonicus* fecal microbiome reveals a link with polyhydroxybutyrate producers in host growth gaps. *Sci. Rep.* **2016**, *6*, 21631. [CrossRef]

93. Dong, P.; Guo, H.; Huang, L.; Zhang, D.; Wang, K. Glucose addition improves the culture performance of Pacific white shrimp by regulating the assembly of Rhodobacteraceae taxa in gut bacterial community. *Aquaculture* **2023**, *567*, 739254. [CrossRef]

 *fishes*

Article

Hepatic Gene Expression Changes of Zebrafish Fed Yeast Prebiotic, Yeast Probiotic, Black Soldier Fly Meal, and Butyrate

Nancy Gao [1], Junyu Zhang [1], Umesh K. Shandilya [1], John S. Lumsden [2], Amir Behzad Barzrgar [3], David Huyben [1] and Niel A. Karrow [1,*]

[1] Department of Animal Biosciences, University of Guelph, Guelph, ON N1G 2W1, Canada; ngao01@uoguelph.ca (N.G.); jzhang48@uoguelph.ca (J.Z.); ushand@uoguelph.ca (U.K.S.); huybend@uoguelph.ca (D.H.)
[2] Department of Pathobiology, OVC, University of Guelph, Guelph, ON N1G 2W1, Canada; jsl@ovc.uoguelph.ca
[3] Ontario Agricultural College, University of Guelph, Guelph, ON N1G 2W1, Canada
* Correspondence: nkarrow@uoguelph.ca

Abstract: As global fish consumption rises, improving fish health through immunomodulatory feed ingredients shows promise while also supporting growth performance. This study investigated the effects of yeast prebiotics, probiotics, a postbiotic (butyrate), and black soldier fly larvae (BSFL) meal on fish immune responses. Zebrafish were fed diets containing these ingredients for 63 days and then exposed to either *Pseudomonas aeruginosa* lipopolysaccharide (LPS) or live *Flavobacterium psychrophilum* to assess hepatic candidate gene expression and weight gain. No mortalities were observed post-immune challenges, and weight gains were not significantly different across treatments. Liver samples were collected for mRNA analysis, and real-time qPCR was used to evaluate the expression of immune-related genes such as *TNF-α*, *IL-1β*, *hepcidin*, and *NF-κB/p65*. *NF-κB/p65* was upregulated in response to immune challenges, indicating a reaction to both LPS and pathogen exposure. Fish on the BSFL diet showed decreased *NF-κB/p65* expression after the pathogen challenge, while probiotic-fed fish had reduced *angiopoietin-like 4 (angptl4)* levels following LPS exposure. Butyrate supplementation had the most significant impact, downregulating pro-inflammatory cytokines and other immune-related genes, suggesting a protective effect. These findings support the health benefits of BSFL and sodium butyrate during an immune challenge.

Keywords: zebrafish; *Saccharomyces cerevisiae*; sodium butyrate; lipopolysaccharide; *Flavobacterium psychrophilum*; immunomodulation

Key Contribution: Supplementation of yeast prebiotic, yeast probiotic, black soldier fly meal, and butyrate influenced the hepatic gene expression following an immune challenge. Butyrate appeared to exert the greatest anti-inflammatory response compared to the other experimental diets.

Citation: Gao, N.; Zhang, J.; Shandilya, U.K.; Lumsden, J.S.; Barzrgar, A.B.; Huyben, D.; Karrow, N.A. Hepatic Gene Expression Changes of Zebrafish Fed Yeast Prebiotic, Yeast Probiotic, Black Soldier Fly Meal, and Butyrate. *Fishes* **2024**, *9*, 495. https://doi.org/10.3390/fishes9120495

Academic Editors: Qiyou Xu, Jianhua Ming, Fei Song, Changle Qi, Chuanpeng Zhou and Marina Paolucci

Received: 25 September 2024
Revised: 29 November 2024
Accepted: 29 November 2024
Published: 2 December 2024

1. Introduction

Aquaculture production has increased over the past three decades to meet the increasing demand for food fish, which has resulted in more intensive production systems, increased risk of infection and the spread of disease among fish [1]. Current disease management strategies frequently rely on antimicrobials or antibiotics, which can contribute to the development of antimicrobial resistance [2]. Recent reports have found antimicrobial resistance in several Gram-negative bacterial species, such as *Pseudomonas aeruginosa* and *Flavobacterium psychrophilum*, which significantly impact the production and cost of cultured fish species due to large mortality events caused by ulcerative syndrome and bacterial cold-water disease, respectively [3,4]. Developing strategies to reduce the dependence on antimicrobials is needed to improve fish disease resistance and produce more food for a growing human population.

Bacterial endotoxins, such as lipopolysaccharide (LPS), that make up the cell membrane of Gram-negative bacteria are often used in animal models to trigger an inflammatory response similar to an acute bacterial infection [5]. The LPS response and mechanism in zebrafish have been well documented, and hence, certain genes such as *TNF-α* and *IL-1β* are known to be induced [6]. However, LPS itself does not cause an infection, as it is not a living microbe. In contrast, the administration of live pathogens (e.g., *F. psychrophilum*) also imposes an immune challenge, and since they are viable, their ability to replicate and interact with the host immune system may better represent the complete pathogen–host interaction. For these reasons, studies can be conducted using LPS or live pathogen challenges at the end of a feeding trial with feed additives to assess their capacity to improve disease resistance [7].

Recent research has highlighted an important relationship between the gut microbiota and the immune system, indicating that supplementation, or alteration of the diet to support the gut environment, may also affect the immune system [5]. Hence, a search for immunomodulatory ingredients that can aid in stress resilience and disease prevention is underway, fueled by the need for innovative and affordable approaches to improve aquaculture production and animal welfare. A few common immunomodulatory ingredients that are suggested to improve the gut and immune health of fish include yeast prebiotics and probiotics, short-chain fatty acids (SCFAs) such as butyrate, and black soldier fly larvae (BSFL) meal [8–11]. In terms of BSFL, it includes several components such as chitin, antimicrobial peptides (AMPs), and lauric acid that modulate the immune system. Besides yeast prebiotics and probiotics, research on the effects of BSFL and butyrate is still limited, although they are currently being utilized in aquaculture, which further supports the need to investigate their effects on fish health [12].

The zebrafish (*Danio rerio*) is a freshwater fish species with a well-characterized genome that is widely recognized as a valuable model organism for studying human diseases and is increasingly being utilized in research to model aquaculture fish species, including salmonids [13]. The advantages of the zebrafish model include optical transparency during its early stages of life, rapid reproduction, a continuous and large amount of egg production, and fast development, which are not characteristics of cultured fish species such as salmonids [13,14].

This study aimed to investigate the immunomodulatory effects of diets containing yeast prebiotic, yeast probiotic, butyrate, or BSFL meal on the expression of hepatic immune and stress-relevant genes following an immune challenge with *P. aeruginosa* LPS or live *F. psychrophilum*. We hypothesized that supplementation of prebiotics, probiotics, butyrate, and BSFL meal would improve the survival of zebrafish following the immune challenges in comparison to the control by downregulating the expression of pro-inflammatory genes.

2. Materials and Methods

2.1. Diet Preparation

All 5 diets, including the control diet, were formulated to be isonitrogenous and isoenergetic, and all diets were prepared at the Department of Animal Biosciences, University of Guelph (Guelph, ON, Canada). Five dietary treatments, including fishmeal-based control, yeast prebiotic (Alltech; Nicholasville, KY, USA) and probiotic (Alltech; Nicholasville, KY, USA), sodium butyrate (Sigma-Aldrich, St. Louis, MO, USA), and defatted BSFL meal (Enterra Feed Corp., Langley, BC, Canada) diets, were tested in the present study. The diet formulations are presented in Table 1. Dry ingredients were mixed for 20 min, followed by an additional 20 min of mixing after adding in the wet ingredients (oils and water). They were then pelleted with a meat grinder with a diameter of 1 mm, oven-dried overnight, ground, and sieved to 1mm.

Table 1. Formulated dietary composition (g/kg on a wet matter basis) for the control, black soldier fly larvae (BSFL), yeast probiotic and prebiotic, and butyrate diets fed to juvenile and adult zebrafish.

Ingredients (g/kg)	Control	BSFL Meal	Probiotic	Yeast Prebiotic	Butyrate
Fishmeal, 68% CP	300	223	300	300	300
Wheat Flour	180	180	180	180	180
Soybean meal, 48% CP	170	170	170	170	170
Poultry by-product meal, 62% CP	100	100	100	100	100
BSF larvae, defatted, Enterra		100			
Probiotic yeast, Alltech			10		
Prebiotic yeast, Alltech				2	
Sodium butyrate (SCFA), Sigma					0.5
Blood meal, porcine	70	70	70	70	70
Starch, corn	98	79	98	98	98
Fish oil, herring	35	35	35	35	35
Canola oil	34	30	34	34	34
Vitamin and mineral premix (DSM)	5	5	5	5	5
Vitamin E	3	3	3	3	3
Choline chloride	3	3	3	3	3
Limestone	1	1	1	1	1
NaCl	1	1	1	1	1
DL-Met	0.5	1	0.5	0.5	0.5

2.2. Zebrafish Culture

Zebrafish adults were housed in a Tecniplast system in the Hagen Aqualab at the University of Guelph at approximately 28 °C on a 12 h light/dark cycle in a recirculation system. Adult zebrafish were set up in 1.7 L Tecniplast Sloping Breeding Tanks with a barrier to separate the males and females at 3 p.m., and then they were allowed to interact at 9 a.m. the next day following the removal of the barrier. Viable eggs were subsequently collected between 2 and 3 p.m. and treated with 50 mg/mL Gentamicin reagent for 1 h at room temperature, then disinfected with 0.004% diluted bleach for 5 min and rinsed. Embryos were then maintained at a density of approximately 100 eggs per plate in egg water (60 mg/L Instant Ocean Sea Salts) at approximately 28.5 °C in an incubator. Once fish reached 5 d post fertilization (dpf), at which point yolk absorption was almost complete, the larvae were transferred to a larger tank where they were fed a diet of GEMMA Micro 75 (Skretting; Westbrook, ME, USA) three times daily. At 30 dpf, the diet was switched to GEMMA 300 (Skretting; Westbrook, ME, USA), where the fish were then fed twice daily.

2.3. Feeding Trial

Once the fish reached 3 to 4 months of age, they were transferred to a Level 2 biosecure facility within the Hagen Aqualab at the University of Guelph at approximately 27 °C on a 12 h light/dark cycle using a flow-through system. Each dietary treatment group comprised three tanks that were positioned into three water baths to ensure the tank temperatures within the room were maintained at 27 °C. Water quality was inspected daily, and ammonia levels were maintained at 0 ppm, pH levels between 7.4 to 7.8, and dissolved oxygen levels at 6 to 8 ppm. In total, 3 separate baths held 5 tanks, each representing a different dietary treatment, with 15 tanks in total. Thus, there were 3 replicate tanks per dietary treatment, containing 50 fish each (Figure 1). The fish were fed their respective diets twice a day for

63 d until satiation, and 10 weighings of different groups of five fish from each replicate tank were recorded on days 0, 10, 20, and 63 of the feeding trial.

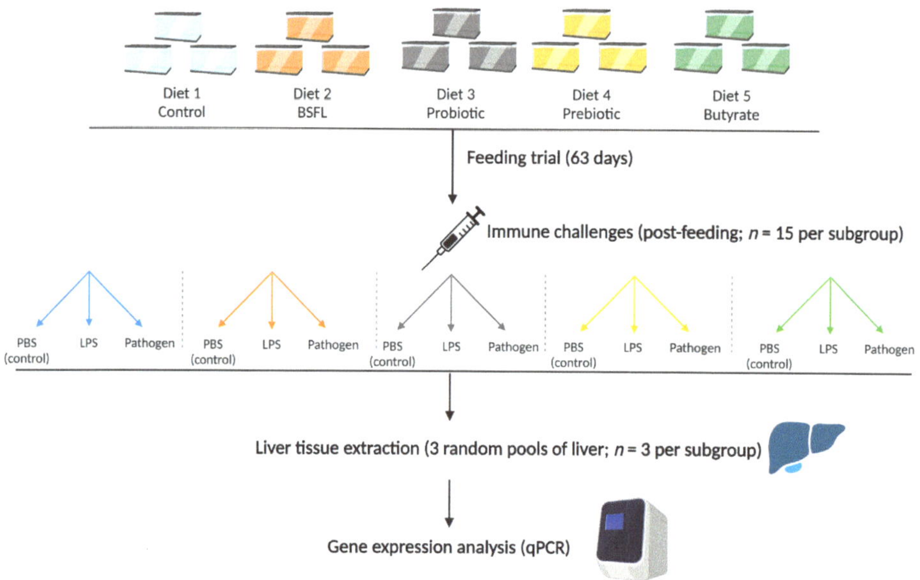

Figure 1. Experimental timeline illustrating the 63-day zebrafish feeding trial with five diets (control, black soldier fly larvae meal (BSFL), yeast probiotic or prebiotic, and sodium butyrate) followed by a 24 h immune challenge with 25 mg/mL *Pseudomonas aeruginosa* lipopolysaccharide (LPS) or 1.5×10^8 CFU/mL of the live pathogen *Flavobacterium psychrophilum*. After the immune challenge, qPCR was conducted on liver tissue samples to assess changes in the expression of hepatic *TNF-α*, *IL-1β*, *hepcidin*, *MyD88*, *caspase-b*, *PGRP*, *serum amyloid A (SAA)*, *angptl4*, *NF-κB (p65 subunit)*, and *cxcl8*. The sample size was 3 replicates (*n* = 3) of 3–4 pooled livers per subgroup (PBS, LPS or pathogen) per dietary treatment. Created in BioRender.com.

2.4. Immune Challenge with Either P. aeruginosa Lipopolysaccharide (LPS) Endotoxin or F. psychrophilum

The LPS from *P. aeruginosa* was purchased from Sigma-Aldrich. A stock solution of 100 mg/mL of LPS was prepared in sterile PBS and diluted to working concentrations of 1 and 25 mg/mL with PBS. The *F. psychophilum* was generously provided by Dr. John Lumsden (Department of Pathobiology, University of Guelph) and cultured on Petri dishes with Cytophaga broth made from Cytophaga Broth Base (Criterion, cat. No. C7790). The pathogen challenge level was prepared by adding *F. psychophilum* bacterial colonies until the suspension reached a turbidity of a 0.5 McFarland standard, which is approximately 1.5×10^8 CFU/mL [15]. The challenge dose was chosen based on a previous research study that experimentally infected rainbow trout with *F. psychophilum* [16].

Before the immune challenges, fish were fasted for 24 h and then anesthetized by immersion with 70 mg/L tricaine methanesulfonate (MS-222) until there was a loss of equilibrium and gill movements ceased. The fish were then placed in a partially cut soft sponge soaked in water to stabilize the body during injections. Fish were intraperitoneally injected with either 10 μL of 25 mg/mL *P. aeruginosa* LPS, 1.5×10^8 CFU/mL *F. psychrophilum*, or an equivalent volume of PBS (pH 7.4) using a Hamilton Model 701 RN Syringe (0.375 in, point style 4 at 12°, 6/PK; Part-#: 207434-10 from Hamilton Company, Reno, NV, USA). After injection, the fish were placed in individually labeled plastic containers containing tank water and equipped with an air stone for 24 h to maintain dissolved oxygen levels, at which time they were then euthanized with 200 mg/L of MS-222. Livers were extracted under a dissection scope, immediately placed in a 3 L tank containing liquid nitrogen, then stored in a −80 °C freezer until RNA extraction could be performed.

2.5. RNA Extraction and cDNA Synthesis

Tissue homogenization was carried out on each sample with 1 mL of TRIzol added in 2.0 mL autoclaved tubes after the tissue samples were thawed using a PowerGen 500 Homogenizer (Fisher Scientific; Waltham, MA, USA). Total RNA was extracted from Trizol cell lysate using an RNeasy Mini Kit (Qiagen; Hilden, Germany) as per the manufacturer's instructions. The RNA concentration was measured using a Cytation 5 spectrophotometer (BioTek; Shoreline, WA, USA). From each sample, 500 ng of total RNA was reverse-transcribed to cDNA using a High-Capacity cDNA Reverse Transcription Kit (Applied Biosystems; Waltham, MA, USA).

2.6. Proximate Composition Analysis

The proximate composition of the diets was analyzed using standard methods at the Department of Animal Biosciences of the University of Guelph (Guelph, ON, Canada), and the data are presented in Table 2. Diets were ground to <1 mm. Crude lipids were determined using petroleum ether and the ANKOM XT15 Extraction System (ANKOM Technology, Macedon, NY, USA) according to the manufacturer's instructions. Nitrogen content was determined using the LECO FP828 System (LECO Corporation, Saint Joseph, MI, USA) and then multiplied by the protein factor of 6.25 to calculate the % crude protein. Gross energy was determined using the IKA C 6000 Isoperibol oxygen bomb calorimeter (IKA Works, Inc, Wilmington, NC, USA). Ash content was determined by placing samples in a muffle furnace at 550 °C for 10 h. Moisture and dry matter were determined by placing samples in a drying oven at 105 °C for 20–24 h.

Table 2. Proximate composition of diets based on the control, black soldier fly larvae (BSFL) meal, yeast prebiotic and probiotic, and butyrate diet.

Proximate Composition	Unit	Control	BSFL Meal	Probiotic	Yeast Prebiotic	Butyrate
Dry matter	%	96.6	96.6	96.8	96.5	96.6
Crude protein (CP)	%	46.8	46.3	46.9	46.9	45.4
Crude lipid (CL)	%	12.1	12.8	12.5	11.9	12.6
Ash	%	9.8	9.1	9.7	9.8	10.3
Carbohydrate	%	28.0	28.5	27.7	27.9	28.2
Gross energy (GE)	GE	21.0	21.0	20.8	21.0	20.8

2.7. Real-Time Quantitative PCR (qPCR)

The primer sequences used for gene expression analysis were selected from previously published studies [9,17]. A list of primer sequences is provided in Table 3. Relative quantitative expression of 10 target genes, namely *TNF-α, IL-1β, hepcidin, MyD88, caspase-b, PGRP, serum amyloid A (SAA), angptl4, NF-κB (p65 subunit),* and *cxcl8,* were determined using the StepOnePlus™ Real-Time PCR System (Applied Biosystems, Burlington, ON, Canada). All qPCR reactions contained a total reaction volume of 10 µL containing 5 µL SYBR Green qPCR SuperMix, 0.5 µL of forward primer, 0.5 µL of reverse primer, 2 µL of nuclease-free water, and 2 µL of cDNA template. The PCR thermal conditions consisted of 50 °C for 2 min, 95 °C for 2 min, 40 cycles of 95 °C for 15 s, 60 °C for 30 s, and 72 °C for 30 s. Dissociation curves were generated at the end of PCR amplification to ensure the presence of single amplified products. The qPCR cycle threshold (Ct) values for each sample were obtained using StepOne Plus software (v2.3). *18s rRNA* was used as a housekeeping gene to normalize the target genes because it was the most stable in comparison to the other tested housekeeping gene, β-actin. The fold changes were determined with the comparative CT method.

Table 3. Forward and reverse primer sequences of zebrafish housekeeping and immune- and stress-related target genes. The annealing temperature for all primers was 60 °C.

Gene	Primer	Sequence 5′-3′	Accession No.
18S ribosomal RNA (18s rRNA/housekeeping gene)	F	TCGCTAGTTGGCATCGTTTATG	BX296557.35
	R	CGGAGGTTCGAAGACGATCA	
Tumor necrosis factor alpha (TNF-α)	F	AGGCAATTTCACTTCCAAGG	NM_212859
	R	AGGTCTTTGATTCAGAGTTGTATCC	
Interleukin 1 beta (IL-1β)	F	ATGCTCATGGCGAACGTC	NM_212844
	R	TGGTTTTAGTGTAAGACGGCACT	
Hepcidin	F	CACAGCCGTTCCCTTCATAC	NM_205583.2
	R	TCAGATGTTGGTTCTCCTGC	
Myeloid differentiation primary response 88 (MyD88)	F	TCCACAGGGACTGACACCTGAGA	NM_212814
	R	GCTGAGTCTTCAGCACAGCAGAT	
Caspase-b	F	ATGGAGGATATTACCCAG	NM_152884
	R	TCACAGTCCAGGAAAC	
Peptidoglycan recognition protein (PGRP)	F	ATGGGACGGTGTATGAAGGC	NM_001044321.1
	R	AGCATTGAGGTTGCCCATGA	
Serum Amlyloid A (SAA)	F	CGGGGTCCTGGGGGCTATTG	NM_001005599.1
	R	GTTGGGGTCTCCGCCGTTTC	
Nuclear factor kappa B p65 subunit (NF-κB/p65)	F	CCTACGGCTAAACGAACTCTAC	XM_005170796.4
	R	GACATGGCCTGCAGATACTT	
CXC motif chemokine ligand 8 (CXCL8/IL-8)	F	CCAGCTGAACTGAGCTCCTC	XM_001342570
	R	GGAGATCTGTCTGGACCCCT	
Angitopoietin-like 4 (Angptl4)	F	CGAGCGCATCAAGCAACA	JN606312–JN606321
	R	TCGCTCGTTTTTCATCGTAATCT	

2.8. Statistical Analysis

All control diets and their respective qPCR data were analyzed using a fixed-effect model. Initially, a one-way Analysis of Variance (ANOVA) was performed to assess the effects of the immune challenges (LPS or pathogen controls versus PBS control) in terms of the fold change in gene expression. Next, a one-way ANOVA was performed to assess the effects of the dietary treatments by comparing the fold changes in gene expression between the experimental diets and control diet within the immune challenge group. Residual analysis was performed to evaluate the assumptions of normality, homogeneity of variances, and independence. Appropriate transformations such as the log transformation were used when necessary to normalize the distribution of residual error. Differences among all treatments with adjustment for multiple comparisons were determined using Tukey's post hoc test. For the survival analysis, a Shapiro–Wilk test was performed to determine the normal distribution of the residuals, and a Levene test was used to determine normal homogeneity of the variance. Log transformations were performed on non-normal data, while ANOVA was performed to determine the significance of survival. Following this, TukeyHSD was performed to determine the differences between treatments. The qPCR data analysis was conducted using RStudio (version 2023.03.0+386), and the survival analysis was conducted using RStudio (version 2023.12.1+402). The qPCR data are presented as the least squares means +/− standard error of the mean (SEM), and the statistical sample size was $n = 3$. The threshold for significant differences was set at a p-value < 0.05, and trends are noted between p-values 0.10 and 0.05.

3. Results

The survival of the zebrafish ranged from 77 to 91%, but was not significant among the diets (p = 0.355). The individual survival rates for the control, BSFL meal, probiotic, yeast prebiotic, and butyrate diets were 85.3, 91, 89.3, 77.0, and 83.3%, respectively. There were no significant changes in weight observed over the 63-day feeding period amongst the diets; however, fish receiving the BSFL diet presented slightly higher final weight gain in comparison to the control diet (Figure 2). As the male and female zebrafish were not separated, and females tend to weigh more than their male counterparts, this may have contributed to the non-significance in the weight data.

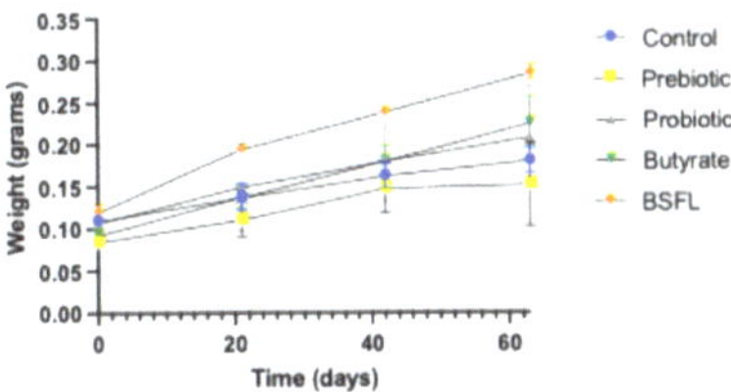

Figure 2. Average weights of adult zebrafish fed the control, prebiotic yeast, yeast probiotic, butyrate, and black soldier fly larvae (BSFL) meal dietary treatments measured on days 0, 20, 40, and 63 (p > 0.05). Each point is an average weight of the 3 replicate tanks for each diet. Data are expressed as mean ± SD, n = 3.

3.1. Effect of Immune Challenge Between Control Groups

Nuclear factor-κB (NF-κB/p65) was significantly upregulated in the control diet within the LPS and pathogen challenge groups in comparison to the PBS-injected group (p = 0.017), indicating that fish responded to the immune challenges (Figure 3). None of the other target genes, including *TNF-α, IL-1β, hepcidin, MyD88, caspase-b, PGRP, serum amyloid A (SAA), angptl4*, and *cxcl8*, were significantly impacted at this sampling time point.

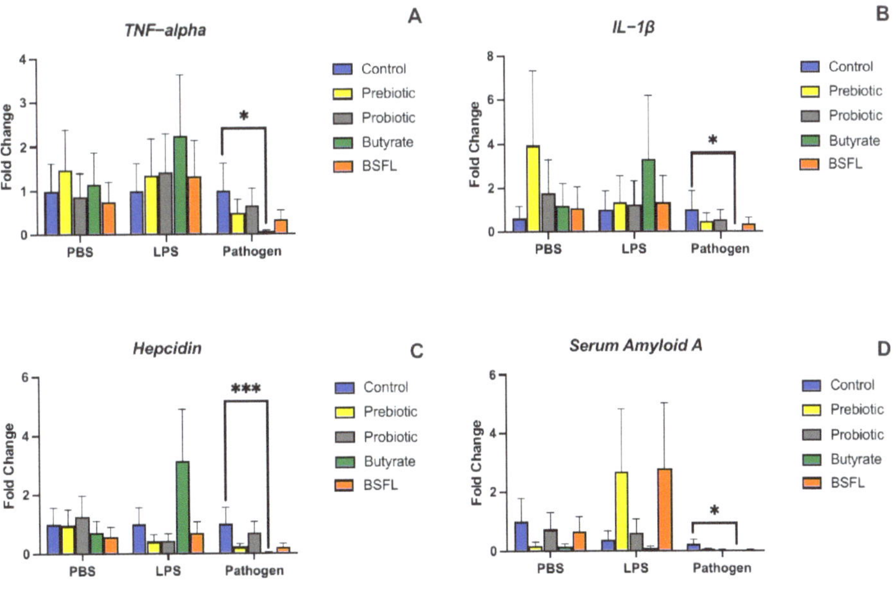

Figure 3. *Cont.*

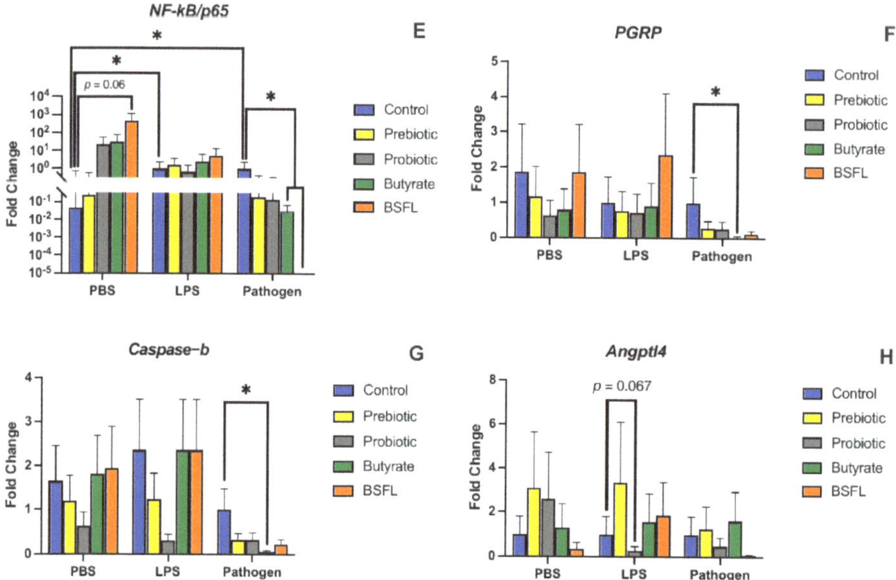

Figure 3. Expression of immune-relevant cytokine and protein genes relative to the housekeeping gene, 18s rRNA, following the 63-day zebrafish feeding trial of five diets (control, black soldier fly larvae (BSFL) meal, yeast probiotic, prebiotic yeast, and sodium butyrate) after a 24 h immune challenge with *Pseudomonas aeruginosa* lipopolysaccharide (LPS) or live *Flavobacterium psychrophilum* (pathogen). (**A,B**) The relative expression levels of pro-inflammatory cytokines, TNF-alpha and *IL-1β*. (**C,D**) The relative expression of the acute phase proteins, hepcidin and serum amyloid A in liver cells. (**E**) The relative expression level of the transcription factor, *NF-κB/p65*. (**F**) The relative expression level of the pattern recognition receptor protein, *PGRP* (peptidoglycan recognition protein). (**G**) The relative expression level of the protease, *caspase-b*. (**H**) The relative expression level of the protein, *angptl4* (angiopoietin-like 4). The sample size comprised 3 replicates (n = 3) of 3–4 pooled livers per subgroup (PBS, LPS or pathogen) per dietary treatment. Data are expressed as least square means ± SEM. In terms of significance, trends between p-values 0.10 and 0.05 are specifically noted, p-values < 0.05 are indicated by *, and p-values < 0.001 are indicated by ***.

3.2. Effect of Dietary Treatments Within the PBS Subgroup

Of all the target genes, only a trend for *NF-κB/p65* was noted (Figure 3). This gene was slightly upregulated in the BSFL group in comparison to the control diet (p = 0.063). The gene expression changes of *TNF-α*, *IL-1β*, *hepcidin*, *MyD88*, *caspase-b*, *PGRP*, *serum amyloid A (SAA)*, *angptl4*, *NF-κB (p65 subunit)*, and *cxcl8* did not statistically differ between the yeast prebiotic, probiotic, and butyrate diet and the control diet groups within the PBS subgroup.

3.3. Effect of Dietary Treatments Within the LPS Challenge Subgroup

Of all the target genes, only a trend for *angptl4* was observed (Figure 3), whereby *angptl4* was slightly downregulated in the probiotic group in comparison to the LPS control (p = 0.070). The gene expression changes of *TNF-α*, *IL-1β*, *hepcidin*, *MyD88*, *caspase-b*, *PGRP*, *serum amyloid A (SAA)*, *angptl4*, *NF-κB (p65 subunit)*, and *cxcl8* did not statistically differ between the yeast probiotic, BSFL meal, and butyrate diets and the control diet within the LPS subgroup.

3.4. Effect of Dietary Treatments Within the Pathogen Challenge Subgroup

In comparison to the pathogen control, the butyrate group displayed significant downregulation of *TNF-α*, *caspase-b*, *IL-1b*, *hepcidin*, *NF-κB/p65*, and *PGRP* (p < 0.050), and the BSFL group displayed significant downregulation of *NF-κB/p65* (p < 0.001) (Figure 3). The gene expression changes of *TNF-α, IL-1β, hepcidin, MyD88, caspase-b, PGRP, serum amyloid A (SAA), angptl4, NF-κB (p65 subunit)*, and *cxcl8* from the yeast prebiotic and probiotic diet groups were not statistically different from the control diet within the pathogen group.

4. Discussion

This study aimed to investigate the effects of supplementary yeast prebiotic, yeast probiotic, BSFL meal, and butyrate on the hepatic immune responses of zebrafish by identifying the changes in immune- and stress-related genes following immune challenges with *P. aeruginosa* LPS and live *F. psychrophilum* pathogen.

Nutrition and diet can considerably influence health by modulating the immune system and the intimately associated gut microbiome [18]. In aquaculture and fish nutrition research, the search for fishmeal alternatives is constantly being pursued due to its unsustainability and high cost. This feeding shift is leading many researchers to explore promising ingredients such as BSFL that can be grown on organic waste and do not directly compete with human food, unlike fishmeal [19]. Additionally, as an effort to limit the acceleration of antimicrobial resistance, in recent years, there has been increasing interest in providing nutrients to production species that support optimal immune function. For example, enriching the gut microbiome with favorable yeast species such as *Saccharomyces cerevisiae* via probiotic and postbiotic supplementation has been shown to increase the immune response and disease resistance of fish species [8]. Sodium butyrate has also shown promising results in reducing LPS-induced zebrafish endotoxemia [9]. This study investigated the immunomodulatory properties of prebiotic yeast, probiotic yeast, butyrate, and BSFL meal in response to immune challenges using *P. aeruginosa*-sourced LPS and live *F. psychrophilum*. This study aimed to assess the host response within each dietary group by comparing the hepatic expression of immune genes to their corresponding positive control groups using qPCR, which included *TNF-α, IL-1β, hepcidin, MyD88, caspase-b, PGRP, SAA, angptl4, NF-κB (p65 subunit)*, and *cxcl8 (IL-8)*. The functional feed additives used in our study, including yeast prebiotic, probiotic yeast, and butyrate, were applied post-pelleting via lipid coating to preserve their stability and efficacy, a widely accepted practice in the aquaculture and animal feed industries [20]. Minor differences in total diet weight were due to the addition of small amounts of these additives, which constituted less than 1% of the diet and had negligible effects on proximate composition and nutrient balance [21]. Top-coating sensitive additives is a recommended approach for maintaining their functional integrity, as supported by product guidelines and previous studies [22]. This methodology aligns with industry standards and ensures that the observed functional effects are attributable to the presence of the additives rather than minor variations in inclusion levels.

4.1. LPS and F. psychrophilum Cause a Significant Immune Response

The control diet was compared between PBS and LPS and pathogen-challenged fish to assess whether these immune challenges were effective in eliciting an immune response. As expected, *NF-κB (p65 subunit)* transcription factor, which mediates inflammatory signaling, was significantly upregulated within the LPS and pathogen challenge groups (Figure 3E) [23]. This study focused on the expression of NF-κB/p65 subunit (also known as RelA), which is involved in the canonical NF-kB signaling pathway [24,25]. The canonical NF-kB pathway is receptive to a large variety of immune-related receptors, including pattern recognition receptors (PRRs) such as TLRs and PGRP, but is also known to be involved in chronic inflammatory diseases such as rheumatoid arthritis and inflammatory bowel disease through overproduction of pro-inflammatory cytokines and chemokines [5,24,25]. Although downstream inflammatory genes were not induced in this study by either immune challenge, possibly because of the chosen sampling time point or tissue, induction of *NF-κB (p65 subunit)* demonstrated that fish responded to the immune challenges.

The results from the LPS and pathogen challenges within the present study are in partial agreement with previous fish studies. Previous research, for example, has shown that a *P. aeruginosa* LPS challenge elicited the upregulation of *NF κB/p65* and *PGRP* in larval zebrafish and the upregulation of complement genes in adult zebrafish according to whole-body tissue analysis [9,26–28].

In zebrafish, the liver has been found to play an important role in the immune response, especially regarding NF-κB activity [29]. Previous studies have demonstrated NF-κB activation of zebrafish liver cells after in vitro infection with snakehead rhabdovirus (SHRV), or *Edwardsiella tarda* [30]. However, the expression of the other inflammatory genes, such as *TNF-α, IL-1β, MyD88, PGRP, SAA*, and *caspase-b*, did not show upregulation in this study, differing from other research results (Figure 3) [9,26,31]. These differences may be attributed to variations in bacterial strains, tissues interrogated, sampling time points, and antigen forms. Although the sample size (n = 3) may have contributed to this study's observed variability, we took great effort to control for this by pooling 3 to 4 livers from individual fish for each sample.

4.2. BSFL Upregulated Immune Genes Following PBS Injection

In the PBS-injected fish, there was a slight upregulation of *NF-κB/p65* in the BSFL meal group compared to the control group, which can be possibly attributed to the chitin in insect-derived ingredients (Figure 3E) [11,32]. A recent meta-analysis revealed that BSFL meal products significantly influenced immune response parameters such as lysozyme activity and *TNF-α* and *IL-10* expression in some fish species, including rainbow trout and zebrafish [33]. Previous research has found that 1% chitin inclusion in the diet resulted in increased lysozyme activity as well as upregulation of *IFN-γ2, IL-10, TNF-α*, and *NF-κB/p65* mRNA expressions in common carp (*Cyprinus carpio*) and grass carp (*Ctenopharyngodon idella*) [34,35]. However, *TNF-α* was not significantly upregulated by the BSFL diet group within the PBS subgroup in the present study.

4.3. Effects of Diet on LPS-Challenged Fish

To investigate the influence of yeast prebiotic, yeast probiotic, butyrate, and BSFL meal on the immune responses of zebrafish, expressions of key hepatic immune-related genes induced by *P. aeruginosa*-derived LPS were analyzed. In the LPS-injected fish, the probiotic diet showed a downward trend in *angptl4* expression (Figure 3H). As far as we know, this is the first study to assess *angptl4* expression in zebrafish fed a probiotic diet followed by an LPS challenge. Angiopoietin-like protein 4 (angptl4) regulates lipid metabolism by inhibiting the enzyme lipoprotein lipase [36]. Interestingly, *angptl4* expression levels have also been correlated with social stress and anxiety in young adult zebrafish [37]. Alnassar et al. (2023), for example, observed that isolated zebrafish displaying signs of anxiety had decreased *angptl4* expression, which was reversed upon reintroduction to a social group [37]. Unfortunately, we did not consider assessing fish behavior in the present study. In addition, peroxisome proliferator-activator receptor (PPAR) (PPAR)-γ, which plays a role in maintaining the homeostasis of the immune system by regulating the differentiation and polarization of various immune cells, has been identified to be an activator of *angptl4* expression [36,38]. Specifically, LPS exposure reduces *PPAR-γ* expression via an NF-κB-dependent mechanism where, upon activation of TLR4, NF-κB drives down *PPAR-γ* expression [5,39]. Even though zebrafish TLR4 receptors do not recognize LPS, a significant reduction in *PPARγ* expression has also been observed in zebrafish exposed to LPS [40]. Thus, it is possible that the downregulated *angptl4* in the current study may be attributed to LPS-mediated reduction in PPAR-γ, although we did not consider *PPAR-γ* when formulating our gene list.

It is unclear whether the decrease in *angptl4* expression was due to LPS-induced stress or prebiotic yeast supplementation. Yeast autolysate and probiotics' impact on stress is novel, but one study showed that zebrafish fed mannan oligosaccharide (MOS) (4 g/kg) had lower stress and anxiety without growth being affected [41]. Provided that social stress and anxiety can potentially be linked with *angptl4* expression, it is possible that yeast cell wall components, such as MOS, can help alleviate stress and anxiety in zebrafish. However, this does not explain the lack of *angptl4* expression changes in the prebiotic yeast group. *S. cerevisiae* has also been observed to influence other stress tolerance- and immune-related parameters in Nile tilapia (*Oreochromis niloticus*) [42,43]. In one study, researchers observed that the yeast-fed juvenile Nile tilapia exhibited greater tolerance to acute heat and hypoxia following stress challenges [43]. A different study also demonstrated downregulation of *TNF-α* and *IL-1β* expression in the liver tissue of Nile tilapia that were fed fermented *S. cerevisiae* [42]. However, we did not observe any significant changes in the expression of *TNF-α* and *IL-1β* expression within our yeast probiotic diet group. It is possible that these reported changes may have been due to differences in the prebiotic and probiotic form, inclusion amount, and sampling time. Overall, this study suggests that probiotic yeast has a potential influence on lipid metabolism and stress in zebrafish upon receiving LPS through the downregulation of *angptl4* expression.

4.4. BSFL and Butyrate Downregulated Immune Genes After F. psychrophilum Challenge

In zebrafish, *F. psychrophilum* infection activates the innate immune system; prevents myeloid cell differentiation; and causes damage to the fins, muscles, and caudal peduncle, similar to signs in commercial fish species like salmonids [44,45]. In the pathogen-challenged diet groups, the BSFL diet group in the current study exhibited significant downregulation of *NF-kB/p65* expression ($p < 0.0001$; Figure 3). The butyrate diet group also displayed significantly downregulated expression of *TNF-α*, *caspase-b*, *IL-1β*, *hepcidin*, *NF-kB/p65*, and *PGRP* ($p < 0.05$), with a trending decrease in *SAA* expression.

The 10% inclusion of defatted BSFL significantly attenuated the expression of *NF-κB/p65* mRNA following a live *F. psychrophilum* challenge in comparison to the PBS subgroup, which experienced significantly elevated levels. Hence, it appears that in the absence of an immune challenge, the BSFL diet activated the *NF-κB* signaling pathway during the 63-day feeding trial period, which was then attenuated upon pathogen introduction. A recent study involving gilthead seabream (*Sparus aurata*) revealed that 15% inclusion of de-fatted BSFL meal significantly upregulated *IL-1β* and generally downregulated *TNF-α* at the end of a 67-day feeding trial [46]. While *IL-1β* and *TNF-α* also appeared to be downregulated by the BSFL diet in the present study, these changes were not statistically significant. Other insect meals, such as yellow mealworm (*Tenebrio molitor*) at 24 and 36% inclusion in the diet, have been found to significantly upregulate the expression of *p65 (RelA)* and *MyD88* in the intestine of largemouth bass after an *Edwardsiella tarda* challenge [47]. While the expression of *NF-κB/p65* from this study varied from others, perhaps due to timing, the overall results agree with BSF's immune-stimulating properties. However, there is still uncertainty regarding the mechanism behind the BSFL diet's influence on the animal's response to *F. psychrophilum*, or how *F. psychrophilum* resulted in the downregulation of *NF-κB/p65* within the BSFL diet group. In mammals, chitin is recognized by the immune system as a microbial-associated molecular pattern (MAMP) by membrane-bound pattern-recognition receptors (PRRs) on innate immune cells such as macrophages [48]. Ligation to this MAMP in turn triggers various signaling cascades to stimulate cytokine production, specifically through TLR2 and dectin-1 signaling [48]. Similarly, in zebrafish, TLR2 plays a role in the protection against bacteria such as *Mycobacterium marinum* and the recognition of other MAMPs such as flagellin [31,49]. In addition to TLR2, C-type lectin receptors (CLRs) are also utilized by zebrafish to recognize other types of MAMPs such as zymosans, which is a mixture of β-glucans, MOS, proteins, and chitin [50,51]. Hence, it is possible that zebrafish chitin recognition in the present study leading to activation of NF-κB is mediated by TLR2 and/or CLR [50]. This may explain the upregulation of NF-κB/p65 expression in the present study in response to the control PBS injection. Provided that chitin can activate TLR2 signaling, which is also involved in LPS recognition, it is possible that chitin exposure also leads to cross-tolerance to LPS through the chemokine receptor CXCR4, which has been previously reported to occur in response to other MAMPs [6,48,52]. Since *F. psychrophilum* and possibly chitin both act on TLR2, which subsequently activates the NF κB signaling pathway, the development of chitin cross tolerance may explain the downregulation of *NF-κB/p65* expression following the pathogen challenge.

Aside from chitin, BSFL also contains antimicrobial peptides (AMPs) and fatty acids such as lauric acid (also known as dodecanoic acid), which modulate the immune system [11,53]. AMPs including attacin, cecropin, defensin, and diptericin have all been identified in black soldier flies [54,55]. Defensins have been shown to inhibit the NF-κB signaling pathway by downregulating the expression of *NF-κB/p65* in immune-challenged mice models, while cecropin has been observed to significantly upregulate *NF-κB/p65* [56,57]. Moreover, lauric acid has also been studied for its antimicrobial, antitumor, and anti-inflammatory activity over the years [53]. Notably, it has been observed to downregulate *IL-6* and *TNF-α* expression and increase serum levels of anti-inflammatory cytokines such as TGF-β1 and IL-10 [58,59]. These components may explain BSFL's influence on the inflammatory response, but further studies are needed to investigate their effects on NF-κB activation.

When examining the immune effects associated with the butyrate diet, the findings from this study demonstrate its potency in attenuating pro-inflammatory responses to *F. psychrophilum*. A recent study that conducted intestinal gene expression profiling of *F. psychrophilum-infected* rainbow trout revealed higher expression levels of *TNF-α* and *IL-1β* [60,61]. Additionally, *F. psychrophilum* infection incites heavy infiltration of neutrophils surrounding the eroded tissues, thereby causing unregulated production of reactive oxygen species (ROS), furthering inflammatory tissue injury [60,62]. The consensus of a profile of an infected salmonid appears to consist of excessive gene expression of *IL-1β*, *TNF-α*, and *IL-10* in spleen tissue, coupled with repressed levels of *transforming growth factor β (TGF-β)* and tight junction-associated proteins in spleen and intestinal tissue in infected rainbow trout [60,63]. Consequently, *F. psychrophilum* infection eventually manifests itself in severe outcomes such as high mortality rates, skin erosion, anemia, gill hemorrhages, necrotic myositis, and septicemia despite antibiotic treatment [62,64]. In recent times, butyrate supplementation in fish has been recognized to ameliorate those pathological immune gene profiles in several studies, which is also supported by the findings from this study [9,65,66].

Butyrate immersion of zebrafish has been shown to exert anti-inflammatory effects by reducing the recruitment of neutrophils and TNF-α-positive M1 macrophage polarization at wound sites via the ortholog of the mammalian hydrocarboxylic acid receptor (*hcar1*) [67]. In turn, this suggests that butyrate reduces excessive ROS produced by neutrophils and the secretion of pro-inflammatory cytokines. The butyrate diet group in this study also exhibited decreased mRNA expression of *TNF-α* and *IL-1β*, which are two markers of pro-inflammatory M1 macrophages. Previous research has indicated that butyrate induces macrophage differentiation into the M2 phenotype, which is mainly involved in anti-inflammatory responses in comparison to M1 macrophages [68,69]. In future studies, it would be interesting to specifically look at macrophage polarization in butyrate-exposed zebrafish. Overall, butyrate supplementation and exposure may elicit anti-inflammatory and anti-microbial properties without exacerbating ROS-associated tissue damage, which is especially important, as fin erosion is common in *F. psychrophilum* pathogenesis. Furthermore, expression of *hepcidin*, the primary regulator of systemic iron homeostasis, was significantly downregulated [70]. *F. psychrophilum* infection is commonly characterized by anemia, which is heavily associated with impaired erythropoiesis, accelerated hemorrhage, or blood loss in teleost fish, as well as increased *hepcidin* expression [71,72]. Hepcidin is a peptide hormone primarily expressed in the liver that plays a central role in iron metabolism, and during infection, it binds to the iron exporter, ferroportin, to block the release of iron from macrophages, hepatocytes, and enterocytes, presumably to restrict iron availability to pathogens [72]. Numerous pathogenic bacteria, including *F. psychrophilum*, have evolved various iron acquisition mechanisms to obtain free iron from the host, which can impair iron absorption and erythropoiesis in the host, leading to a condition known as anemia of inflammation [73,74]. In a recent study, butyrate supplementation was shown to mitigate inflammation-induced anemia in mice by upregulating ferroportin, a molecular target of hepcidin that works to release iron from storage, as well as reduce *TNF-α* production in macrophages [75]. Thus, butyrate potentially improves iron availability by both increasing iron release through upregulation of ferroportin and decreasing iron sequestration by downregulating hepcidin. This study demonstrates that the inclusion of 0.05% sodium butyrate in the zebrafish diet significantly attenuates pro-inflammatory marker expression in response to live *F. psychrophilum* when compared to the challenged control diet group.

5. Conclusions

The findings from this study provide insight on the immunomodulatory properties of *S. cerevisiae* probiotic, BSFL meal, and butyrate when supplemented in zebrafish. The BSFL meal diet demonstrated both immunostimulatory and anti-inflammatory effects, as seen with the changes in *NF-κB/p65* expression between the PBS and pathogen subgroups, which may possibly be attributed to chitins, AMPs, and lauric acid that are found in BSFL. For the LPS challenge, the probiotic diet exhibited downregulated *angptl4*, suggesting an influence on lipid metabolism and stress levels. Additional studies should be conducted to validate the role of *angptl4* in social stress and its connection to probiotic supplementation. Lastly, sodium butyrate supplementation significantly downregulated various pro-inflammatory genes during an *F. psychrophilum* infection, indicating that it may have therapeutic potential in terms of BCWD or rainbow trout fry syndrome. Therefore, future studies should be carried out to investigate the mechanism behind the anti-inflammatory effects of butyrate, how BSFL or its immunomodulatory components influence pro-inflammatory markers depending on the immune status of the animal, and the immunometabolism effect of probiotics and prebiotic yeast.

Author Contributions: Conceptualization, D.H., J.S.L., U.K.S. and N.A.K.; formal analysis, U.K.S., A.B.B. and N.G.; investigation, N.G. and J.Z.; methodology, N.G., U.K.S. and J.Z.; writing—original draft, N.G.; writing—review and editing, D.H., U.K.S. and N.A.K. All authors have read and agreed to the published version of the manuscript.

Funding: This work was financially supported by Dr. Niel Karrow's NSERC Discovery grant (400232) and by Dr. David Huyben's OMAFRA (UG-T1-2021-101077) and NSERC Alliance grants (ALLRP-568553-21).

Institutional Review Board Statement: All animal utilization protocols involved in this study were approved by the Animal Care Committee of the University of Guelph, Ontario, Canada (protocol code AUP # 4076).

Data Availability Statement: The datasets used and analyzed during the current study are available from the corresponding author upon request.

Acknowledgments: We would like to thank Matt Cornish and Mike Davies at the Hagen Aqualab as well as Kayla Price and Philip Lyons from Alltech for donating the prebiotic and probiotic yeast. Also, thanks to Maddie Borland, Carmi Riesenbach, Cody Anderson, and Megan Woo for their help sampling and weighing fish.

Conflicts of Interest: The authors declare no conflicts of interest.

References

1. Naylor, R.; Fang, S.; Fanzo, J. A Global View of Aquaculture Policy. *Food Policy* **2023**, *116*, 102422. [CrossRef]
2. Schar, D.; Klein, E.Y.; Laxminarayan, R.; Gilbert, M.; Van Boeckel, T.P. Global Trends in Antimicrobial Use in Aquaculture. *Sci. Rep.* **2020**, *10*, 21878. [CrossRef] [PubMed]
3. Knupp, C.; Wiens, G.D.; Faisal, M.; Call, D.R.; Cain, K.D.; Nicolas, P.; Van Vliet, D.; Yamashita, C.; Ferguson, J.A.; Meuninck, D.; et al. Large-Scale Analysis of *Flavobacterium psychrophilum* Multilocus Sequence Typing Genotypes Recovered from North American Salmonids Indicates That Both Newly Identified and Recurrent Clonal Complexes Are Associated with Disease. *Appl. Environ. Microbiol.* **2019**, *85*, e02305–e02318. [CrossRef] [PubMed]
4. Hussein, M.A.; El-tahlawy, A.S.; Abdelmoneim, H.M.; Abdallah, K.M.E.; El Bayomi, R.M. *Pseudomonas aeruginosa* in Fish and Fish Products: A Review on the Incidence, Public Health Significance, Virulence Factors, Antimicrobial Resistance, and Biofilm Formation. *J. Adv. Vet. Res.* **2023**, *13*, 1464–1468.

5. Necela, B.M.; Su, W.; Thompson, E.A. Toll-like Receptor 4 Mediates Cross-talk between Peroxisome Proliferator-activated Receptor γ and Nuclear factor-κB in Macrophages. *Immunology* **2008**, *125*, 344–358. [CrossRef]

6. Novoa, B.; Bowman, T.V.; Zon, L.; Figueras, A. LPS Response and Tolerance in the Zebrafish (*Danio rerio*). *Fish Shellfish Immunol.* **2009**, *26*, 326–331. [CrossRef]

7. Landolt, M.L. The Relationship between Diet and the Immune Response of Fish. *Aquaculture* **1989**, *79*, 193–206. [CrossRef]

8. Del Valle, J.C.; Bonadero, M.C.; Fernández-Gimenez, A.V. Saccharomyces Cerevisiae as Probiotic, Prebiotic, Synbiotic, Postbiotics and Parabiotics in Aquaculture: An Overview. *Aquaculture* **2023**, *569*, 739342. [CrossRef]

9. Wang, M.X.; Shandilya, U.K.; Wu, X.; Huyben, D.; Karrow, N.A. Assessing Larval Zebrafish Survival and Gene Expression Following Sodium Butyrate Exposure and Subsequent Lethal Bacterial *Lipopolysaccharide* (LPS) Endotoxin Challenge. *Toxins* **2023**, *15*, 588. [CrossRef]

10. Abdel-Latif, H.M.R.; Abdel-Tawwab, M.; Khalil, R.H.; Metwally, A.A.; Shakweer, M.S.; Ghetas, H.A.; Khallaf, M.A. Black Soldier Fly (*Hermetia illucens*) Larvae Meal in Diets of European Seabass: Effects on Antioxidative Capacity, Non-Specific Immunity, Transcriptomic Responses, and Resistance to the Challenge with *Vibrio alginolyticus*. *Fish Shellfish Immunol.* **2021**, *111*, 111–118. [CrossRef]

11. Koutsos, E.; Modica, B.; Freel, T. Immunomodulatory Potential of Black Soldier Fly Larvae: Applications beyond Nutrition in Animal Feeding Programs. *Transl. Anim. Sci.* **2022**, *6*, txac084. [CrossRef]

12. Boyd, C.E.; D'Abramo, L.R.; Glencross, B.D.; Huyben, D.C.; Juarez, L.M.; Lockwood, G.S.; McNevin, A.A.; Tacon, A.G.J.; Teletchea, F.; Tomasso, J.R.; et al. Achieving Sustainable Aquaculture: Historical and Current Perspectives and Future Needs and Challenges. *J. World Aquac. Soc.* **2020**, *51*, 578–633. [CrossRef]

13. Jørgensen, L.V.G. Zebrafish as a Model for Fish Diseases in Aquaculture. *Pathogens* **2020**, *9*, 609. [CrossRef] [PubMed]

14. Meyers, J.R. Zebrafish: Development of a Vertebrate Model Organism. *CP Essent. Lab. Tech.* **2018**, *16*, e19. [CrossRef]

15. Leber, A.L. (Ed.) *Clinical Microbiology Procedures Handbook*, 4th ed.; ASM Press: Washington, DC, USA, 2016; ISBN 978-1-55581-880-7.

16. Jarau, M.; MacInnes, J.I.; Lumsden, J.S. Erythromycin and Florfenicol Treatment of Rainbow Trout *Oncorhynchus mykiss* (Walbaum) Experimentally Infected with *Flavobacterium psychrophilum*. *J. Fish Dis.* **2019**, *42*, 325–334. [CrossRef]

17. Camp, J.G.; Jazwa, A.L.; Trent, C.M.; Rawls, J.F. Intronic Cis-Regulatory Modules Mediate Tissue-Specific and Microbial Control of Angptl4/Fiaf Transcription. *PLoS Genet.* **2012**, *8*, e1002585. [CrossRef]

18. Childs, C.E.; Calder, P.C.; Miles, E.A. Diet and Immune Function. *Nutrients* **2019**, *11*, 1933. [CrossRef]

19. Oteri, M.; Di Rosa, A.R.; Lo Presti, V.; Giarratana, F.; Toscano, G.; Chiofalo, B. Black Soldier Fly Larvae Meal as Alternative to Fish Meal for Aquaculture Feed. *Sustainability* **2021**, *13*, 5447. [CrossRef]

20. El-Saadony, M.T.; Alagawany, M.; Patra, A.K.; Kar, I.; Tiwari, R.; Dawood, M.A.O.; Dhama, K.; Abdel-Latif, H.M.R. The Functionality of Probiotics in Aquaculture: An Overview. *Fish Shellfish Immunol.* **2021**, *117*, 36–52. [CrossRef]

21. Huyben, D.; Chiasson, M.; Lumsden, J.S.; Pham, P.H.; Chowdhury, M.A.K. Dietary Microencapsulated Blend of Organic Acids and Plant Essential Oils Affects Intestinal Morphology and Microbiome of Rainbow Trout (*Oncorhynchus mykiss*). *Microorganisms* **2021**, *9*, 2063. [CrossRef]

22. Pech-Canul, A.D.L.C.; Ortega, D.; García-Triana, A.; González-Silva, N.; Solis-Oviedo, R.L. A Brief Review of Edible Coating Materials for the Microencapsulation of Probiotics. *Coatings* **2020**, *10*, 197. [CrossRef]

23. Schlein, L.J.; Thamm, D.H. Review: NF-kB Activation in Canine Cancer. *Vet. Pathol.* **2022**, *59*, 724–732. [CrossRef]

24. Ouyang, G.; Liao, Q.; Zhang, D.; Rong, F.; Cai, X.; Fan, S.; Zhu, J.; Wang, J.; Liu, X.; Liu, X.; et al. Zebrafish NF-κB/P65 Is Required for Antiviral Responses. *J. Immunol.* **2020**, *204*, 3019–3029. [CrossRef]

25. Lawrence, T. The Nuclear Factor NF-B Pathway in Inflammation. *Cold Spring Harb. Perspect. Biol.* **2009**, *1*, a001651. [CrossRef] [PubMed]

26. Zhou, J.; Gu, X.; Fan, X.; Zhou, Y.; Wang, H.; Si, N.; Yang, J.; Bian, B.; Zhao, H. Anti-Inflammatory and Regulatory Effects of Huanglian Jiedu Decoction on Lipid Homeostasis and the TLR4/MyD88 Signaling Pathway in LPS-Induced Zebrafish. *Front. Physiol.* **2019**, *10*, 1241. [CrossRef] [PubMed]

27. Ko, E.-Y.; Cho, S.-H.; Kwon, S.-H.; Eom, C.-Y.; Jeong, M.S.; Lee, W.; Kim, S.-Y.; Heo, S.-J.; Ahn, G.; Lee, K.P.; et al. The Roles of NF-κB and ROS in Regulation of pro-Inflammatory Mediators of Inflammation Induction in LPS-Stimulated Zebrafish Embryos. *Fish Shellfish Immunol.* **2017**, *68*, 525–529. [CrossRef]

28. Wang, Z.; Han, Y. Response of Gene Expression to LPS Challenge Manifests the Ontogeny and Maturation of the Complement System in Zebrafish Larvae. *J. Mar. Biol. Ass.* **2013**, *93*, 1965–1971. [CrossRef]

29. Sullivan, C.; Kim, C.H. Zebrafish as a Model for Infectious Disease and Immune Function. *Fish Shellfish Immunol.* **2008**, *25*, 341–350. [CrossRef]

30. Phelan, P.E.; Mellon, M.T.; Kim, C.H. Functional Characterization of Full-Length TLR3, IRAK-4, and TRAF6 in Zebrafish (*Danio rerio*). *Mol. Immunol.* **2005**, *42*, 1057–1071. [CrossRef]

31. Yang, D.; Zheng, X.; Chen, S.; Wang, Z.; Xu, W.; Tan, J.; Hu, T.; Hou, M.; Wang, W.; Gu, Z.; et al. Sensing of Cytosolic LPS through Caspy2 Pyrin Domain Mediates Noncanonical Inflammasome Activation in Zebrafish. *Nat. Commun.* **2018**, *9*, 3052. [CrossRef]

32. Nunes, C.S.; Philipps-Wiemann, P. Chitinases. In *Enzymes in Human and Animal Nutrition*; Elsevier: Amsterdam, The Netherlands, 2018; pp. 361–378, ISBN 978-0-12-805419-2.

33. Xiao, Y.; Zhu, L.; Liang, R.; Su, J.; Yang, J.; Cao, X.; Lu, Y.; Yu, Y.; Hu, J. A Meta-analysis of the Effects of Black Soldier Fly Meal on Fish Immune Response and Antioxidant Capacity. *Comp. Immunol. Rep.* **2024**, *7*, 200162. [CrossRef]

34. Gopalakannan, A.; Arul, V. Immunomodulatory Effects of Dietary Intake of Chitin, Chitosan and Levamisole on the Immune System of *Cyprinus carpio* and Control of *Aeromonas hydrophila* Infection in Ponds. *Aquaculture* **2006**, *255*, 179–187. [CrossRef]
35. Harikrishnan, R.; Devi, G.; Van Doan, H.; Balasundaram, C.; Thamizharasan, S.; Hoseinifar, S.H.; Abdel-Tawwab, M. Effect of Diet Enriched with *Agaricus bisporus* Polysaccharides (ABPs) on Antioxidant Property, Innate-Adaptive Immune Response and pro-Anti Inflammatory Genes Expression in *Ctenopharyngodon idella* against *Aeromonas hydrophila*. *Fish Shellfish Immunol.* **2021**, *114*, 238–252. [CrossRef] [PubMed]
36. Mattijssen, F.; Alex, S.; Swarts, H.J.; Groen, A.K.; Van Schothorst, E.M.; Kersten, S. Angptl4 Serves as an Endogenous Inhibitor of Intestinal Lipid Digestion. *Mol. Metab.* **2014**, *3*, 135–144. [CrossRef] [PubMed]
37. Alnassar, N.; Hillman, C.; Fontana, B.D.; Robson, S.C.; Norton, W.H.J.; Parker, M.O. Angptl4 Gene Expression as a Marker of Adaptive Homeostatic Response to Social Isolation across the Lifespan in Zebrafish. *Neurobiol. Aging* **2023**, *131*, 209–221. [CrossRef]
38. Christofides, A.; Konstantinidou, E.; Jani, C.; Boussiotis, V.A. The Role of Peroxisome Proliferator-Activated Receptors (PPAR) in Immune Responses. *Metabolism* **2021**, *114*, 154338. [CrossRef]
39. Simonin, M.-A.; Bordji, K.; Boyault, S.; Bianchi, A.; Gouze, E.; Bécuwe, P.; Dauça, M.; Netter, P.; Terlain, B. PPAR-γ Ligands Modulate Effects of LPS in Stimulated Rat Synovial Fibroblasts. *Am. J. Physiol. Cell Physiol.* **2002**, *282*, C125–C133. [CrossRef]
40. Zhang, Y.; Wang, C.; Jia, Z.; Ma, R.; Wang, X.; Chen, W.; Liu, K. Isoniazid Promotes the Anti-Inflammatory Response in Zebrafish Associated with Regulation of the PPARγ/NF-κB/AP-1 Pathway. *Chem.-Biol. Interact.* **2020**, *316*, 108928. [CrossRef]
41. Forsatkar, M.N.; Nematollahi, M.A.; Rafiee, G.; Farahmand, H.; Martínez-Rodríguez, G. Effects of Prebiotic Mannan Oligosaccharide on the Growth, Survival, and Anxiety-like Behaviors of Zebrafish (*Danio rerio*). *J. Appl. Aquac.* **2017**, *29*, 183–196. [CrossRef]
42. Abu-Elala, N.M.; Younis, N.A.; AbuBakr, H.O.; Ragaa, N.M.; Borges, L.L.; Bonato, M.A. Influence of Dietary Fermented Saccharomyces Cerevisiae on Growth Performance, Oxidative Stress Parameters, and Immune Response of Cultured *Oreochromis niloticus*. *Fish Physiol. Biochem.* **2020**, *46*, 533–545. [CrossRef]
43. Abass, D.A.; Obirikorang, K.A.; Campion, B.B.; Edziyie, R.E.; Skov, P.V. Dietary Supplementation of Yeast (*Saccharomyces cerevisiae*) Improves Growth, Stress Tolerance, and Disease Resistance in Juvenile Nile Tilapia (*Oreochromis niloticus*). *Aquacult. Int.* **2018**, *26*, 843–855. [CrossRef]
44. Avendaño-Herrera, R.; Benavides, I.; Espina, J.A.; Soto-Comte, D.; Poblete-Morales, M.; Valdés, J.A.; Feijóo, C.G.; Reyes, A.E. Zebrafish (*Danio rerio*) as an Animal Model for Bath Infection by *Flavobacterium psychrophilum*. *J. Fish Dis.* **2020**, *43*, 561–570. [CrossRef] [PubMed]
45. Huyben, D.; Jarau, M.; MacInnes, J.; Stevenson, R.; Lumsden, J. Impact of Infection with *Flavobacterium psychrophilum* and Antimicrobial Treatment on the Intestinal Microbiota of Rainbow Trout. *Pathogens* **2023**, *12*, 454. [CrossRef]
46. Moutinho, S.; Oliva-Teles, A.; Fontinha, F.; Martins, N.; Monroig, Ó.; Peres, H. Black Soldier Fly Larvae Meal as a Potential Modulator of Immune, Inflammatory, and Antioxidant Status in Gilthead Seabream Juveniles. *Comp. Biochem. Physiol. Part B Biochem. Mol. Biol.* **2024**, *271*, 110951. [CrossRef]
47. Ge, C.; Liang, X.; Wu, X.; Wang, J.; Wang, H.; Qin, Y.; Xue, M. Yellow Mealworm (*Tenebrio molitor*) Enhances Intestinal Immunity in Largemouth Bass (*Micropterus salmoides*) via the NFκB/Survivin Signaling Pathway. *Fish Shellfish Immunol.* **2023**, *136*, 108736. [CrossRef]
48. Elieh Ali Komi, D.; Sharma, L.; Dela Cruz, C.S. Chitin and Its Effects on Inflammatory and Immune Responses. *Clin. Rev. Allergy Immunol.* **2018**, *54*, 213–223. [CrossRef]
49. Hu, W.; Yang, S.; Shimada, Y.; Münch, M.; Marín-Juez, R.; Meijer, A.H.; Spaink, H.P. Infection and RNA-Seq Analysis of a Zebrafish Tlr2 Mutant Shows a Broad Function of This Toll-like Receptor in Transcriptional and Metabolic Control and Defense to *Mycobacterium marinum* Infection. *BMC Genom.* **2019**, *20*, 878. [CrossRef] [PubMed]
50. Glass, E.; Robinson, S.L.; Rosowski, E.E. Zebrafish Use Conserved CLR and TLR Signaling Pathways to Respond to Fungal PAMPs in Zymosan. *bioRxiv* **2024**. [CrossRef] [PubMed]
51. Takeda, K.; Akira, S. Microbial Recognition by Toll-like Receptors. *J. Dermatol. Sci.* **2004**, *34*, 73–82. [CrossRef]
52. Novoa, B.; Figueras, A. Zebrafish: Model for the Study of Inflammation and the Innate Immune Response to Infectious Diseases. In *Current Topics in Innate Immunity II*; Lambris, J.D., Hajishengallis, G., Eds.; Advances in Experimental Medicine and Biology; Springer: New York, NY, USA, 2012; Volume 946, pp. 253–275, ISBN 978-1-4614-0105-6.
53. Ameena, M.; Arumugham, M.; Ramalingam, K.; Shanmugam, R. Biomedical Applications of Lauric Acid: A Narrative Review. *Cureus* **2024**, *16*, e62770. [CrossRef]
54. Vogel, H.; Müller, A.; Heckel, D.G.; Gutzeit, H.; Vilcinskas, A. Nutritional Immunology: Diversification and Diet-Dependent Expression of Antimicrobial Peptides in the Black Soldier Fly *Hermetia illucens*. *Dev. Comp. Immunol.* **2018**, *78*, 141–148. [CrossRef] [PubMed]
55. Vogel, M.; Shah, P.N.; Voulgari-Kokota, A.; Maistrou, S.; Aartsma, Y.; Beukeboom, L.W.; Salles, J.F.; Van Loon, J.J.A.; Dicke, M.; Wertheim, B. Health of the Black Soldier Fly and House Fly under Mass-Rearing Conditions: Innate Immunity and the Role of the Microbiome. *J. Insects Food Feed* **2022**, *8*, 857–878. [CrossRef]
56. Zeng, L.; Tan, J.; Xue, M.; Liu, L.; Wang, M.; Liang, L.; Deng, J.; Chen, W.; Chen, Y. An Engineering Probiotic Producing Defensin-5 Ameliorating Dextran Sodium Sulfate-Induced Mice *Colitis* via Inhibiting NF-kB Pathway. *J. Transl. Med.* **2020**, *18*, 107. [CrossRef]

57. Zhou, X.; Li, X.; Wang, X.; Jin, X.; Shi, D.; Wang, J.; Bi, D. Cecropin B Represses CYP3A29 Expression through Activation of the TLR2/4-NF-κB/PXR Signaling Pathway. *Sci. Rep.* **2016**, *6*, 27876. [CrossRef]
58. Wang, Y.; Abdullah; Zhong, H.; Wang, J.; Feng, F. Dietary Glycerol Monolaurate Improved the Growth, Activity of Digestive Enzymes and Gut Microbiota in Zebrafish (*Danio rerio*). *Aquac. Rep.* **2021**, *20*, 100670. [CrossRef]
59. Sypniewski, J.; Kierończyk, B.; Benzertiha, A.; Mikołajczak, Z.; Pruszyńska-Oszmałek, E.; Kołodziejski, P.; Sassek, M.; Rawski, M.; Czekała, W.; Józefiak, D. Replacement of Soybean Oil by *Hermetia illucens* Fat in Turkey Nutrition: Effect on Performance, Digestibility, Microbial Community, Immune and Physiological Status and Final Product Quality. *Br. Poult. Sci.* **2020**, *61*, 294–302. [CrossRef]
60. Deng, F.; Wang, D.; Yu, Y.; Lu, T.; Li, S. Systemic Immune Response of Rainbow Trout Exposed to *Flavobacterium psychrophilum* Infection. *Fish Shellfish Immunol.* **2024**, *144*, 109305. [CrossRef] [PubMed]
61. Muñoz-Atienza, E.; Távara, C.; Díaz-Rosales, P.; Llanco, L.; Serrano-Martínez, E.; Tafalla, C. Local Regulation of Immune Genes in Rainbow Trout (*Oncorhynchus mykiss*) Naturally Infected with *Flavobacterium psychrophilum*. *Fish Shellfish. Immunol.* **2019**, *86*, 25–34. [CrossRef]
62. Nilsen, H.; Johansen, R.; Colquhoun, D.; Kaada, I.; Bottolfsen, K.; Vågnes, Ø.; Olsen, A. Flavobacterium *Psychrophilum* Associated with *Septicaemia* and *Necrotic myositis* in Atlantic Salmon Salmo Salar: A Case Report. *Dis. Aquat. Org.* **2011**, *97*, 37–46. [CrossRef]
63. Orieux, N.; Douet, D.-G.; Le Hénaff, M.; Bourdineaud, J.-P. Prevalence of *Flavobacterium psychrophilum* Bacterial Cells in Farmed Rainbow Trout: Characterization of *Metallothionein* A and Interleukin1-β Genes as Markers Overexpressed in Spleen and Kidney of Diseased Fish. *Vet. Microbiol.* **2013**, *162*, 127–135. [CrossRef]
64. Nematollahi, A.; Decostere, A.; Pasmans, F.; Haesebrouck, F. *Flavobacterium psychrophilum* infections in Salmonid Fish. *J. Fish Dis.* **2003**, *26*, 563–574. [CrossRef] [PubMed]
65. Mirghaed, A.T.; Yarahmadi, P.; Soltani, M.; Paknejad, H.; Hoseini, S.M. Dietary Sodium Butyrate (Butirex® C4) Supplementation Modulates Intestinal Transcriptomic Responses and Augments Disease Resistance of Rainbow Trout (*Oncorhynchus mykiss*). *Fish Shellfish Immunol.* **2019**, *92*, 621–628. [CrossRef] [PubMed]
66. Tian, L.; Zhou, X.-Q.; Jiang, W.-D.; Liu, Y.; Wu, P.; Jiang, J.; Kuang, S.-Y.; Tang, L.; Tang, W.-N.; Zhang, Y.-A.; et al. Sodium Butyrate Improved Intestinal Immune Function Associated with NF-κB and p38MAPK Signalling Pathways in Young Grass Carp (*Ctenopharyngodon idella*). *Fish Shellfish Immunol.* **2017**, *66*, 548–563. [CrossRef] [PubMed]
67. Cholan, P.M.; Han, A.; Woodie, B.R.; Watchon, M.; Kurz, A.R.; Laird, A.S.; Britton, W.J.; Ye, L.; Holmes, Z.C.; McCann, J.R.; et al. Conserved Anti-Inflammatory Effects and Sensing of Butyrate in Zebrafish. *Gut Microbes* **2020**, *12*, 1824563. [CrossRef]
68. Ji, J.; Shu, D.; Zheng, M.; Wang, J.; Luo, C.; Wang, Y.; Guo, F.; Zou, X.; Lv, X.; Li, Y.; et al. Microbial Metabolite Butyrate Facilitates M2 Macrophage Polarization and Function. *Sci. Rep.* **2016**, *6*, 24838. [CrossRef]
69. Schulthess, J.; Pandey, S.; Capitani, M.; Rue-Albrecht, K.C.; Arnold, I.; Franchini, F.; Chomka, A.; Ilott, N.E.; Johnston, D.G.W.; Pires, E.; et al. The Short Chain Fatty Acid Butyrate Imprints an Antimicrobial Program in Macrophages. *Immunity* **2019**, *50*, 432–445.e7. [CrossRef] [PubMed]
70. Pagani, A.; Nai, A.; Silvestri, L.; Camaschella, C. Hepcidin and Anemia: A Tight Relationship. *Front. Physiol.* **2019**, *10*, 1294. [CrossRef]
71. Clauss, T.M.; Dove, A.D.M.; Arnold, J.E. Hematologic Disorders of Fish. *Vet. Clin. N. Am. Exot. Anim. Pract.* **2008**, *11*, 445–462. [CrossRef]
72. Neves, J.V.; Caldas, C.; Ramos, M.F.; Rodrigues, P.N.S. Hepcidin-Dependent Regulation of *Erythropoiesis* during Anemia in a Teleost Fish, *Dicentrarchus labrax*. *PLoS ONE* **2016**, *11*, e0153940. [CrossRef]
73. Vaibarová, V.; Čížek, A. Supposed Virulence Factors of *Flavobacterium psychrophilum*: A Review. *Fishes* **2024**, *9*, 163. [CrossRef]
74. Liu, M.; Hu, R.; Li, W.; Yang, W.; Xu, Q.; Chen, L. Identification of Antibacterial Activity of Hepcidin From Antarctic Notothenioid Fish. *Front. Microbiol.* **2022**, *13*, 834477. [CrossRef] [PubMed]
75. Xiao, P.; Cai, X.; Zhang, Z.; Guo, K.; Ke, Y.; Hu, Z.; Song, Z.; Zhao, Y.; Yao, L.; Shen, M.; et al. Butyrate Prevents the Pathogenic Anemia-Inflammation Circuit by Facilitating Macrophage Iron Export. *Adv. Sci.* **2024**, *11*, 2306571. [CrossRef] [PubMed]

Article

Taurochenodeoxycholic Acid Improves Growth, Physiology, Intestinal Microbiota, and Muscle Development in Red Swamp Crayfish (*Procambarus clarkii*)

Xiaodi Xu [1,2], Xiaochuan Zheng [1,2], Changyou Song [1,2], Xin Liu [1,2], Qunlan Zhou [1,2], Cunxin Sun [1,2], Aimin Wang [3], Aiming Zhu [4] and Bo Liu [1,2,*]

[1]   Wuxi Fisheries College, Nanjing Agricultural University, Wuxi 214081, China; 2021213005@stu.njau.edu.cn (X.X.); zhengxiaochuan@ffrc.cn (X.Z.); songchangyou@ffrc.cn (C.S.); 2023213007@stu.njau.edu.cn (X.L.); zhouqunlan@ffrc.cn (Q.Z.); suncunxin@ffrc.cn (C.S.)
[2]   Freshwater Fisheries Research Center, Chinese Academy of Fishery Sciences, Wuxi 214081, China
[3]   College of Marine and Biology Engineering, Yancheng Institute of Technology, Yancheng 224007, China; blueseawam@ycit.cn
[4]   Yancheng Academy of Fishery Science, Yancheng 224008, China; zam--3@163.com
*   Correspondence: liub@ffrc.cn

Academic Editor: Sung Hwoan Cho

Received: 26 December 2024

Revised: 19 January 2025

Accepted: 20 January 2025

Published: 22 January 2025

Citation: Xu, X.; Zheng, X.; Song, C.; Liu, X.; Zhou, Q.; Sun, C.; Wang, A.; Zhu, A.; Liu, B. Taurochenodeoxycholic Acid Improves Growth, Physiology, Intestinal Microbiota, and Muscle Development in Red Swamp Crayfish (*Procambarus clarkii*). Fishes **2025**, *10*, 38. https://doi.org/10.3390/fishes10020038

Abstract: Taurochenodeoxycholic acid (TCDCA), one of the bile acids, is thought to be involved in the regulation of muscle nutrient metabolism and gut microbial homeostasis. However, the effect of dietary addition of TCDCA on *Procambarus clarkii* is unclear. Therefore, in this study, an 8-week feeding experiment was conducted to explore the potential regulatory mechanisms of TCDCA on *P. clarkii* growth, physiology, muscle quality and gut microbes. The results indicated that dietary addition of TCDCA not only improved growth performance (final weight; weight gain; and specific growth rate) but also increased muscle elasticity and protein content. In addition, dietary TCDCA promotes muscle growth and development by increasing myofiber length, which is consistent with the activation of the expression of genes related to protein utilization (TOR and AKT) and muscle proliferation and differentiation (MyHC, MLC1, MEF2A, MEF2B). Importantly, 16s rRNA sequencing demonstrated that dietary TCDCA had no significant effect on gut microbial composition (alpha diversity) but significantly increased microbial abundance at the genus level. Functional prediction analysis of differential microbes revealed that dietary TCDCA may promote metabolism by altering gut microbes, thereby promoting muscle quality. In conclusion, our study demonstrates that the dietary addition of TCDCA promotes *P. clarkii* growth and muscle quality and protein deposition by altering gut microbes.

Keywords: TCDCA; muscle quality; protein deposition; gut microbiota; *Procambarus clarkii*

Key Contribution: This paper uncovered the function of TCDCA in the growth, muscle development, muscle nutrient quality, and gut microbiota of *P. clarkii*.

1. Introduction

Aquaculture is considered one of the sustainable sources of animal protein, with crustaceans being an economically important aquatic product. Healthy culture and disease prevention and control are key to the sustainable development of aquaculture [1,2]. With an increasing awareness of food safety and environmental protection, the research for safe and efficient feed additives has become a new trend in the industry [3]. *Procambarus clarkii* is a freshwater species which has become one of the important species in aquaculture industry due to its high nutritional value of muscle and soft and tender meat [4]. In 2023, *P. clarkii* production in China reached nearly 3.1 million tons, accounting for about 9.259% of the country's total freshwater aquaculture production [5]. With the expansion of *P. clarkii* culture, a variety of feed additives have been widely used to improve its growth, immunity, and antioxidant capacity [6–8] and thus improve the efficiency of farming.

Bile acids (BAs), as a major component of bile, are initially synthesized from cholesterol in the liver and further into the intestine [9]. Bile acids have a variety of physiological functions, including improving growth performance [10], accelerating intestinal digestion and absorption of fat-soluble vitamins [11], and regulating glucose metabolism [12] and immune function [13]. Therefore, the dietary addition of bile acids is widely used in aquatic animals [14–16]. Although bile acids have beneficial effects in animals, it has not been reported whether crustaceans are able to synthesize bile acids themselves [17]. Taurochenodeoxycholic acid (TCDCA), as one of the bile acids, is thought to be involved in nutritional regulation and to have an adjunctive therapeutic role in metabolic or immune disorders [18,19]. Studies have shown that TCDCA treatment suppresses SGIV infection in *Epinephelu* spp. [20]. In mammals, TCDCA not only modulates immunity and inflammation but also treats apoptosis-related diseases [18,21]. However, TCDCA has been limitedly studied in aquaculture, and it is worth investigating whether the regulatory mechanisms of TCDCA in aquatic animals, including crustaceans, are similar to those in mammals.

Studies have shown that changes in gut microbes may increase susceptibility to disease and simultaneously enhance immune system function [22,23], such as diabetes [24] and insulin sensitivity [25]. When bile acids enter the intestine, they interact with the intestinal flora [26]. Studies have indicated that bile acids are able to inhibit the growth of harmful bacteria and remodel the structure and function of the intestinal flora. In addition, coupling bile acids further increases the diversity of the gut microbiota [27,28]. Our previous study suggested that intestine microbes and their specific bile acids contribute to the process of intestinal barrier damage in *Macrobrachium rosenbergii* [29]. A study in mammals noted that altered gut flora may trigger skeletal muscle atrophy via the bile acid-FXR pathway [30]. Conversely, studies have shown that dietary bile acids can significantly improve feed conversion and muscle quality in *Larimichthys crocea* [15], as well as increase protein deposition in *Ctenopharyngodon idella* [31]. However, in crustaceans, the role of dietary addition of TCDCA in regulating muscle metabolism as well as gut microbes has not been reported.

Therefore, this study will investigate the effects of dietary TCDCA on the growth, physiology, muscle nutrition, and development of *P. clarkii*. Furthermore, we used 16s sequencing to investigate the regulatory mechanisms of dietary TCDCA on intestinal flora. This study provides a reference for the development of novel exogenous additives for *P. clarkii* aquaculture and also lays the foundation for the study of the molecular mechanism of the effect of TCDCA on *P. clarkii*.

2. Materials and Methods

2.1. Ethics Statement

The animal experiments involved in this study were approved by the Animal Care and Use Committee of Nanjing Agricultural University (Nanjing, China). All relevant procedures were conducted in accordance with the Guidelines for the Care and Use of Laboratory Animals in China.

2.2. Experimental Design of TCDCA Feeding

Prawns were provided by Jiangsu Jinfeng Agricultural Technology Co., Ltd., (Yancheng, China) for this experiment. In terms of the TCDCA feeding validation trial, 360 prawns with a uniform initial weight (4.94 ± 0.02 g) were randomly assigned to two groups with four replicates and 45 individuals in each treatment (cement pond: length × width × water depth, 2.5 m × 2.0 m × 0.4 m). The experiment was designed to include two groups: the mixed-protein-source group was the control group, and the group supplemented with 300 mg/kg TCDCA (provided by Shanghai yuanye Bio-Technology Co., Ltd, Shanghai, China) was the TCDCA group. During the culture period, the CON and TCDCA groups were fed two times a day (6:00 and 21:00) until apparent satiety for a total of 8 weeks. The amount of remaining bait was observed after 30 min of feeding and the feeding rate was adjusted accordingly. The ingredients and proximal composition of the two diets are shown in Table 1. The starch used in this experiment was pre-gelatinized starch. All the raw materials were ground through a 60 mm mesh. After being carefully weighed, the fine powder was mixed with 6% oil (w/w) and 30% water (w/w) and further blended. The feed was pelleted by laboratory pelletizer (Guangzhou Huagong Optical Mechanical & Electrical Technology Co. Ltd., Guangzhou, China) into 2 mm diameter pellets. After drying the feed for 24 h at 65 °C, the feed was stored at −20 °C for feeding experiment. To reduce cannibalism, shelters were set up to avoid disturbance for *P. clarkii*, especially when they were under molting. Prawn aquaculture conditions were as follows: water temperature 28–32 °C, pH 7.2–7.8, DO > 6.0 mg/L, with a natural photoperiod.

Table 1. Components and proximate analysis of CON and TCDCA diets.

	CON	TCDCA
Components (% dry matter)		
Fish meal	5.0	5.0
Soybean meal	25.0	25.0
Rapeseed meal	18.0	18.0
Pork powder	3.0	3.0
Peanut meal	5.0	5.0
Rice bran	4.0	4.0
Salt	0.3	0.3
Spray-dried blood cell powder	5.0	5.0
Squid paste	3.0	3.0
Shrimp meal	3.0	3.0
Starch	18.79	18.76
Soybean oil	3.0	3.0
Ecdysone (2%)	0.01	0.01
Vitamin premix [a]	1.0	1.0
Mineral premix [b]	1.0	1.0
Choline chloride (50%)	0.5	0.5
Calcium dihydrogen phosphate	2.0	2.0
Carboxymethyl cellulose	0.5	0.5
Bicarbonate	1.5	1.5
Microcrystalline methionine	0.4	0.4
TCDCA	0.0	0.03
Total	100.0	100.0
Proximate analysis (%)		
Dry matter (DM)	83.03	82.27
Crude protein, CP (%)	32.78	33.21

Table 1. *Cont.*

	CON	TCDCA
Crude lipid (Ether extract), EE (%)	5.68	6.08
Gross energy (MJ/kg)	15.89	16.09

Note: [a] Premix supplied the following vitamins (g/kg): vitamin A, 4 g; vitamin D3, 0.02 g; vitamin E, 10 g; vitamin K3, 10 g; thiamin, 10 g; riboflavin, 10 g; calcium pantothenate, 20 g; pyridoxine, 20 g; cyanocobalamin, 0.01 g; biotin, 0.2 g; folic acid, 0.5 g; niacin, 40 g; inositol, 400 g; vitamin C, 20 g; microcrystalline cellulose, 455.27 g. [b] Premix supplied the following minerals (g/kg): potassium iodate, 0.6 g; sodium selenite pentahydrate, 0.08 g; potassium dihydrogen phosphate, 320 g; magnesium sulfate, 200 g; manganese sulfate monohydrate, 20 g; copper chloride dihydrate, 2 g; zinc sulfate heptahydrate, 60 g; ferrous sulfate heptahydrate, 50 g; sodium chloride, 100 g; cobalt chloride hexahydrate, 2 g; microcrystalline cellulose, 245.32 g.

2.3. Sample Collection

After the feeding experiment, the prawns were fasted for 24 h. All the prawns in each pond were counted and weighed to calculate the growth parameters. Nine prawns were taken from each group, muscle and intestinal samples were quickly collected, and the tissues were immediately frozen in liquid nitrogen and stored at $-80\ ^\circ$C until analysis in subsequent experiments. Meanwhile, dorsal muscle was collected for muscle texture measurement. In addition, muscle tissue was fixed in glutaraldehyde and 4% paraformaldehyde for further analysis of muscle tissue structure. The remaining muscle tissue was frozen at $-20\ ^\circ$C to measure muscle proximate composition.

2.4. Growth Performance

Growth indices were calculated according to the following formula:

$$\text{Weight gain rate (WGR, \%)} = 100 \times (W_t - W_0)/W_0$$

$$\text{Specific growth rate (SGR, \%/day)} = 100 \times (\ln W_t - \ln W_0)/t$$

$$\text{Feed intake (FI, g)} = W_f/\text{number}$$

$$\text{Feed conversion ratio (FCR)} = (W_d/\text{number})/(W_t - W_0).$$

$$\text{Hepatosomatic index (HSI, \%)} = 100 \times W_h/W_t$$

W_t: final average weight (g); W_0: initial average weight (g); W_d: total weight of feed intake (g); t: cultivation cycle (d); W_h: liver weight; W_f: weight of feed.

2.5. Proximate Composition Analysis of Muscle

In each group, 9 muscle samples were randomly taken and measured for texture, moisture content, ash, crude protein, and ether extract. In detail, the muscle samples were dried at 65 $^\circ$C for 12 h and 105 $^\circ$C for 4 h, and the moisture content was calculated by measuring the difference in mass of the samples before and after drying. By burning the muscle samples at high temperature (550 $^\circ$C), the organic matter was completely oxidized and the remaining inorganic matter was the ash content. The nitrogen content of the muscle samples was determined using the Kjeldahl method, and the crude protein content was estimated from the average proportion of nitrogen in protein. Muscle samples were repeatedly heated and extracted in an organic solvent using petroleum ether Soxhlet extraction, and the residue was weighed to determine the ether extract content.

2.6. Histological Analysis of Muscle

The sampled muscles were fixed in 4% paraformaldehyde for at least 24 h, dehydrated with an ethanol gradient, embedded in paraffin, and cut into 5 μm sections (Leica, RM2255, Nussloch, Germany). The sections were stained with hematoxylin–eosin (H&E)

and sealed with neutral resin, observed under a microscope, and photographed (Leica, DM4B, Germany) [32].

2.7. Transmission Electron Microscopy (TEM) Observation of Muscle

Muscle samples were cut into 4 mm^3 tissue pieces, fixed with 2.5% glutaraldehyde at 4 °C for 24 h, then rinsed with 0.1 M phosphoric acid rinse solution 3 times, fixed with 1% osmium acid for 2 h, rinsed with PBS 3 times, and then subjected to gradient dehydration in different concentrations of ethanol (50, 70, 90, 95 and 100%). The samples were then permeabilized in 1:1 and 1:3 acetone and epoxy resin, respectively, and the samples were put into capsules or embedding plates, with the embedding agent epoxy resin added, and polymerized in a temperature chamber at 60 °C for 48 h. Ultra-thin sections of 70 nm were cut using an ultramicrotome (Leica ultracut UCT25, Los Angeles, CA, USA). Finally, the sections were double-stained with uranium–lead and dried overnight. Photographs were taken for observation using a transmission electron microscope (Hitachi, Tokyo, Japan).

2.8. 16s rRNA Sequencing

To explore the effect of TCDCA on the diversity and composition of *P. clarkii* gut microbes, six intestine samples from each group were selected for 16s rRNA sequencing according to the method previously described by Liu [33]. Simplistically, DNA was extracted from the intestine using the FastDNA kit (Mpbio, Santa Ana, CA, USA). Subsequently, the DNA was linearly amplified and coded using the bacterial 16s rRNA variable region V3–V4 primers (515F: 5′-GTGCCAGCMGCCGCGG-3′ and 907R: 5′-CCGTCAATTCMTTTRAGTTT-3′. Finally, high-throughput sequencing based on the Illumina Novaseq6000 platform (QIIME2-2021.11) was performed to obtain the raw sequence for 16s analysis.

The PE reads obtained from Illumina PE sequencing were first merged according to the overlap relationship (pandaseq (V2.11)), and the quality of the sequences was also quality-controlled and filtered (PRINSEQ (V0.20.4)). Finally, high-quality sequences were obtained from each sample. Then, de-priming, mass filtering, and denoise were performed by the DADA2 method using the software QIIME2 (qiime2-2021.11). Data from each sample were processed for random draw leveling and statistically sequenced quantities for ASV species annotation. Based on the abundance and annotation information of ASV, the number of sequences per sample at each taxonomic level (Phylum, Class, Order, Family, Genus, Species) was counted as a proportion of the total number of sequences.

The PCA diagrams, as well as the dilution curves for the Alpha Diversity Index, were plotted using R (V3.6.2). Braycurtis, Weighted Unifrac, and Unweighted Unifrac distances were calculated using the software QIIME2 and heatmaps of diversity indices were plotted using R (V3.6.2). Representative ASV sequences were compared to the KEGG PATHWAY database (https://www.kegg.jp/kegg/pathway.html, accessed on 10 June 2024) and the COG database (https://www.ncbi.nlm.nih.gov/research/, accessed on 10 September 2024) using the software PICRUSt2 (V2.1.2). cog-project/) for the functional prediction of microbes.

2.9. RT-PCR Analysis

Total RNA was extracted using the TRIzol method (Invitrogen, Carlsbad, CA, USA), and the RNA OD260/280 (1.8–2.0) ratio was detected to determine the quality of RNA. Reverse-transcription PCR was performed using the PrimeScript TM FAST RT reagent kit with gDNA Eraser (Takara, Dalian, China). RT-qPCR was performed on a BioRad CFX96 system (Bio-Rad, Shanghai, China) in conjunction with TB Green™Premix Ex Taq™II reagent. EIF was used as an internal reference gene to correct the expression level of the target gene. All primer sequences are shown in Table 2.

Table 2. Primer sequences for real-time PCR.

Gene	Forward (5′-3′)	Reverse (5′-3′)	PL (bp)	Reference
EIF	GGAATAAGGGGACGAAGACC	GCAAACACACGCTGGGAT	126	[34]
TOR	GAAGGCATGCTGCGGTATTG	CGCAGGCTTTGGGTCTCTTA	122	[34]
S6K	ACAGCCGAGAATCGCAAGAA	ATCACCATTATCGGGTCCGC	153	[34]
4E-BP1	ACCTGCCAGTGATACCAGGA	TGGCTCCTCTGAAATCGTTCC	80	[34]
AKT	CCTTGGGGCGTCTACTCCTA	TCCTCATAATCCTCACTTTCCT	176	database
FOXO	ACGCGCTAACACCATGGAAG	GACTCTCACTCAGCGACGAA	158	[35]
MyHC	AAGCCAACCGTACCCTCAA	AGTAGCACGTTCTCTGCATTCA	174	database
MLC1	TGAGAAGGTCGGAGGCAAG	TGCCATTCTCAGATTTGTCGT	155	[36]
MEF2A	CATCTTCCAACCATCCTGGG	GTTTGCTCAACGGGGTATCA	125	[36]
MEF2B	ACCAGCACCACCTTCACATT	GAAGATGGACCCAAATGTGAA	133	[36]
MSTN	AGCAACAGCAACAACAAGGA	GCAGGAAGGGACATTTACCG	136	[36]
LC3	TGAGTAGTCCGTCTCGGTGT	CCATGTAGAGGAACCCGTCG	169	[37]
ATG2	GTACTTCCCGTGGTCGGATG	CCATCCACGAACCTGAGAGG	175	[37]
ATG3	GCCAAGACAACCACCATAGC	AGAGCCGAGGTGTCTGGTAG	201	database
ATG9	TCATACATCCAGGGTTCGCC	GGGCAAAGGAACAAACGTCC	189	database
ATG12	TGGAGGGGAAGGACTTACGG	AGCTTTCCCTTAGCAGTCTTC	203	[37]
ATG16	AGATGGATGGCACAGAAGGC	GTTCACTTGCTTGGGCTCAC	178	database
ATG18	CGTGTTGTAGTGGAGGAGCA	CGTGGCTGCTTTTGAATCGT	194	database
ub	TCCAGCCTCTCCTGCCTT	CCTTCCTTATCCTGAATCTTTGCC	172	[38]
Psma1	CTTTACCTCATTGACCCATCT	CACAACCATAGTATCCATTACACAT	149	database
psmd1	ACTCATACAGCAAACAGAATCC	CGTCCACCAGCATCAATAA	147	database
psmd6	AGCTTTTGCTAAAACCTACG	TCCCAATCTCCTCCCTCT	159	database
Psmc1	TGTCTCCATTCTCTCCTTTGT	TTGAGGTGCCTTCTCTAGCT	148	database

Note: the database was the RNA-seq results conducted by the aquatic animal disease and nutrition lab of FFRC.

2.10. Statistical Analysis

Data were first tested for normality and homogeneity and analyzed using SPSS software (version 26.0) and expressed as mean $\pm$ standard error of the mean (SEM). Gene expression levels were calculated by $2^{-\Delta\Delta ct}$. Statistical differences between the two groups were analyzed by Student's *t*-test.

3. Results

3.1. Growth Performance

The growth performance and biometric indices are detailed in Table 3. FW, WGR and SGR were significantly higher in the experimental group than in the control group ($p < 0.05$). However, FI, FCR, and HSI exhibited no statistical differences ($p > 0.05$). The results indicate that TCDCA supplementation improves the growth performance of *P. clarkii*.

Table 3. The effects of TCDCA on the growth performance of *P. clarkii*.

	CON	TCDCA	*p* Value
Initial weight, IW (g)	4.93 $\pm$ 0.012	4.94 $\pm$ 0.031	0.795
FW (g)	33.16 $\pm$ 1.033	37.09 $\pm$ 0.983	0.044
WGR (%)	572.24 $\pm$ 19.25	651.01 $\pm$ 22.067	0.043
SGR (%/day)	3.67 $\pm$ 0.054	3.88 $\pm$ 0.057	0.043
FI (g)	25.69 $\pm$ 0.235	29.32 $\pm$ 3.431	0.098
FCR	0.91 $\pm$ 0.033	0.91 $\pm$ 0.076	0.985
HSI (%)	5.72 $\pm$ 0.216	6.24 $\pm$ 0.299	0.161

Note: Data are expressed as means with SEM (IW, FW, WGR, SGR, FI, and FCR were detected for all the *P. clarkii* that were collected; HSI was calculated for 9 *P. clarkii* collected). *p* values less than 0.05 represent significant differences according to Student's *t*-test.

3.2. Effects of TCDCA on Muscle Texture and Nutrient Content of P. clarkii

Subsequently, we further evaluated the effects of TCDCA addition on muscle texture and nutrient composition in *P. clarkii*. TCDCA addition did not increase the meat rate (Figure 1A), while it did affect muscle texture. The shearing force in the TCDCA group was higher than that in the control group, but adhesiveness was lower (Figure 1B). The findings suggest that TCDCA can render muscles more resilient. The differences in moisture, ether extract, and ash content between the control and TCDCA groups of *P. clarkii* muscles were not significant (Figure 1C,E,F, $p > 0.05$). Conversely, the crude protein levels in the TCDCA group were remarkably elevated (Figure 1D, $p < 0.05$), suggesting TCDCA addition may contribute to an increase in crude protein content in *P. clarkii* muscles. Together, TCDCA addition improved the muscle texture and nutrient content of *P. clarkii*.

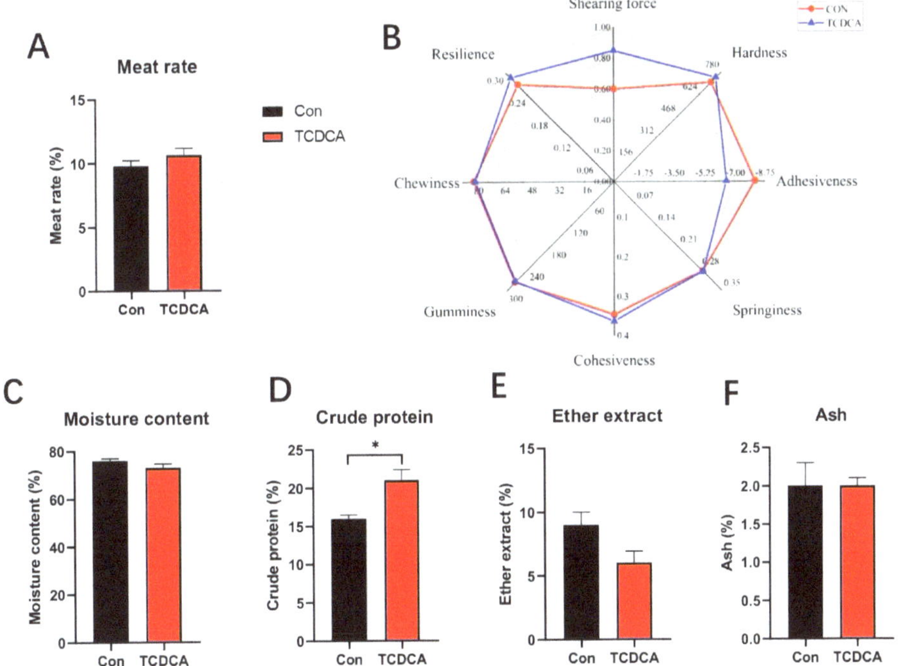

Figure 1. The effects of TCDCA on muscle texture and nutrient content of *P. clarkii*. (**A**), meat rate; (**B**), radar map on muscle texture; (**C**), moisture content; (**D**), crude protein; (**E**), ether extract (crude lipid); (**F**), ash. Data were analyzed by Student's *t*-test; * represent $p < 0.05$, results are indicated as mean $\pm$ SEM, $n = 9$.

3.3. Effect of TCDCA on Muscle Tissue Morphology of P. clarkii

Further investigation on morphology found muscle fiber density and length were considerably increased in the TCDCA group (Figure 2A,C). In detail, the TCDCA group exhibited a significant increase in muscle fiber diameter of over 80 μm (Figure 2B,C, $p < 0.05$). Ultra-microstructural observation revealed that TCDCA promoted the growth of *P. clarkii* sarcomeres (Figure 2B,C, $p < 0.05$). This finding suggests that TCDCA supplementation exerted a measurable impact on muscle structure.

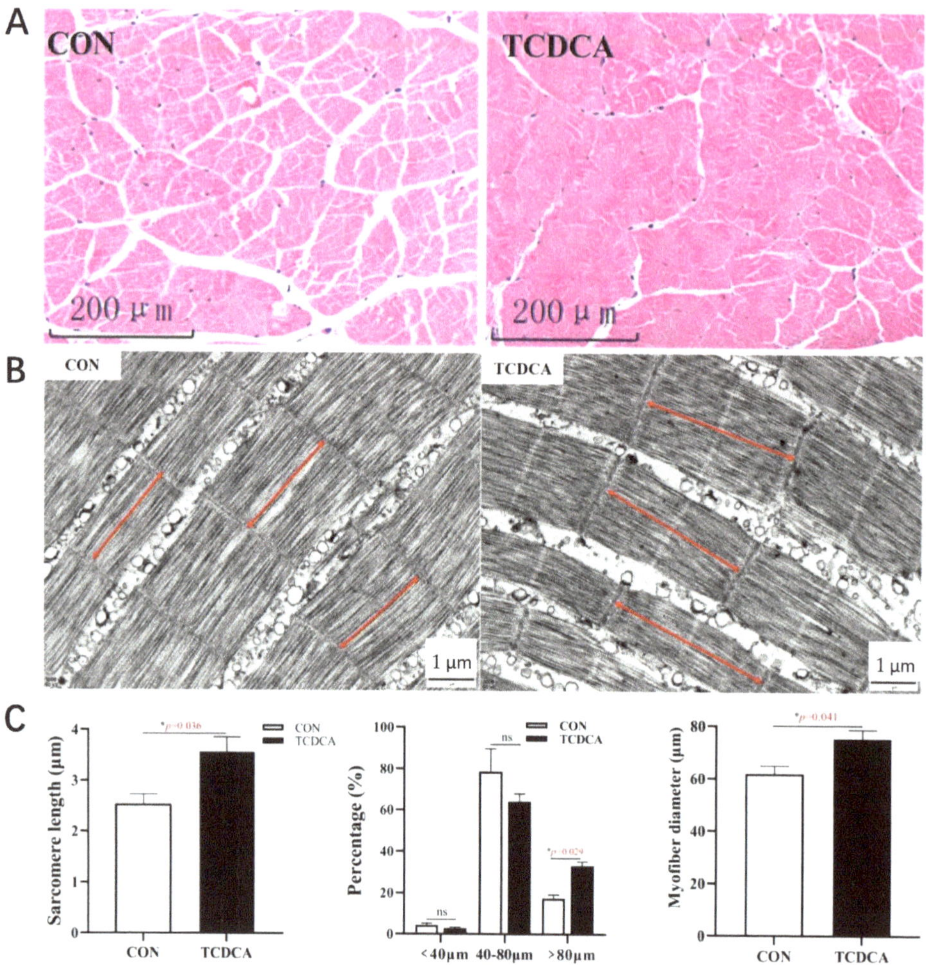

Figure 2. The effect of TCDCA on the muscle tissue morphology of *P. clarkii.* (**A**), H&E staining of muscle fiber; (**B**), TEM morphology of muscle fiber, red arrows represent the sarcomere length; (**C**) statistics of sarcomere length, muscle fiber diameter distribution, and myofiber diameter comparison between Con and TCDCA. Data were analyzed by Student's *t*-test; * represent $p < 0.05$, results are indicated as mean $\pm$ SEM, $n = 9$.

3.4. Effects of TCDCA on Transcription Levels of Genes Related to Muscle Development in P. clarkii

Based on the above results, we further evaluated the expression of key genes in the signaling pathways on protein metabolism and muscle proliferation and differentiation. The genes related to protein synthesis (Figure 3A), such as TOR and AKT, were up-regulated in the TCDCA group, while 4E-BP1 and FOXO were down-regulated. Meanwhile, genes related to muscle proliferation and differentiation (Figure 3B), such as MyHC, MLC1, MEF2A, MEF2B were up-regulated, while MSTN was down-regulated in TCDCA. Accordingly, genes related to autophagy (Figure 3C), such as LC3 and ATG3, were suppressed in the TDCDA group. Additionally, ubiquitination-related genes (Ub) were inhibited in the TCDCA group (Figure 3D). These data reveal that TCDCA activates protein utilization and muscle proliferation and differentiation, affecting the muscle development of *P. clarkii.*

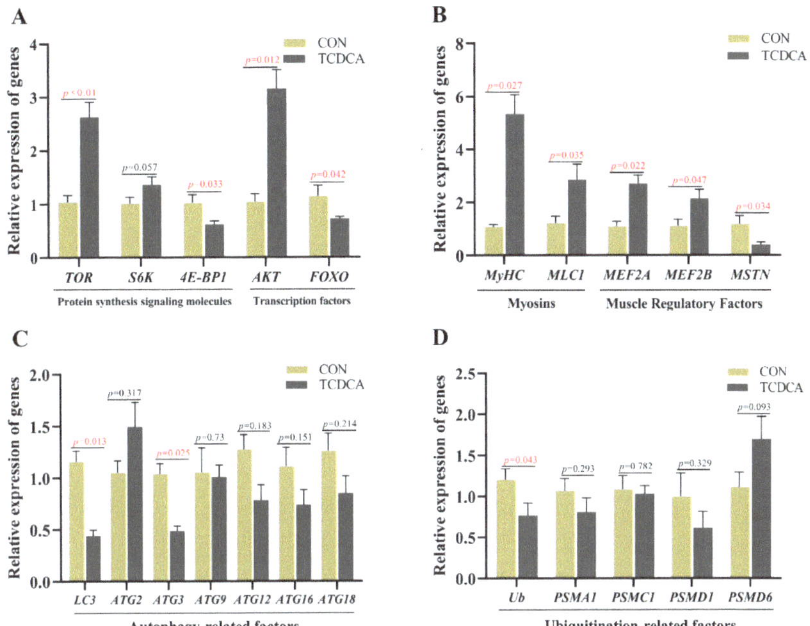

Figure 3. Effects of TCDCA on the transcription levels of genes related to muscle development in *P. clarkii*. (**A**), relative expression of genes related to protein synthesis signaling molecules and transcription factors; (**B**), relative expression of genes related to myosin and muscle regulatory factors; (**C**), relative expression of genes related to autophagy factors; (**D**), relative expression of genes related to ubiquitination factors. Red data mean *p* < 0.05.

3.5. Analysis of Differential Intestinal Microbes Between Con and TCDCA Groups

TCDCA, as a bile acid, may influence muscle development by microorganisms. Therefore, we explored the role of the gut flora with 16s rRNA sequencing. Alpha diversity is mainly used to reflect species richness and evenness, as well as sequencing depth. The results indicated TCDCA supplementation did not affect alpha diversity, including ace, chao1, Shannon, simpson and coverage index (Figure 4A). Meanwhile, a PCA based on braycurtis was used to describe the alpha diversity (Figure 4B). The results indicated that the control and TCDCA groups were not well separated. These results suggest TCDCA had no effect on the general composition of the intestinal flora.

To further corroborate the shift in intestinal flora, LEfSe was used to analyze the microbial communities under TCDCA diets, as shown in Figure 4C,D. The most abundant phylotypes in the TCDCA group, from phylum to genus level, were *c_Chloroflexales*, *o_Lactobacillales*, *o_Chloroflexales*, *o_Gemmatales*, *f_Chloroflexaceae*, *f_Lactobacillaceae*, *f_Gemmataceae*, *f_Microbacteriaceae*, *g_Chloronema*, *g_Lactobacillus*, *g_Fimbriiglobus*, and *g_Aurantimicrobium* in TCDCA group (Figure 4C). The most abundant phylotypes in the Con group included *g_Fusibacter* and *g_Fusibacteraceae* (Figure 4C). Additionally, differences in microbial composition at the genus level were analyzed to further understand the effects of TCDCA; the results are shown in Figure 4D. TCDCA significantly increased bacterial abundance at the genus level. *g_Chloronema*, *g_Tabrizicola*, *g_PeM15*, *g_Tropicimonas*, *g_Hyphomicrobium*, *g_Fimbriiglobus*, and *g_KD4-96* were considered key differential flora due to their higher relative abundance.

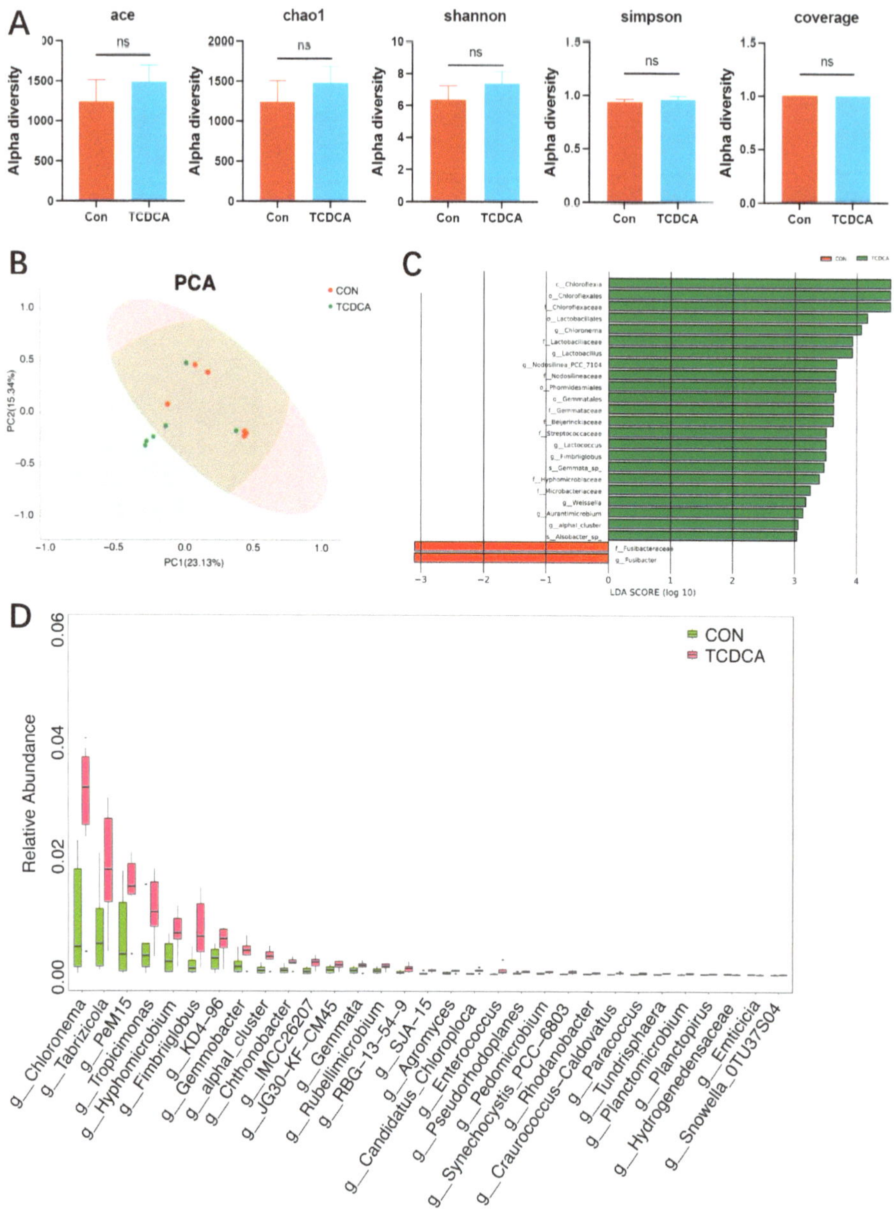

Figure 4. Intestinal microbes between Con and TCDCA groups in *P. clarkia*. (**A**), alpha diversity analysis, ns represent *p* > 0.05; (**B**), Principal Component Analysis (PCA); (**C**), microbial community biomarkers: LEfSe analysis identified the taxa with the most remarkable differences in abundance among the three taxa; (**D**), microbial comparation analysis between Con and TCDCA groups at the genus level. In (**C**,**D**), "c_, o_, f_, g_, s_" are the prefixes of the bacteria.

3.6. Effect of TCDCA on Intestinal Microbial Functions

The prediction of the function of intestinal microorganisms by KEGG function analysis is shown in Figure 5. The function prediction showed that the TDCDA group increased the abundance of metabolism pathways (protein metabolism, fatty acid metabolism, biosynthesis of metabolites, metabolism of cofactors and vitamins, xenobiotic biodegradation and metabolism, and nucleic acid metabolism) and excretory system. However, TCDCA decreased the abundance of drug resistance (beta-Lactam resistance). Therefore, TCDCA modified the gut flora to boost metabolism.

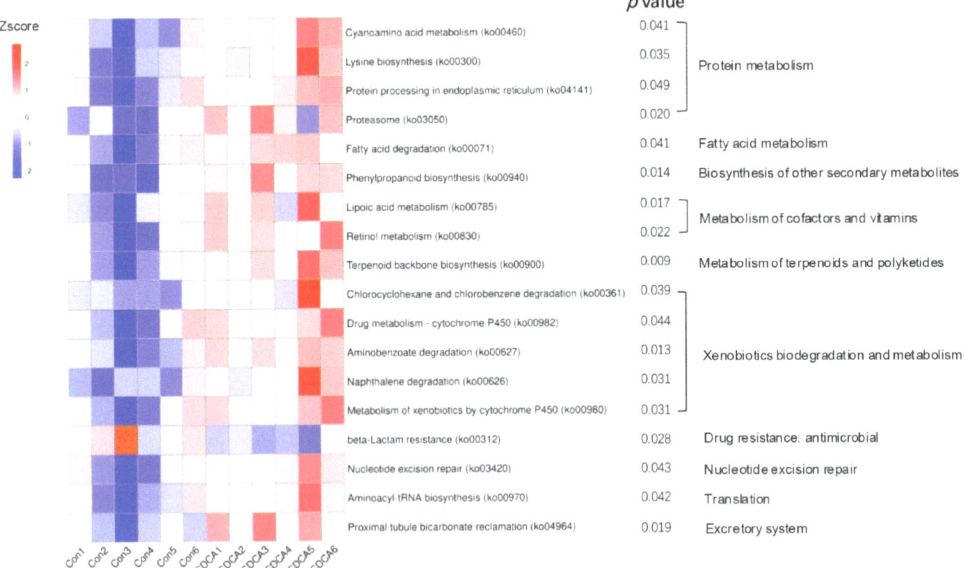

Figure 5. Functional prediction of intestinal microbes under TCDCA. KEGG functional annotation. Red color represents up-regulated functions, and blue color represents down-regulated functions.

4. Discussion

Exogenous BA supplementation to aquafeeds has been demonstrated to enhance growth performance and feed utilization [39]. The results of our study indicated that the appropriate addition of TCDCA enhanced the growth performance of *P. clarkii*, especially body weight. Notably, there was no significant difference in HSI. According to Du et al. [40], adding BAs generally reduced liver fat deposits and enhanced the growth performance of aquatic animals. Consequently, we proposed that TCDCA supplementation could improve body weight in multiple manners rather than promoting hepatic metabolism.

The primary edible tissue of aquatic species is muscle, which makes up 30–80% of the entire body weight. Nutrient accumulation in aquatic animals often manifests as increased muscle mass [41]. Even though there was no discernible change in meat rate in the present experiment, the higher trend indicated TCDCA encouraged nutrient accumulation in muscle. Further muscle composition results showed that protein content in the TCDCA group increased significantly, which confirms that TCDCA may promote protein deposition in muscle to increase body weight. Jiang et al. [42] demonstrated that a moderate administration of bile acids improved the deposition of crude proteins in *Oreochromis niloticus* muscle. A study in *Litopenaeus vannamei* showed that exogenous bile acids could increase crude protein deposition to alleviate the reduction in muscle protein content caused by plant protein feeds [35]. Additionally, skeletal muscle protein deposition, which dictates muscle development, critically depends on the equilibrium between protein synthesis and degradation [43]. The mTOR signaling pathway plays an essential role in protein synthesis. The mTOR signaling pathway is regulated by transcription factors (AKT and FOXO) to impact downstream effector molecules, including 4E-BP1, governing the synthesis of proteins [44]. The two main mechanisms for intracellular protein degradation are autophagy and ubiquitination [45]. Autophagy degraded intracellular proteins and organelles to preserve cellular homeostasis and functionality [46]. Ubiquitination is an important post-translational modification of proteins which covalently binds ubiquitin molecules to target proteins and directs proteins for proteasomal degradation [47]. In this experiment, protein metabolism-related gene expression revealed that TCDCA promoted protein synthesis and inhibited protein catabolism, further elucidating that TCDCA supplementation improved muscle protein deposition in *P. clarkii*.

Important inherent characteristics of texture, including hardness, adhesion, bonding, elasticity, plasticity, flexibility, and shear, have a significant impact on the quality of muscles [48]. Muscle fibers, intramuscular connective tissue (IMCT), and intramuscular fat (IMF) are the primary components that affect the textural characteristics of muscle [49]. However, a complex interplay of internal and environmental variables impacts these qualities [50]. For *P. clarkii*, an appropriate increase in muscle hardness and elasticity and a decrease in adhesion are more sought after by consumers [51]. In this experiment, TCDCA addition greatly increased the shear force and decreased the adhesiveness of *P. clarkii* muscle, suggesting that TCDCA supplementation improved muscle quality. However, another study in *Larimichthys crocea* found that supplemental mixed bile acids increased muscle cohesiveness and gumminess [52]. Since various species have distinct metabolisms, bile acids may have varied impacts on muscle texture. Moreover, it has been shown that myofiber diameter is positively correlated with muscle shear and negatively correlated with adhesiveness [53]. Chen et al. [54] showed that bile acid supplementation significantly increased the mass and myofiber diameter of breast muscles in chickens on a high-fat diet and promoted the growth of muscle tissues, which is in agreement with our results. Therefore, we propose that the primary mechanism by which TCDCA modulates the structure of *P. clarkii* muscle tissue might involve muscle proliferation spurred on by an increase in myofiber diameter. In addition, changes in sarcomeres are one of the main causes of muscle hypertrophy [55]. The addition of TCDCA increased the sarcomere length in *P. clarkii*, according to ultra-microstructures, indicating that TCDCA also had a positive impact on muscle hypertrophy. Myogenic regulators such as Myocyte Enhancer Factor (MEF) and muscle growth inhibitor (MSTN), which suppresses the development of skeletal muscle, collaborate in concert to control muscle proliferation and hypertrophy [56]. Research has demonstrated that MSTN adversely affects myocyte proliferation and differentiation, as well as muscle growth and development. According to Acosta et al. [57], zebrafish with suppression of MSTN expression grow embryos more quickly and have bigger bodies. Under the cooperative control of elements like myosin light chain (MLC) and myosin heavy chain (MyHc), mononuclear myoblasts differentiate and fuse to produce multinucleated myotubes, which ultimately form myofibers [58]. In this experiment, TCDCA promoted muscle hyperplasia and hypertrophy in *P. clarkii* by affecting important regulatory factors, improving muscle quality.

An increasing amount of data points to the intimate interaction between the gut flora and BAs [59]. BAs impact host health and metabolism-related enzymes and pathways by selectively inhibiting the intestinal flora, which encourages the growth of BA-tolerant bacteria while inhibiting bile acid-sensitive bacteria [60]. Our results suggest that TCDCA supplementation had no effect on the α diversity of the gut flora. However, at the genus level, TCDCA supplementation increased the relative abundance of Chloronema, Tabrizicola, PeM15, Tropicimonas, and Hyphomicrobium. Chloronema is a filamentous anaerobic photosynthetic bacterium that can participate in protein metabolism due to its chloroplast antennae structure [61]. Tabrizicola have a complex protein synthesis capacity and encodes a wide range of proteins to support physiological functions [62]. It has been proposed that PeM15, an uncultured actinobacterium (Actinobacteria), might be one of the probiotics that aid aquatic animals withstand environmental changes [63]. Tropicimonas belongs to the Proteobacteria family and is engaged in degrading branched-chain alkanes and several host metabolic processes [64]. It has been discovered that primary bile salts in grass carp raised the relative abundance of Proteobacteria, impacting host health [65]. Hyphomicrobium is a proteobacterium that can utilize mono-carbons, such as methanol, as a carbon source through the serine pathway [66].

Therefore, we suggest that TCDCA supplementation may affect the metabolic function of the organism by regulating the intestinal flora abundance. Further functional predictions demonstrated that changes in intestinal flora predominantly impacted metabolic changes, in which protein metabolism was significantly enhanced. Therefore, we hypothesize that TCDCA enhanced intestinal protein metabolism by the changes in gut flora abundance, thus promoting muscle protein deposition and development to improve muscle quality.

5. Conclusions

In conclusion, dietary TCDCA improves growth performance, muscle development, and nutrient quality and ameliorates muscular autophagy and the gut microbiota relating to fatty acid and protein metabolism, as well as immunity. These data provide a theoretical basis for the application of TCDCA in *P. clarkii* aquaculture.

Author Contributions: Conceptualization, X.X. and B.L.; methodology, X.X. and C.S. (Changyou Song); software, X.Z.; validation, X.X. and X.L.; formal analysis, Q.Z.; investigation, X.X.; resources, A.W. and A.Z.; data curation, X.X. and C.S. (Cunxin Sun); writing—original draft preparation, X.X.; writing—review and editing, X.X. and C.S. (Changyou Song); visualization, X.X.; supervision, B.L.; project administration, B.L.; funding acquisition, C.S. (Changyou Song), X.Z. and B.L. All authors have read and agreed to the published version of the manuscript.

Funding: This work was supported by the National Key R&D Program of China, grant number 2023YFD2402000; the Jiangsu Province Agricultural Science and Technology Independent Innovation Fund, grant number CX(24)3065; the earmarked fund for Agriculture Research System of China, grant number CARS-48; the project of High Quality Fishery Development in Yancheng, China, grant number 2022yc003; the Central Public-interest Scientific Institution Basal Research Fund of Freshwater Fisheries Research Center, CAFS, grant number 2024JBFR06; and the Natural Science Foundation of Jiangsu Province for Youths, grant number BK20230178.

Institutional Review Board Statement: Ethics Committee Name: Animal Care and Use Committee of the Committee on the Ethics of Animal Experiments of the Freshwater Fisheries Research Center. Approval Code: LAECFFRC-2024-03-19. Approval Date: 19 March 2024.

Data Availability Statement: All data are available upon request.

Acknowledgments: The authors acknowledge Yancheng Shangshui Environmental Biotechnology Engineering Co., Ltd., Yancheng, China, for their kind help during the experiment.

Conflicts of Interest: The authors declare no conflicts of interest.

References

1. Thornber, K.; Verner-Jeffreys, D.; Hinchliffe, S.; Rahman, M.M.; Bass, D.; Tyler, C.R. Evaluating antimicrobial resistance in the global shrimp industry. *Rev. Aquac.* **2020**, *12*, 966–986. [CrossRef]
2. Farag, M.A.; Mansour, S.T.; Nouh, R.A.; Khattab, A.R. Crustaceans (shrimp, crab, and lobster): A comprehensive review of their potential health hazards and detection methods to assure their biosafety. *J. Food Saf.* **2023**, *43*, 10326. [CrossRef]
3. Dawood, M.A.O.; Koshio, S.; Esteban, M.A. Beneficial roles of feed additives as immunostimulants in aquaculture: A review. *Rev. Aquac.* **2018**, *10*, 950–974. [CrossRef]
4. Qin, L.R.; He, J.H.; Rong, K.; Guo, C.; Liu, J.S.; Zhang, T.L.; Li, W. Changes in secondary sexual characteristics of female red swamp crayfish (*Procambarus clarkii*) and relationship to ovarian development: Implications for intensive breeding of seedlings. *Aquaculture* **2024**, *592*, 741156. [CrossRef]

5. Ou, J.T.; Wang, X.; Luan, X.Q.; Yu, S.; Chen, H.; Dong, H.Z.; Zhang, B.H.; Xu, Z.Q.; Liu, Y.; Zhao, W.H. Comprehensive analysis of the mRNA and miRNA transcriptome implicated in the immune response of *Procambarus clarkii* to *Spiroplasma eriocheiris*. *Microb. Pathog.* **2024**, *196*, 106928. [CrossRef] [PubMed]

6. Lin, Y.B.; Li, S.X.; Li, Y.L.; Fang, L.; Zhang, H.; Wang, Q.; Ruan, G.L. Effects of luteolin supplementation on growth, histology, antioxidant capacity, non-specific immunity and intestinal microbiota of the red swamp crayfish (*Procambarus clarkii*). *Anim. Feed Sci. Tech.* **2024**, *313*, 115986. [CrossRef]

7. Kong, D.H.; Ji, Y.X.; Zhang, B.Y.; Li, K.C.; Liao, Z.Y.; Wang, H.; Zhou, J.X.; Wang, Q.J. Effects of hydroxy methionine zinc on growth performance, immune response, antioxidant capacity, and intestinal microbiota of red claw crayfish (*Procambarus clarkii*). *Fish Shellfish Immun.* **2024**, *144*, 109231. [CrossRef]

8. Yang, D.; Sun, W.B.; Hou, M.D.; Xiao, C.B.; Jin, H.H.; Lin, Y.; Wang, D.P.; Ye, H.; Luo, H. *Eucommia ulmoides* Oliv. leaf extract: Effects on growth, antioxidant capacity, and lipid metabolism in Red swamp crayfish (*Procambarus clarkii*). *Aquac. Rep.* **2024**, *39*, 102425. [CrossRef]

9. Di Ciaula, A.; Garruti, G.; Baccetto, R.L.; Molina-Molina, E.; Bonfrate, L.; Wang, D.Q.H.; Portincasa, P. Bile Acid Physiology. *Ann. Hepatol.* **2017**, *16*, S4–S14. [CrossRef]

10. Adam, A.H.; Verdegem, M.; Soliman, A.A.; Zaki, M.; Khalil, R.H.; Nour, A.M.; Khaled, A.A.; Basuini, M.F.E.; Khalil, H.S. Effect of dietary bile acids: Growth performance, immune response, genes expression of fatty acid metabolism, intestinal, and liver morphology of striped catfish (*Pangasianodon hypophthalmus*). *Aquac. Rep.* **2023**, *29*, 101510. [CrossRef]

11. Copple, B.L.; Li, T.G. Pharmacology of bile acid receptors: Evolution of bile acids from simple detergents to complex signaling molecules. *Pharmacol. Res.* **2016**, *104*, 9–21. [CrossRef] [PubMed]

12. Fiorucci, S.; Distrutti, E.; Carino, A.; Zampella, A.; Biagioli, M. Bile acids and their receptors in metabolic disorders. *Prog. Lipid Res.* **2021**, *82*, 101094. [CrossRef] [PubMed]

13. Wang, L.; Sagada, G.; Wang, C.Y.; Liu, R.C.; Li, Q.; Zhang, C.; Yan, Y.Z. Exogenous bile acids regulate energy metabolism and improve the health condition of farmed fish. *Aquaculture* **2023**, *562*, 738852. [CrossRef]

14. Li, L.; Liu, T.Y.; Li, J.R.; Yang, Y.C.; Liu, H.Y.; Zhang, P.Y. Effectiveness of bile acids as a feed supplement to improve growth performance, feed utilization, lipid metabolism, digestive enzymes, and hepatic antioxidant status in aquaculture animals: A meta-analysis. *Aquac. Rep.* **2024**, *36*, 102121. [CrossRef]

15. Li, J.B.; Wang, Z.; Cao, X.F.; Wang, J.M.; Gong, Y.; Wang, X.E.; Lai, W.C.; Bu, X.Y.; Zheng, J.C.; Mai, K.S.; et al. Effects of supplemental mixed bile acids on growth performance, body composition, digestive enzyme activities, skin color, and flesh quality of juvenile large yellow croaker (*Larimichthys crocea*) in soybean oil based diet. *Front. Mar. Sci.* **2023**, *10*, 1149887. [CrossRef]

16. Peng, X.R.; Feng, L.; Jiang, W.D.; Wu, P.; Liu, Y.; Jiang, J.; Kuang, S.Y.; Tang, L.; Zhou, X.Q. Supplementation exogenous bile acid improved growth and intestinal immune function associated with NF-κB and TOR signalling pathways in on-growing grass carp (*Ctenopharyngodon idella*): Enhancement the effect of protein-sparing by dietary lipid. *Fish Shellfish Immun.* **2019**, *92*, 552–569. [CrossRef]

17. Kumar, V.; Sinha, A.K.; Romano, N.; Allen, K.M.; Bowman, B.A.; Thompson, K.R.; Tidwell, J.H. Metabolism and Nutritive Role of Cholesterol in the Growth, Gonadal Development, and Reproduction of Crustaceans. *Rev. Fish. Sci. Aquac.* **2018**, *26*, 254–273. [CrossRef]

18. Liu, Y.C.; Niu, K.M.; Wang, R.X.; Liang, X.X.; Lin, C.; Wu, X.; Zhai, Z.Y. Taurochenodeoxycholic acid inhibits intestinal epithelial cell proliferation and induces apoptosis independent of the farnesoid X receptor. *Food Funct.* **2023**, *14*, 5277–5289. [CrossRef]

19. Bao, L.G.; Hao, D.C.; Wang, X.; He, X.L.; Mao, W.; Li, P.F. Transcriptome investigation of anti-inflammation and immuno-regulation mechanism of taurochenodeoxycholic acid. *BMC Pharmacol. Toxicol.* **2021**, *22*, 23. [CrossRef]

20. Zhu, F.Y.; Xu, W.H.; Wang, W.J.; Liao, J.M.; Huang, Y.H.; Huang, X.H.; Qin, Q.W. Taurochenodeoxycholic acid exerts anti-viral activities upon SGIV infection via anti-inflammatory response. *Aquaculture* **2024**, *578*, 740124. [CrossRef]

21. Wang, X.; Zhang, Z.Y.; He, X.L.; Mao, W.; Zhou, L.; Li, P.F. Taurochenodeoxycholic acid induces NR8383 cells apoptosis via PKC/JNK-dependent pathway. *Eur. J. Pharmacol.* **2016**, *786*, 109–115. [CrossRef] [PubMed]

22. Song, C.Y.; Wen, H.B.; Liu, G.X.; Ma, X.Y.; Lv, G.H.; Wu, N.Y.; Chen, J.X.; Xue, M.M.; Li, H.X.; Xu, P. Gut Microbes Reveal Pseudomonas Medicates Ingestion Preference Protein Utilization and Cellular Homeostasis Under Feed Domestication in Freshwater Drum, *Aplodinotus grunniens*. *Front. Microbiol.* **2022**, *13*, 861705, Corrigendum in *Front. Microbiol.* **2023**, *14*, 1259988. [CrossRef] [PubMed]

23. de Bruijn, I.; Liu, Y.Y.; Wiegertjes, G.F.; Raaijmakers, J.M. Exploring fish microbial communities to mitigate emerging diseases in aquaculture. *FEMS Microbiol. Ecol.* **2018**, *94*, 161. [CrossRef] [PubMed]

24. Zhang, X.Y.; Shen, D.Q.; Fang, Z.W.; Jie, Z.Y.; Qiu, X.M.; Zhang, C.F.; Chen, Y.L.; Ji, L.N. Human Gut Microbiota Changes Reveal the Progression of Glucose Intolerance. *PLoS ONE* **2013**, *8*, e71108. [CrossRef]

25. Chambers, E.S.; Byrne, C.S.; Morrison, D.J.; Murphy, K.G.; Preston, T.; Tedford, C.; Garcia-Perez, I.; Fountana, S.; Serrano-Contreras, J.I.; Holmes, E.; et al. Dietary supplementation with inulin-propionate ester or inulin improves insulin sensitivity in adults with overweight and obesity with distinct effects on the gut microbiota, plasma metabolome and systemic inflammatory responses: A randomised crossover trial. *Gut* **2019**, *68*, 1430–1438. [CrossRef]
26. Lin, H.R.; Xu, F.Z.; Chen, D.Y.; Xie, K.L.; Yang, Y.D.; Hu, W.; Li, B.Y.; Jiang, Z.L.; Liang, Y.H.; Tang, X.Y.; et al. The gut microbiota-bile acid axis mediates the beneficial associations between plasma vitamin D and metabolic syndrome in Chinese adults: A prospective study. *Clin. Nutr.* **2023**, *42*, 887–898. [CrossRef]
27. Goodrich, J.K.; Davenport, E.R.; Beaumont, M.; Jackson, M.A.; Knight, R.; Ober, C.; Spector, T.D.; Bell, J.T.; Clark, A.G.; Ley, R.E. Genetic Determinants of the Gut Microbiome in UK Twins. *Cell Host Microbe* **2016**, *19*, 731–743. [CrossRef]
28. Dai, H.Y.; Shan, Z.Y.; Shi, L.; Duan, Y.H.; An, Y.C.; He, C.H.; Lyu, Y.; Zhao, Y.G.; Wang, M.L.; Du, Y.H.; et al. Mulberry leaf polysaccharides ameliorate glucose and lipid metabolism disorders via the gut microbiota-bile acids metabolic pathway. *Int. J. Biol. Macromol.* **2024**, *282*, 136876. [CrossRef]
29. Zheng, X.C.; Xu, X.D.; Liu, M.Y.; Yang, J.; Yuan, M.; Sun, C.X.; Zhou, Q.L.; Chen, J.M.; Liu, B. Bile acid and short chain fatty acid metabolism of gut microbiota mediate high-fat diet induced intestinal barrier damage in *Macrobrachium rosenbergii*. *Fish Shellfish Immun.* **2024**, *146*, 109376. [CrossRef]
30. Mancin, L.; Wu, G.D.; Paoli, A. Gut microbiota-bile acid-skeletal muscle axis. *Trends Microbiol.* **2023**, *31*, 254–269. [CrossRef]
31. Zhou, J.S.; Chen, H.J.; Ji, H.; Shi, X.C.; Li, X.X.; Chen, L.Q.; Du, Z.Y.; Yu, H.B. Effect of dietary bile acids on growth, body composition, lipid metabolism and microbiota in grass carp (*Ctenopharyngodon idella*). *Aquac. Nutr.* **2018**, *24*, 802–813. [CrossRef]
32. Cheng, L.; Zhou, J.L.; Cheng, J.P. Bioaccumulation, tissue distribution and joint toxicity of erythromycin and cadmium in Chinese mitten crab (*Eriocheir sinensis*). *Chemosphere* **2018**, *210*, 267–278. [CrossRef] [PubMed]
33. Liu, B.; Song, C.Y.; Gao, Q.; Liu, B.; Zhou, Q.L.; Sun, C.X.; Zhang, H.M.; Liu, M.; Tadese, D.A. Maternal and environmental microbes dominate offspring microbial colonization in the giant freshwater prawn *Macrobrachium rosenbergii*. *Sci. Total Environ.* **2021**, *790*, 148062. [CrossRef] [PubMed]
34. Wen, C.; Ma, S.; Tian, H.Y.; Jiang, W.B.; Jia, X.Y.; Zhang, W.X.; Jiang, G.Z.; Li, X.F.; Chi, C.; He, C.F.; et al. Evaluation of the protein-sparing effects of carbohydrates in the diet of the crayfish, *Procambarus clarkii*. *Aquaculture* **2022**, *556*, 738275. [CrossRef]
35. Yang, H.J.; Mo, A.J.; Yi, L.Y.; Wang, J.H.; He, X.G.; Yuan, Y.C. Selenium attenuated food borne cadmium-induced intestinal inflammation in red swamp crayfish (*Procambarus clarkii*) via regulating PI3K/Akt/NF-κBpathway. *Chemosphere* **2024**, *349*, 140814. [CrossRef]
36. Cai, M.L.; Zhang, Y.; Zhu, J.Q.; Li, H.H.; Tian, H.Y.; Chu, W.Y.; Hu, Y.; Liu, B.; Wang, A.M. Intervention of re-feeding on growth performance, fatty acid composition and oxidative stress in the muscle of red swamp crayfish (*Procambarus clarkii*) subjected to short-term starvation. *Aquaculture* **2021**, *545*, 737110. [CrossRef]
37. Zhu, M.R.; Zhan, M.; Xi, C.J.; Gong, J.; Shen, H.S. Molecular characterization and expression of the autophagy-related gene Atg14 in WSSV-infected *Procambarus clarkii*. *Fish Shellfish. Immunol.* **2022**, *125*, 200–211. [CrossRef]
38. Liu, J.; Du, H.; Liu, T.; Chen, C.; Yan, Y.; Liu, T.Q.; Liu, L.; Wang, E.L. Accurate reflection of hepatopancreas antioxidation and detoxification in *Procambarus clarkii* during virus infection and drug treatment: Reference gene selection, evaluation and expression analysis. *Aquaculture* **2022**, *556*, 738283. [CrossRef]
39. Wang, Y.Y.; Xu, Z.F.; Li, M.L.; Shuai, K.; Lei, L.; Li, X.Q.; Leng, X.J. Supplemental bile acids in low fishmeal diet improved the growth, nutrient utilization of Pacific white shrimp. *Aquac. Rep.* **2023**, *28*, 101452. [CrossRef]
40. Du, Y.H.; Wang, G.J.; Yu, E.M.; Xie, J.; Xia, Y.; Li, H.Y.; Zhang, K.; Gong, W.B.; Li, Z.F.; Xie, W.P.; et al. Dietary deoxycholic acid decreases fat accumulation by activating liver farnesoid X receptor in grass crap (*Ctenopharyngodon idella*). *Aquaculture* **2024**, *578*, 740123. [CrossRef]
41. Weatherley, A.H.; Gill, H.S. *The Biology of Fish Growth*; Academic Press: Cambridge, MA, USA, 1987.
42. Jiang, J.Y.; Lu, X.; Dong, L.X.; Tian, J.; Zhang, J.M.; Guo, Z.B.; Luo, Y.J.; Cui, Z.B.; Wen, H.; Jiang, M. Enhancing growth, liver health, and bile acid metabolism of tilapia (*Oreochromis niloticus*) through combined cholesterol and bile acid in diets. *Anim. Nutr.* **2024**, *17*, 335–346. [CrossRef] [PubMed]
43. McCarthy, J.J.; Esser, K.A. Anabolic and catabolic pathways regulating skeletal muscle mass. *Curr. Opin. Clin. Nutr.* **2010**, *13*, 230–235. [CrossRef] [PubMed]
44. Saxton, R.A.; Sabatini, D.M. mTOR Signaling in Growth, Metabolism, and Disease. *Cell* **2017**, *168*, 960–976. [CrossRef] [PubMed]
45. Nedelsky, N.B.; Todd, P.K.; Taylor, J.P. Autophagy and the ubiquitin-proteasome system: Collaborators in neuroprotection. *BBA-Mol. Basis Dis.* **2008**, *1782*, 691–699. [CrossRef]
46. Eskelinen, E.L. Autophagy: Supporting cellular and organismal homeostasis by self-eating. *Int. J. Biochem. Cell B* **2019**, *111*, 1–10. [CrossRef]
47. Magnani, N.D.; Dada, L.A.; Sznajder, J.I. Ubiquitin-proteasome signaling in lung injury. *Transl. Res.* **2018**, *198*, 29–39. [CrossRef]
48. Cheng, J.H.; Sun, D.W.; Han, Z.; Zeng, X.A. Texture and Structure Measurements and Analyses for Evaluation of Fish and Fillet Freshness Quality: A Review. *Compr. Rev. Food Sci. F* **2014**, *13*, 52–61. [CrossRef]

49. Roy, B.C.; Bruce, H.L. Contribution of intramuscular connective tissue and its structural components on meat tenderness-revisited: A review. *Crit. Rev. Food Sci.* **2024**, *64*, 9280–9310. [CrossRef]

50. Matarneh, S.K.; Silva, S.L.; Gerrard, D.E. New Insights in Muscle Biology that Alter Meat Quality. *Annu. Rev. Anim. Biosci.* **2021**, *9*, 355–377. [CrossRef]

51. Zhou, M.Z.; Shi, G.P.; Deng, Y.; Wang, C.; Qiao, Y.; Xiong, G.Q.; Wang, L.; Wu, W.J.; Shi, L.; Ding, A.Z. Study on the physicochemical and flavor characteristics of air frying and deep frying shrimp (crayfish) meat. *Front. Nutr.* **2022**, *9*, 1022590. [CrossRef]

52. Li, X.; Qu, K.; Liu, Y.; Dong, X.; Chi, S.; Yang, Q.; Tan, B.; Zhang, S.; Xie, S. Effects of Dietary Bile Acids Supplementation on Growth Performance, Sterol Metabolism, Bile Acids Enterohepatic Circulation, and Apoptosis of Juvenile Pacific White Shrimp (*Litopenaeus Vannamei*). *SSRN* **2023**. [CrossRef]

53. Bouton, P.E.; Harris, P.V.; Shorthose, W.R. Possible relationships between shear, tensile, and adhesion properties of meat and meat structure. *J. Texture Stud.* **1975**, *6*, 297–314. [CrossRef]

54. Chen, L.; Shi, Y.H.; Li, J.B.; Shao, C.M.; Ma, S.; Shen, C.; Zhao, R.Q. Dietary bile acids improve breast muscle growth in chickens through FXR/IGF2 pathway. *Poult. Sci.* **2024**, *103*, 103346. [CrossRef] [PubMed]

55. Nicol, R.L.; Frey, N.; Pearson, G.; Cobb, M.; Richardson, J.; Olson, E.N. Activated MEK5 induces serial assembly of sarcomeres and eccentric cardiac hypertrophy. *EMBO J.* **2001**, *20*, 2757–2767. [CrossRef] [PubMed]

56. Johnston, I.A. Muscle development and growth: Potential implications for flesh quality in fish. *Aquaculture* **1999**, *177*, 99–115. [CrossRef]

57. Acosta, J.; Carpio, Y.; Borroto, I.; González, O.; Estrada, M.P. Myostatin gene silenced by RNAi show a zebrafish giant phenotype. *J. Biotechnol.* **2005**, *119*, 324–331. [CrossRef]

58. Dumont, N.A.; Rudnicki, M.A. Characterizing satellite cells and myogenic progenitors during skeletal muscle regeneration. *Methods Mol. Biol.* **2017**, *1560*, 179–188. [CrossRef]

59. Jia, W.; Xie, G.X.; Jia, W.P. Bile acid-microbiota crosstalk in gastrointestinal inflammation and carcinogenesis. *Nat. Rev. Gastro Hepatol.* **2018**, *15*, 111–128. [CrossRef]

60. Barut, I.; Kaya, S. The Diagnostic Value of C-Reactive Protein in Bacterial Translocation in Experimental Biliary Obstruction. *Adv. Clin. Exp. Med.* **2014**, *23*, 197–203. [CrossRef]

61. Furt, F.; Lemoi, K.; Tüzel, E.; Vidali, L. Quantitative analysis of organelle distribution and dynamics in Physcomitrella patens protonemal cells. *BMC Plant Biol.* **2012**, *12*, 70. [CrossRef]

62. Tarhriz, V.; Eyvazi, S.; Shakeri, E.; Hejazi, M.S.; Dilmaghani, A. Antibacterial and Antifungal Activity of Novel Freshwater Bacterium *Tabrizicola aquatica* as a Prominent Natural Antibiotic Available in Qurugol Lake. *Pharm. Sci.* **2020**, *26*, 88–92. [CrossRef]

63. Cai, X.; Yu, Y.; Wang, Y.; Zhang, Z.; Liao, M.; Li, B.; Liu, X.; Zhu, H.; Rong, X. Effects of potassium monopersulfate on water environment index and microbial community structure of *Litopenaeus vannamei* in pond culture system. *J. Shanghai Ocean Univ.* **2022**, *31*, 452–461. (In Chinese)

64. Harwati, T.U.; Kasai, Y.; Kodama, Y.; Susilaningsih, D.; Watanabe, K. *Tropicimonas isoalkanivorans* gen. nov., sp. nov., a branched-alkane-degrading bacterium isolated from Semarang Port in Indonesia. *Int. J. Syst. Evol. Micr.* **2009**, *59*, 388–391. [CrossRef]

65. Xiong, F.; Wu, S.G.; Zhang, J.; Jakovlic, I.; Li, W.X.; Zou, H.; Li, M.; Wang, G.T. Dietary Bile Salt Types Influence the Composition of Biliary Bile Acids and Gut Microbiota in Grass Carp. *Front. Microbiol.* **2018**, *9*, 2209. [CrossRef]

66. Attwood, M.M.; Harder, W. The Oxidation and Assimilation of C_2 Compounds by *Hyphomicrobium* sp. *Microbiology* **1974**, *84*, 350–356. [CrossRef]

 *fishes*

Article

Dietary Alpha-Lipoic Acid Alleviated Hepatic Glycogen Deposition and Improved Inflammation Response of Largemouth Bass (*Micropterus salmoides*) Fed on High Dietary Carbohydrates

Zishuo Fang [1,†], Xianwei Pan [2,†], Ye Gong [1], Nihe Zhang [1], Shiwen Chen [1], Ning Liu [2,*], Naisong Chen [1] and Songlin Li [1,3,*]

[1] Research Centre of the Ministry of Agriculture and Rural Affairs on Environmental Ecology and Fish Nutrition, Shanghai Ocean University, Shanghai 201306, China; 42fzs@163.com (Z.F.); y-gong@shou.edu.cn (Y.G.); gfaznh@163.com (N.Z.); jx202201012@gmail.com (S.C.); nschen@shou.edu.cn (N.C.)

[2] International Research Centre for Food and Health, College of Food Science and Technology, Shanghai Ocean University, Shanghai 201306, China; xianweipan2023@163.com

[3] National Demonstration Center on Experiment Teaching of Fisheries Science, Shanghai Ocean University, Shanghai 201306, China

* Correspondence: nliu@shou.edu.cn (N.L.); slli@shou.edu.cn (S.L.)

† These authors contributed equally to this work.

Academic Editor: Francisco J. Moyano

Received: 6 December 2024
Revised: 22 December 2024
Accepted: 26 December 2024
Published: 28 December 2024

Citation: Fang, Z.; Pan, X.; Gong, Y.; Zhang, N.; Chen, S.; Liu, N.; Chen, N.; Li, S. Dietary Alpha-Lipoic Acid Alleviated Hepatic Glycogen Deposition and Improved Inflammation Response of Largemouth Bass (*Micropterus salmoides*) Fed on High Dietary Carbohydrates. *Fishes* 2025, 10, 9. https://doi.org/10.3390/fishes10010009

Abstract: In order to mitigate the adverse effects of high carbohydrates on largemouth bass and to investigate the feasibility of LA as a feed additive, the present study observed the effects of added α-lipoic acid (LA) on growth performance, glucose metabolism and immunity in largemouth bass fed on high dietary carbohydrates (10% α-cassava starch inclusion). A total of 315 juvenile largemouth bass (initial body weight, 5.09 ± 0.10 g) were divided into nine tanks (800 L) (upper radius 0.65 m × lower radius 0.5 m × height 1 m), with each holding 35 fish. Three iso-nitrogenous and iso-lipidic diets supplementing with 0 g/kg, 0.5 g/kg and 1 g/kg LA (LA0, LA500, LA1000) were designed to feed juvenile largemouth bass on a satiation diet twice daily for eight weeks with each diet feeding to triplicate groups. The results indicated that the performance in growth was significantly enhanced by the addition of dietary LA ($p < 0.05$). Meanwhile, hepatic glycogen content was significantly reduced ($p < 0.05$), and the expression of genes relating to insulin pathway and glycolysis significantly increased with LA inclusion ($p < 0.05$). The relative expression of insulin receptor a (*ira*) in the LA500 group was the highest, while the relative expression of glycerol kinase (*gk*), phosphofructokinase liver type (*pfkl*) and phosphoenolpyruvate carboxykinase (*pepck*) was the highest in the LA1000 group ($p < 0.05$). In addition, LA supplementation significantly increased the activity of lysozyme, which reached its maximum value in the LA500 group ($p < 0.05$). LA supplementation also promoted the expression of genes relating to anti-inflammatory and inhibited the expression of pro-inflammatory related genes ($p < 0.05$). Above all, the dietary addition of LA could improve performance in growth, alleviated hepatic glycogen deposition, and improved the immunity function of largemouth bass fed on high dietary carbohydrates. This provides us with ideas to mitigate the adverse effects of high carbohydrates on largemouth bass in actual production and provides a basis for the application of LA in aquatic biology.

Keywords: α-lipoic acid; glycogen deposition; immunity function; largemouth bass

Key Contribution: This study demonstrated that dietary alpha-lipoic acid alleviated hepatic glycogen deposition and improved inflammation response of largemouth bass

(*Micropterus salmoides*) fed on a high carbohydrate diet, which provides a theoretical basis for the application of LA in aquaculture.

1. Introduction

Starch is an economical energy source and a good binder for aquatic feeds [1], which has the advantages of easy accessibility and cost-effectiveness in aquafeed production. However, carnivorous fish could not effectively utilize dietary glucose, and high carbohydrate intake could cause hepatic glycogen deposition, in turn affecting fish growth performance and health status, which has been well demonstrated by studies on largemouth bass (*Micropterus salmoides*) [2,3], whereas decreased insulin secretion or increased insulin resistance has been suggested to be one of the causes for the limited carbohydrate utilization [4]. Meanwhile, insulin pathway activation is also involved in the improvement of glucose metabolism, which has been proven in largemouth bass [3,5,6]. However, high dietary carbohydrates are not effective in promoting insulin secretion and activating insulin pathways in carnivorous fish [7,8]. Additionally, excessive carbohydrates might also exacerbate inflammatory response in carnivorous fish [9,10], which would negatively affect the health of cultured fish. Consequently, finding an effective additive that could both regulate glucose metabolism and mitigate the adverse effects caused by high carbohydrates is important to produce for carnivorous fish.

Immune function is important for fish health and modern aquaculture exposes fish to acute stresses (e.g., crowding and handling), leading to increased morbidity [11,12]. The improvement of immunity could reduce the incidence of disease in fish [13]. In the immune system, stimulation of specific immunity through vaccines and other means is economically inefficient; therefore, improving non-specific immunity is a key area of focus for aquaculture [14]. Anti-inflammatory cytokines (e.g., IL-10) and pro-inflammatory cytokines (e.g., IL-6) also play important roles in fish immune organs and immune functions [15]. In addition, previous studies have shown that p38-MAPK/I-κB/NF-κB signaling pathway regulates inflammatory cytokines in largemouth bass [16–18].

α-Lipoic acid (LA) is a natural compound that can be obtained from microorganisms, plants and animals and has a regulatory role in insulin sensitivity and insulin secretion [19,20]. In mammals, it has been well demonstrated that LA could stimulate the insulin-signaling cascade through its pro-oxide properties [21]. Apart from mammals, in common carp (*Cyprinus carpio*), LA has been shown to increase the utilization of feed carbohydrates [22], in turn improving glycemic control in fish. Meanwhile, LA also has a role in immunomodulation, which could improve immunity through enhancing the activity of lysozyme, inhibiting the expression genes relating to pro-inflammatory cytokine genes and promoting genes relating to the expression of anti-inflammatory cytokines [16,23]. Therefore, LA shows a potential function in improving carbohydrates utilization and maintaining a state of health for carnivorous fish and has the potential to be used as a feed additive and thus mitigate the adverse effects of high carbohydrates on carnivorous fish. However, studies on LA in aquaculture have mainly focused on its effects on growth performance and non-specific immunity of cultured fish, and there is still a paucity of studies on the regulation of sugar metabolism pathways in fish by LA. It remains to be investigated whether LA can regulate the utilization of carbohydrates in carnivorous fish and mitigate the adverse effects caused by high carbohydrates.

Largemouth bass have been extensively cultured in China. Considerable research has proved the limitation of this fish in utilizing carbohydrates, and 10% starch inclusion could impair its growth and health status [2,3]. Above all, considering the problems identified in the actual culture of largemouth bass, as well as the functions exhibited by LA, this study investigated the effects of dietary LA addition on hepatic glycogen deposition and inflammatory responses in largemouth bass consuming high carbohydrates.

2. Materials and Methods

2.1. Preparation for Diets

Previous studies on LA in a large number of other fish species showed that most of the LA additions in fishes were concentrated at 0.5–1 g/kg [24–27]. Therefore, three iso-nitrogenous and iso-lipid experimental diets with 0 g/kg (LA0), 500 mg/kg (LA500) and 1000 mg/kg (LA1000) LA additive concentrations were designed (Table 1). Prior research has demonstrated that 10% dietary starch consistently induces liver glycogen accumulation in largemouth bass, which negatively affects the fish's health [2,3]. Consequently, 10% of dietary starch addition with 0 mg/kg of LA supplementation in the present study was selected as the control group. After the components were thoroughly mixed, an extrusion swelling machine was used to extrude the experimental diets. The low-lipid components of the experimental meals were pulverized, well combined, and then well blended with the lipid ingredients. Afterwards, water was added to harden the dough, and, after that, the dough was extruded by a pelletizer into two sizes (2.5 mm and 4 mm). The manufactured diets were then placed at −20 °C until the experiment began.

Table 1. Formulation and proximate composition of experimental diets (% dry matter).

Ingredients	LA0	LA500	LA1000
White fish meal [2]	40.00	40.00	40.00
Wheat gluten meal [2]	3.00	3.00	3.00
Blood meal [2]	3.00	3.00	3.00
Shrimp meal [2]	3.00	3.00	3.00
Fermented soybean meal [2]	12.00	12.00	12.00
Corn gluten meal [2]	12.00	12.00	12.00
Brewer's yeast meal [2]	3.00	3.00	3.00
Squid viscera meal [2]	2.00	2.00	2.00
Soybean phospholipids [2]	2.00	2.00	2.00
Fish oil [2]	2.00	2.00	2.00
Soybean oil [2]	2.00	2.00	2.00
α-lipoic acid [1]	0.00	0.05	0.10
Microcrystalline cellulose	3.00	2.95	2.90
Vitamin mixture [3]	1.00	1.00	1.00
Mineral mixture [4]	1.00	1.00	1.00
$Ca(H_2PO_4)_2$ [2]	1.00	1.00	1.00
α-cassava Starch [2]	10.00	10.00	10.00
Proximate analysis (Mean values, % dry weight)			
Crude protein	52.17	51.93	52.03
Crude lipid	11.84	10.54	10.73

[1] Acquired from Sanhua Biotechnology Co. (Zhengzhou, China), purity, $\geq$98%. [2] Provided by Xinxin Tian'en Feed Corporation (Zhejiang, China). [3] Vitamin mixture (mg/kg diet): vitamin B_1, 17.80; vitamin A, 16,000 IU; vitamin B_2, 48; vitamin B_6, 29.52; vitamin B_{12}, 0.24; vitamin C, 800; vitamin D_3, 8000 IU; vitamin E, 160; vitamin K_3, 14.72; choline chloride, 1500; folic acid, 6.40; niacinamide, 79.20; inositol, 320; calcium-pantothenate, 73.60; biotin, 0.64. [4] Mineral mixture (mg/kg diet): I [Ca $(IO_3)_2$], 1.63; Mn $(MnSO_4)$, 6.20; Cu $(CuSO_4)$, 2.00; Fe $(FeSO_4)$, 21.10; Se (Na_2SeO_3), 0.18; Co $(CoCl_2)$, 0.24; Zn $(ZnSO_4)$, 34.4.

2.2. Experiment Design

This feeding study was carried out in the joint lab of Shanghai Ocean University and Shanghai Nonghao Feed Co., Ltd. (Shanghai, China). The experiment fish were temporarily raised for three weeks with feeding on the commercial feed supplied by Xinxin Tian'en

Aquafeed Ltd. (Zhejiang, China), with the same water quality parameters as during the experiment. After the temporary rearing, 315 fish with an average initial body weight of (5.09 ± 0.10 g) were screened and distributed into nine 800 L tanks (upper radius 0.65 m × lower radius 0.5 m × height 1 m), each holding 35 fish. The fish in triplicate tanks were each fed an experiment diet until apparent saturation two times daily (07:30 and 17:30), and intake was recorded for 56 days. Over the process, the water exchange was 5% daily, and the water conditions were as follows: pH 7.1 ± 0.3; temperature 25 ± 0.5 °C; ammonia and nitrate content < 0.1 mg/L.

2.3. Sample Collection

After the trial, fish fasted for 24 h for further sampling, while counting the number of fish in the bucket to calculate survival rate (SR) and measuring the total weight of the fish in each bucket to measure the final mean body weight were conducted. Twelve fish from each tank were taken and anesthetized using eugenol (1:10,000) for body length and weight measurements, and three fish from each tank were randomly taken for the analysis of whole-body composition. After that, liver and viscera were isolated from 8 of the 12 fish, liver and viscera weights were determined and visceral index (VSI) and hepatic index (HSI) were analyzed. The liver samples were separated for the examination of the glycogen content, and the livers and kidneys of the remaining four fish were used for gene expression analysis. Muscle samples used for chemical analysis were obtained from the 12 fish. After being immediately frozen in liquid nitrogen, the samples were maintained at −80 °C for further examination.

2.4. Chemical Analysis

The moisture, crude protein and crude lipid content were analyzed following the method described in AOAC [28]. The whole fish, liver and muscle samples were dried to a consistent weight to assess the moisture content, and the crude lipid content was determined using the Soxhlet extraction method (SX-360, Opsis, Furulund, Sweden). The crude protein content was determined using the Dumas combustion technique (FP828, LECO, St. Joseph, MI, USA). The glycogen content of the liver and muscles was measured with potassium hydroxide/anthrone, using a commercial glycogen assay kit, (operated according to the instructions to accurately organize the weight added to the alkaline solution in a boiling water bath for 20 min). Then, a certain amount of double-distilled water was added to make the blood concentration of the test solution, and a certain amount of detection solution was taken and added as a colorant in the boiling water bath for 5 min. Finally, colorimetry at 620 nm wavelength with 1 cm aperture was conducted (A043-1-1, Nanjing Jiancheng Bioengineering Institute, Nanjing, China) [29]. Lysozyme (LZM) activity was determined using a turbidimetric method, using a commercial lysozyme assay kit (the prepared application bacterial solution was put into a 37 °C water bath for more than 10 min, so that the temperature of the bacterial solution reached 37 °C and so that the standard solution and the sample were at the same temperature). And the visible spectrophotometer was set at 530 nm, with 1 cm aperture cuvette and with double-distilled water to adjust the transmittance of 100%. Next, this was placed into the corresponding number of test tubes, and 0.2 mL of the sample was added in order to be tested. Then 2 mL of the application of bacterial solution was quickly rushed into the test tube and was immediately mixed and timed. After that, it was quickly poured it into the cuvette at 530 nm in the visible spectrophotometer, the transmittance value T1 at 15 s was read. The cuvette was not taken out, and the transmittance value T2 at 2 min and 15 s was read, and the difference in the transmittance of the two times was found out and then was calculated according to the instructions (A050-1-1, Nanjing Jiancheng Bioengineering Institute, Nanjing, China) [30].

2.5. Relative Gene Expression Analysis

The total liver RNA of cultured fish was extracted using the TransZol Up (Code#ET111-01) (Takara, Shiga, Japan), following the instruction of the manufacturer. Then, a commercial kit (Cat#RR074A) (Takara, Shiga, Japan) was used to reverse the extracted RNA to single strand cDNA for real-time quantification PCR (RT-qPCR). The primers of genes involved in the present study were listed in Table S1. After that, a melting curve analysis was performed after 40 cycles of 95 °C for 10 s, 57 °C for 10 s and 72 °C for 20 s, as well as RT-qPCR using a quantitative thermal cycler (Mastercycler EP Realplex, Eppendorf, Germany). *β-actin* was selected to be the housekeeping gene, and a $2^{-\Delta\Delta Ct}$ method was used to analyze relative gene expression [31].

2.6. Calculation and Statistical Analysis

The variance test and the chi-square test were used, respectively, to evaluate the data's homogeneity and normality. The dose of LA for optimum fish growth was determined using the orthogonal polynomial regression. Using SPSS 24.0 software, one-way ANOVA analysis was performed to analyze all the data, which were represented as mean ± standard error of the mean (SEM). The multiple comparison test, with a significance threshold of 0.05, was Duncan's multiple range test.

Calculation formula:

$$\text{Survival rate (SR, \%)} = \frac{\text{Nf}}{\text{N0}} \times 100$$

$$\text{Specific growth rate (SGR, \%/d)} = \frac{[\ln(\text{Wf}) - \ln(\text{Wi})]}{\text{days}} \times 100$$

$$\text{Condition factor (CF, \%)} = \frac{\text{BW}}{(\text{BL})3} \times 100$$

$$\text{Feed intake rate (FIR, \%/d·individual)} = \frac{\text{Wd}}{[\text{days} \times (\text{Wf} + \text{Wi}) / 2 \times \text{Nf}]} \times 100$$

$$\text{Hepatosomatic index (HSI, \%)} = \frac{\text{LW}}{\text{BW}} \times 100$$

$$\text{Viscerosomatic index (VSI, \%)} = \frac{\text{VW}}{\text{BW}} \times 100$$

N0: number of individuals; Wi: the mean initial body weight; BW: individual body weight; LW: liver weight; VW: visceral weight; BL: fish body length; Wf: the mean of final body weight; Nf: the number of final individuals; Wd: the weight of the meal ingested.

3. Results

3.1. Growth Performance

The growth performance results are given in Table 2. LA supplementation had no significant effect on SR, condition factor (CF) and feed intake rate (FIR) ($p > 0.05$) but significantly increased specific growth rate (SGR) ($p < 0.05$). The VSI showed no discernible changes among different groups ($p > 0.05$). However, the HSI in the LA500 group was substantially lower than that of the control group ($p < 0.05$). Orthogonal polynomial analysis shows that an LA supplementation of 0.74 g/kg feed resulted in maximum growth in largemouth bass (Figure 1).

Table 2. Effects of adding different doses of LA on growth performance of largemouth bass.

	LA0	LA500	LA1000	*p*-Value
Initial body weight (g)	5.08 ± 0.01	5.10 ± 0.02	5.09 ± 0.01	0.144, 0.552
Final body weight (g)	40.86 ± 0.84 [b]	45.53 ± 0.96 [a]	45.39 ± 1.55 [a]	0.029, 0.033
Specific growth rate (%/d)	3.72 ± 0.04 [b]	3.90 ± 0.04 [a]	3.90 ± 0.06 [a]	0.034, 0.034
Condition factor (%g/cm^3)	2.29 ± 0.02	2.33 ± 0.02	2.33 ± 0.02	0.201, 0.201
Feed intake rate (%/d·individual)	1.91 ± 0.02	1.92 ± 0.03	1.94 ± 0.03	0.796, 0.403
Hepatosomatic index (%)	3.77 ± 0.04 a	3.38 ± 0.16 b	3.52 ± 0.06 ab	0.034, 0.131
Viscerosomatic index (%)	8.32 ± 0.20	8.31 ± 0.20	8.20 ± 0.11	0.990, 0.644

Means standard error of the mean, or SEM, N = 3, indicates that the data inside a row with a comparable superscript letter are not significantly different from the other dietary groups ($p > 0.05$). The calculation formula is shown below.

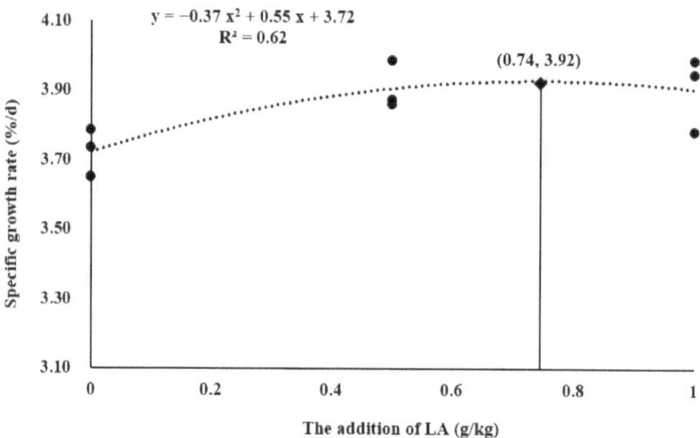

Figure 1. Polynomial regression relationship between dietary supplementation levels of LA and specific growth rate (SGR) of largemouth bass.

3.2. Body Composition and Glycogen Content Analysis

The body composition results are given in Table 3, and the glycogen content analysis results are shown in Figure 2. The supplementation of LA had no effect on the proximate composition of the whole fish body ($p > 0.05$), while liver crude protein content in LA500 was significantly decreased ($p < 0.05$). Furthermore, the LA addition considerably reduced the hepatic glycogen level and raised the muscle glycogen content ($p < 0.05$).

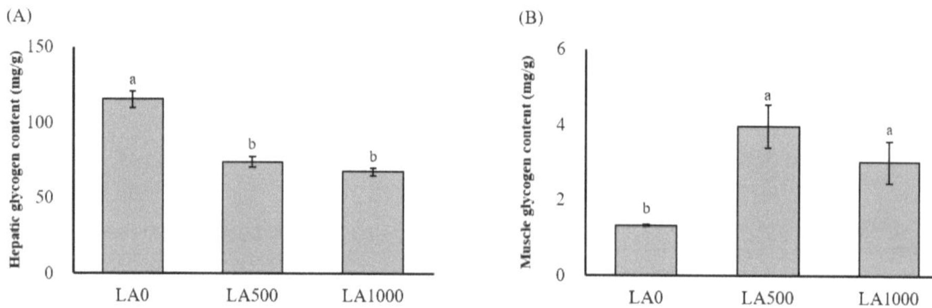

Figure 2. The amount of muscle glycogen (**B**) and hepatic glycogen (**A**) in largemouth bass given the experimental diets (N = 3). Bars bearing the same letter's values (mean standard error of the mean, or SEM) do not significantly differ from one another ($p > 0.05$; Duncan's test).

Table 3. Effects of adding different doses of LA (0 mg/kg, 500 mg/kg and 1000 mg/kg) on the body approximate composition of largemouth bass.

	LA0	LA500	LA1000	*p*-Value
Whole fish				
Moisture (%)	73.91 ± 1.16	74.09 ± 0.58	74.06 ± 0.45	0.877, 0.902
Crude lipid (%)	3.91 ± 0.41	3.92 ± 0.38	4.25 ± 0.37	0.986, 0.554
Crude protein (%)	17.26 ± 0.12	17.20 ± 0.14	17.32 ± 0.13	0.756, 0.730
Ash (%)	3.05 ± 0.13	2.97 ± 0.09	3.02 ± 0.02	0.588, 0.849
Liver				
Moisture (%)	74.38 ± 0.48	75.38 ± 0.50	73.62 ± 0.41	0.344, 0.528
Crude lipid (%)	2.78 ± 0.04	2.78 ± 0.03	2.74 ± 0.05	0.911, 0.543
Crude protein (%)	9.00 ± 0.04 [b]	10.47 ± 0.45 [a]	9.73 ± 0.19 [ab]	0.030, 0.207
Muscle				
Moisture (%)	78.34 ± 0.32	78.00 ± 0.25	78.12 ± 0.17	0.378, 0.555
Crude lipid (%)	1.79 ± 0.14	1.84 ± 0.10	1.48 ± 0.07	0.735, 0.085
Crude protein (%)	19.12 ± 0.36	19.43 ± 0.25	19.39 ± 0.12	0.431, 0.490

Means standard error of the mean, or SEM, N = 3, indicates that the data inside a row with a comparable superscript letter are not significantly different from the other dietary groups ($p > 0.05$).

3.3. Relative Expression of Genes Related to Insulin Pathway and Glucose Metabolism

The relative expression of genes related to insulin pathway results are shown in Figure 2. The relative expression of *ira* and *irb* were significantly enhanced by the addition of dietary LA ($p < 0.05$), with the LA500 group exhibiting the highest value level among all the groups (Figure 3A,B). The relative expression of *akt1* and *pi3kr1* in the groups with LA supplementation was significantly higher than the control group ($p < 0.05$) (Figure 3C,D).

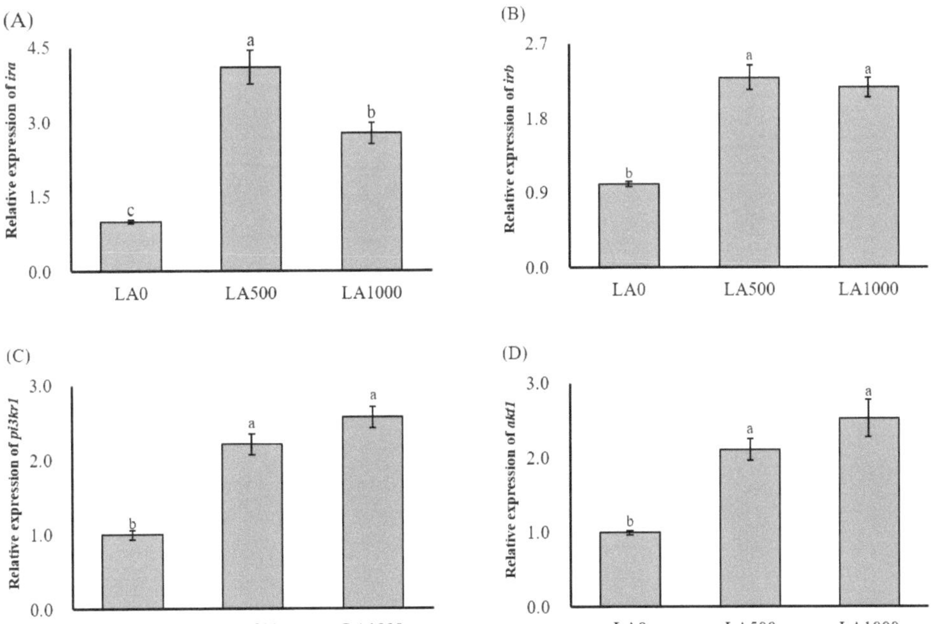

Figure 3. Expression of the genes associated with the insulin pathway in cultured fish: *ira* (**A**), *irb* (**B**), *akt1* (**C**), and *pi3kr1* (**D**). There are no significantly different values (meaning standard error of the mean, SEM) in bars with the same letter among treatments (N = 3; $p > 0.05$; Duncan's test).

The relative expression of genes related to glucose metabolism are shown in Figure 3. The relative expression of *gk* and *pfkl* were considerably increased by dietary LA inclusion, with the LA1000 group exhibiting the highest value among all the groups ($p < 0.05$) (Figure 4A,B). Nevertheless, there was no discernible change in *pk* expression ($p > 0.05$) (Figure 4C). When dietary LA was added, the expression of *g6pc* and *pepck* was dramatically increased and the maximum value of the expression of these genes was found in the LA1000 group ($p < 0.05$) (Figure 4D,E). However, the expression of *fbp1* was not affected by dietary LA supplementation ($p > 0.05$) (Figure 4F).

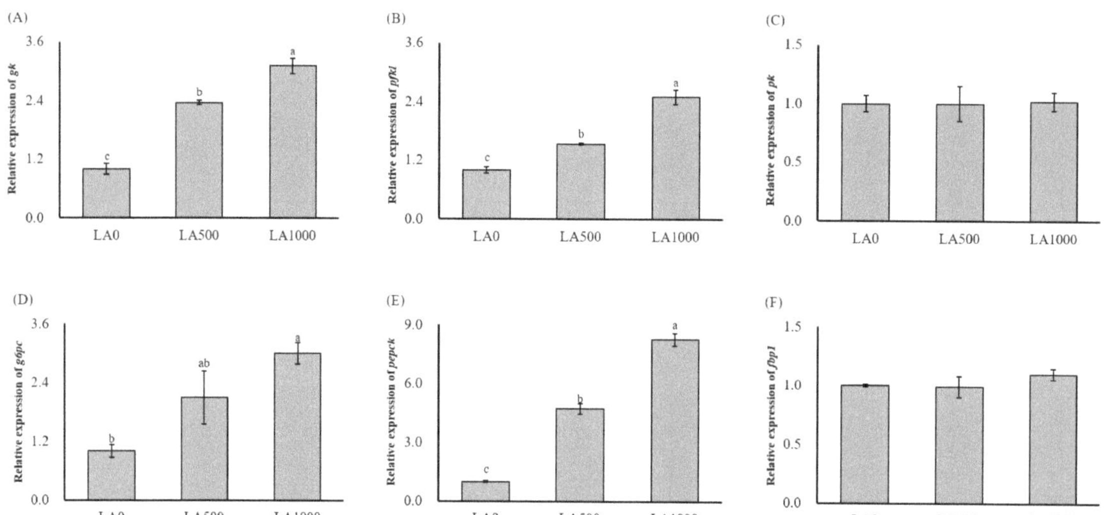

Figure 4. The expression of the genes associated with the gluconeogenesis of the experimental fish, namely *gk* (**A**), *pfkl* (**B**), and *pk* (**C**), *g6pc* (**D**), *pepck* (**E**) and *fbp1* (**F**). Values (standard error of the mean, SEM) in bars with no letter are not significantly different when comparing treatments (N = 3) ($p > 0.05$; Duncan's test).

3.4. Lysozyme Activity and Inflammation Response

The lysozyme activity and inflammation response results are shown in Figure 4. The activity of lysozyme in the LA500 and LA1000 groups was notably enhanced compared to the control group ($p < 0.05$) (Figure 5A). Additionally, dietary LA inclusion significantly decreased the relative expression of *mapk* ($p < 0.05$) (Figure 5B). The relative expression of *iκb* in LA500 and LA1000 groups was significantly higher than the control group ($p < 0.05$) (Figure 5C). The relative expression of *nfκb* in LA500 and LA1000 was significantly lower than the other treatments ($p < 0.05$) (Figure 5D). The relative expression of interleukin 10 (*il-10*) was significantly increased with the addition of LA ($p < 0.05$) (Figure 5E). Among the pro-inflammatory factors, the relative expression of *tnf-α* was significantly decreased in the LA500 group, whereas the relative expression of *il-1β* was significantly decreased in both the LA500 and LA1000 groups (Figure 5F,G).

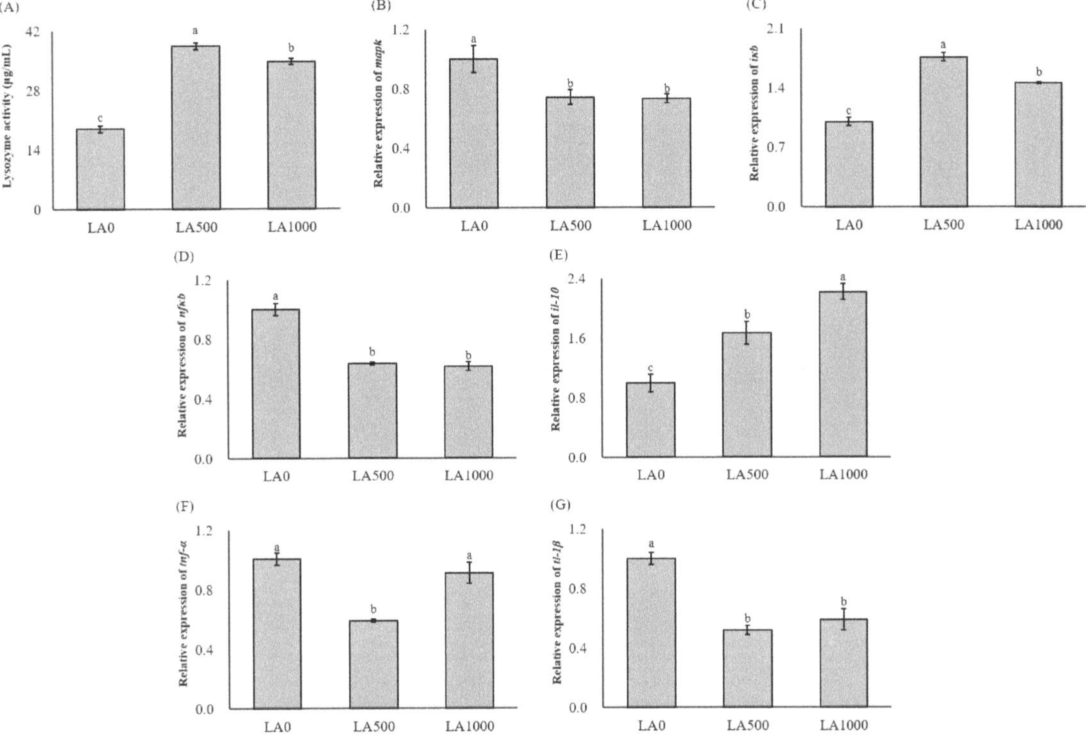

Figure 5. Addition of LA to feed: effects on lysozyme activity (**A**) and expression of the genes *mapk* (**B**), *iκb* (**C**), *nfκb* (**D**), *il-10* (**E**), *tnf-α* (**F**) and *il-1β* (**G**), which are related to immunocompetence in experimental fish. Values (standard error of the mean, SEM) in bars with the same letter are not significantly different when comparing treatments (N = 3) ($p > 0.05$; Duncan's test).

4. Discussion

Due to its variety of applications, LA is a naturally occurring substance that is widely distributed in nature. In recent decades, it has drawn a lot of attention as a frequent feed additive [20,22,24,32,33]. With the addition of LA, the growth performance of largemouth bass increased significantly in this study, but the final body weight in LA1000 decreased slightly compared to LA500. This result suggested that largemouth bass on a high dietary carbohydrate diet would have better growth performance if suitable amounts of LA were added, while excessive addition of LA might inhibit this beneficial effect. This result was consistent with the studies on African catfish [34], juvenile hybrid grouper [26] and northern snakehead [24], which is likely because excess dietary α-LA could increase energy expenditure by modulating hypothalamic AMPK activity and negatively affect growth performance [35].

It has been proved that high carbohydrates could cause excessive glycogen accumulation in the liver of carnivorous fish [7,23]. In recent research, a reduced liver glycogen level indicated the improvement of LA inclusion on the utilization of glucose by largemouth bass. Consistently, in common carp (*Cyprinus carpio*), LA has been shown to improve glycemic control and carbohydrate metabolism [22]. It has been established that the insulin pathway is essential for preserving glucose homeostasis in animals [36]. The insulin receptor substrate protein (IRS) is phosphorylated as a key node, leading to the activation of downstream pathways of two other key nodes, phosphatidylinositol 3 kinase (PI3K) and serine/threonine kinase (AKT), which control the metabolic actions of insulin [37,38].

Recent studies in largemouth bass suggest that inadequate insulin secretion is responsible for their limited glucose utilization [5]. Therefore, the promotion of insulin secretion or activation of the insulin pathway could improve glucose utilization in carnivorous fish. The study on insulin delivery validates the function of the insulin pathway in controlling the metabolism of glucose in some teleost, including hybrid grouper [39] and largemouth bass [5]. Meanwhile, it has been proposed that the enhanced metabolism of glucose in carnivorous fish on a high carbohydrate diet is caused by the activation of the insulin pathway [3,40], whereas, on the one hand, LA has pro-oxidant properties that stimulate of the insulin-signaling cascade, which in turn increases glucose uptake by muscle and adipocytes, as reflected in muscle glycogen content [21]. Consistent with this, the gene expression results in this study suggested an activation of the insulin-signaling pathway in largemouth bass, which was induced by dietary LA inclusion. Therefore, the activation of the insulin pathway might underline the mechanism for how LA could improve glucose metabolism.

Glycolysis and gluconeogenesis are opposing metabolic pathways in the degradation and synthesis of carbohydrates and play important roles in maintaining glucose homeostasis. Although glycerol kinase (GK), phosphofructokinase (PFK) and pyruvate kinase (PK) catalyze unidirectional reactions and play key roles in glycolysis, most of the enzymes in these two pathways are shared. Phosphoenolpyruvate carboxykinase, fructose-bisphosphatase (FBP), glucose-6-phosphatase catalytic region (G6PC) and phosphoenolpyruvate carboxykinase (PEPCK) are key steps in the control of gluconeogenesis [41,42]. The insulin administration test has shown that teleost's glycolysis and gluconeogenesis changes when the insulin pathway is activated [5,8]. Recent studies on freshwater prawns confirm that LA promotes glycolysis in the tricarboxylic acid cycle at the enzyme level and positively regulates carbohydrate metabolism [43]. Thus, in recent research, LA proximally promoted glycolysis through the activation of the insulin pathway. However, in recent research, the addition of LA unexpectedly increased the expression of two genes linked to gluconeogenesis, *pepck* and *g6pc* of largemouth bass. In line with this, the expression of hepatic genes, *pepck* and *g6pc* of zebrafish fed with LA were up-regulated [44]. The presumed reason for this is that LA can influence gluconeogenesis but not through the insulin-signaling pathway.

Excessive carbohydrate intake deteriorates immunity in carnivorous fish such as golden pompano (*Trachinotus ovatus*) [45], hybrid snakehead (♀*Channa maculate* × ♂*Channa argus*) [46] and largemouth bass [47]. LA has a potential role in enhancing immunocompetence in fish in addition to regulating the insulin pathway, which has been proven in several fish species, such as northern snakehead [24] and Nile tilapia (*Oreochromis niloticus*) [48]. In recent research, the addition of LA resulted in a significant increase in lysozyme activity, which confirmed the potential role of LA in enhancing the immunocompetence of the organism. Recent research has proven the effect of the p38-MAPK/I-κB/NF-κB pathway on immune regulation [16–18]. Previous research in mammals has shown that LA could inhibit NF-kB by regulating the upstream kinase, MAPK, and by avoiding the degradation of I-κB [49]. Pro-inflammatory factors, such as IL-1β and TNF-α are the primary middlemen of inflammation in fish, and IL-10, as a major anti-inflammatory gene, inhibits the release and synthesis of inflammatory cytokines, thereby modulating the inflammatory response [50,51]. Meanwhile, the modulatory effect of LA upon immunity has also been demonstrated in some fish species, such as grass carp and snakehead. After the addition of LA to their diets, pro-inflammatory factors were down-regulated, and anti-inflammatory factors were up-regulated, which is consistent with our findings [24,52]. In recent research, the addition of LA also activated the expression of *iκb*, and the expression of *mapk* and *nfκb* was inhibited. Meanwhile, LA suppressed the expression of pro-inflammatory factors *tnf-α* and *il-1β* and promoted the expression of anti-inflammatory factor *il-10*. The above results suggest that LA had a modulating effect on immunity.

Hepatic glycogen deposition, growth retardation and reduced immunity are all adverse effects that tend to occur when fed high amounts of carbohydrates, and this study experimentally illustrates the mitigating effects of LA on these three adverse effects, revealing its potential as a high-carbohydrate feed additive. Furthermore, this experiment only discussed the effects of LA on growth, glucose metabolism and the immunity of fish consuming high levels of α-starch, which still has some limitations. The role of LA in the aquaculture process can be further determined by conducting subsequent experiments on the feeding of different carbohydrate sources.

5. Conclusions

Above all, dietary LA improves growth performance of high carbohydrate-fed largemouth bass. In addition, LA could reduce hepatic glycogen accumulation through the activation of the insulin pathway and attenuate the decreased immunity of largemouth bass caused by high carbohydrates. This provides us with ideas to mitigate the adverse effects of high carbohydrates on carnivorous fish in practical production and lays the foundation for the application of LA in aquatic biology.

Supplementary Materials: The following are available online at https://www.mdpi.com/article/10.3390/fishes10010009/s1, Table S1: Primers used in the present study. Reference [53] is cited in the supplementary material.

Author Contributions: Z.F., Investigation, Data curation, Formal analysis, Writing—Original Draft; X.P., Investigation, Data curation. Y.G., Data curation, Writing—Original Draft; N.Z., Investigation, Methodology; S.C., Investigation, Methodology; N.L., Supervision, Writing—Review & Editing; N.C., Conceptualization, Supervision, Writing—Review and Editing; S.L., Conceptualization, Supervision, Writing—Review & Editing, Funding acquisition. All authors discussed the results and commented on the manuscript. All authors have read and agreed to the published version of the manuscript.

Funding: This work was financially supported by Shanghai Rising-Star Program (Grant 24QA2703500), National Natural Science Foundation of China (31802308), and SciTech Funding by CSPFTZ Lingang Special Area Marine Biomedical Innovation Platform (No. RWS-2024-003).

Institutional Review Board Statement: This animal study was reviewed and approved by the Animal Care and Use Committee of the Shanghai Ocean University. (Protocol number: SHOU-DW-2024–037, Approval date: 6 February 2024).

Informed Consent Statement: Not applicable.

Data Availability Statement: The data of this research is available upon request.

Acknowledgments: The authors thank their respective universities and institutes for their support.

Conflicts of Interest: The authors declare no conflicts of interest.

References

1. NRC. *Carbohydrates and Fibre, in Nutrient Requirements of Fish and Shrimp*; The National Academies Press: Washington, DC, USA, 2011; pp. 135–162.
2. Ma, H.J.; Mou, M.M.; Pu, D.C.; Lin, S.M.; Chen, Y.J.; Luo, L. Effect of dietary starch level on growth, metabolism enzyme and oxidative status of juvenile largemouth bass *Micropterus salmoides*. *Aquaculture* **2019**, *498*, 482–487. [CrossRef]
3. Liu, Y.; Liu, N.; Wang, A.; Chen, N.; Li, S. Resveratrol inclusion alleviated highdietary-carbohydrate-induced glycogen deposition and immune response of largemouth bass, *Micropterus salmoides*. *Br. J. Nutr.* **2022**, *127*, 165–176. [CrossRef] [PubMed]
4. MacDonald, I.A. A review of recent evidence relating to sugars, insulin resistance and diabetes. *Eur. J. Nutr.* **2016**, *55* (Suppl. S2), 17–23. [CrossRef]

5. Chen, P.; Wu, X.; Gu, X.; Han, J.; Xue, M.; Liang, X. FoxO1 in *Micropterus salmoides*: Molecular characterization and its roles in glucose metabolism by glucose or insulin-glucose loading. *Gen. Comp. Endocr.* **2021**, *310*, 113811. [CrossRef]

6. Wang, X.; Gong, Y.; Li, W.; Liu, N.; Fang, Z.; Zhang, N.; Chen, N.; Li, S. The beneficial effects of metformin inclusion on growth performance, glucose utilization, antioxidant capacity and apoptosis of largemouth bass (*Micropterus salmoides*) fed with high dietary carbohydrates. *Aquaculture* **2024**, *588*, 740957. [CrossRef]

7. Li, S.; Sang, C.; Wang, A.; Zhang, J.; Chen, N. The impacts of dietary carbohydrate levels on growth performance, feed utilization, glycogen accumulation and hepatic glucose metabolism in hybrid grouper (*Epinephelus fuscoguttatus* ♀× *E. lanceolatus* ♂). *Aquaculture* **2019**, *512*, 734351. [CrossRef]

8. Capilla, E.; M'edale, F.; Navarro, I.; Panserat, S.; Vachot, C.; Kaushik, S.; Guti'errez, J. Muscle insulin binding and plasma levels in relation to liver glucokinase activity, glucose metabolism and dietary carbohydrates in rainbow trout. *Regul. Pept.* **2003**, *110*, 123–132. [CrossRef]

9. Xu, C.; Liu, W.B.; Zhang, D.D.; Cao, X.F.; Shi, H.J.; Li, X.F. Interactions between dietary carbohydrate and metformin: Implications on energy sensing, insulin signaling pathway, glycolipid metabolism and glucose tolerance in blunt snout bream *Megalobrama amblycephala*. *Aquaculture* **2018**, *483*, 183–195. [CrossRef]

10. Zhao, L.; Liang, J.; Liu, H.; Gong, C.; Huang, X.; Hu, Y.; Liu, Q.; He, Z.; Zhang, X.; Yang, S.; et al. Yinchenhao Decoction ameliorates the high-carbohydrate diet induced suppression of immune response in largemouth bass (*Micropterus salmoides*). *Fish. Shellfish. Immunol.* **2022**, *125*, 141–151. [CrossRef]

11. Tapia-Paniagua, S.T.; Vidal, S.; Lobo, C.; Prieto-Álamo, M.J.; Jurado, J.; Cordero, H.; Cerezuela, R.; Garcia de la Banda, I.; Esteban, M.A.; Balebona, M.C.; et al. The treatment with the probiotic *Shewanella putrefaciens* Pdp11 of specimens of Solea senegalensis exposed to high stocking densities to enhance their resistance to disease. *Fish Shellfish. Immunol.* **2014**, *41*, 209–221. [CrossRef]

12. Eslamloo, K.; Akhavan, S.R.; Fallah, F.J.; Henry, M.A. Variations of physiological and innate immunological responses in goldfish (*Carassius auratus*) subjected to recurrent acute stress. *Fish Shellfish Immunol.* **2014**, *37*, 147–153. [CrossRef] [PubMed]

13. Van Doan, H.; Tapingkae, W.; Moonmanee, T.; Seepai, A. Effects of low molecular weight sodium alginate on growth performance, immunity, and disease resistance of tilapia, *Oreochromis niloticus*. *Fish Shellfish Immunol.* **2016**, *55*, 186–194. [CrossRef] [PubMed]

14. Pérez-Sánchez, T.; Mora-Sánchez, B.; Balcázar, J.L. Biological approaches for disease control in aquaculture: Advantages, limitations and challenges. *Trends Microbiol.* **2018**, *26*, 896–903. [CrossRef] [PubMed]

15. Pan, F.Y.; Feng, L.; Jiang, W.D.; Jiang, J.; Wu, P.; Kuang, S.Y.; Tang, L.; Tang, W.N.; Zhang, Y.A.; Zhou, X.Q.; et al. Methionine hydroxy analogue enhanced fish immunity via modulation of NF-κB, TOR, MLCK, MAPKs and Nrf2 signaling in young grass carp (*Ctenopharyngodon idella*). *Fish Shellfish Immunol.* **2016**, *56*, 208–228. [CrossRef] [PubMed]

16. Tibullo, D.; Li Volti, G.; Giallongo, C.; Grasso, S.; Tomassoni, D.; Anfuso, C.D.; Lupo, G.; Amenta, F.; Avola, R.; Bramanti, V. Biochemical and clinical relevance of alpha lipoic acid: Antioxidant and anti-inflammatory activity, molecular pathways and therapeutic potential. *Inflamm. Res.* **2017**, *66*, 947–959. [CrossRef]

17. Dong, B.; Wu, L.; Chen, Q.; Xu, W.; Li, D.; Han, D.; Zhu, X.; Liu, H.; Yang, Y.; Xie, S.; et al. Tolerance assessment of atractylodes macrocephla polysaccharide in the diet of largemouth bass (*Micropterus salmoides*). *Antioxidants* **2022**, *11*, 1581. [CrossRef]

18. Tao, J.; Wang, S.; Qiu, H.; Xie, R.; Zhang, H.; Chen, N.; Li, S. Modulation of growth performance, antioxidant capacity, non-specific immunity and disease resistance in largemouth bass (*Micropterus salmoides*) upon compound probiotic cultures inclusion. *Fish. Shellfish. Immunol.* **2022**, *127*, 804–812. [CrossRef]

19. Islam, M.T. Antioxidant activities of dithiol alpha-lipoic acid. *Bangladesh J. Med. Sci.* **2009**, *8*, 46. [CrossRef]

20. Capece, U.; Moffa, S.; Improta, I.; Di Giuseppe, G.; Nista, E.C.; Cefalo, C.M.A.; Cinti, F.; Pontecorvi, A.; Gasbarrini, A.; Giaccari, A.; et al. Alpha-lipoic acid and glucose metabolism: A comprehensive update on biochemical and therapeutic features. *Nutrients* **2023**, *15*, 18. [CrossRef]

21. Konrad, D. Utilization of the insulin-signaling network in the metabolic actions of α-lipoic acid—Reduction or oxidation? *Antioxid. Redox Signal.* **2005**, *7*, 1032–1039. [CrossRef]

22. Santos, R.A.; Caldas, S.; Primel, E.G.; Tesser, M.B.; Monserrat, J.M. Effects of lipoic acid on growth and biochemical responses of common carp fed with carbohydrate diets. *Fish Physiol. Biochem.* **2016**, *42*, 1699–1707. [CrossRef] [PubMed]

23. Li, X.; Han, T.; Zheng, S.; Wu, G. Hepatic glucose metabolism and its disorders in fish. *Recent Adv. Anim. Nutr. Metab.* **2022**, 207–236. [CrossRef]

24. Li, M.; Kong, Y.; Lai, Y.; Wu, X.; Zhang, J.; Niu, X.; Wang, G. The effects of dietary supplementation of α-lipoic acid on the growth performance, antioxidant capacity, immune response, and disease resistance of northern snakehead, *Channa argus*. *Fish Shellfish Immunol.* **2022**, *126*, 57–72. [CrossRef] [PubMed]

25. Ou, G.; Xie, R.; Huang, J.; Huang, J.; Wen, Z.; Li, Y.; Jiang, X.; Ma, Q.; Chen, G. Effects of dietary alpha-lipoic acid on growth performance, serum biochemical indexes, liver antioxidant capacity and transcriptome of juvenile hybrid grouper (*Epinephelus fuscoguttatus*♀× *Epinephelus polyphekadion*♂). *Animals* **2023**, *13*, 887. [CrossRef]

26. Kütter, M.T.; Monserrat, J.M.; Primel, E.G.; Caldas, S.S.; Tesser, M.B. Effects of dietary α-lipoic acid on growth, body composition and antioxidant status in the Plata pompano *Trachinotus marginatus* (Pisces, Carangidae). *Aquaculture* **2012**, *368*, 29–35. [CrossRef]

27. Samuki, K.; Setiawati, M.; Jusadi, D.; Agus Suprayudi, M. The evaluation of α-Lipoic acid supplementation in diet on the growth performance of giant gourami (*Osphronemus goramy*) juvenile. *Aquacult. Res.* **2021**, *52*, 1538–1547. [CrossRef]

28. AOAC. *Official Methods of Analysis*, 17th ed.; Association of Official Agricultural Chemists: Washington, DC, USA, 2003.

29. Seifter, S.; Dayton, S.; Novic, B. The estimation of glycogen with the anthrone reagent. *Arch. Biochem. Biophys.* **1950**, *25*, 191–200.

30. Cui, Z.; He, F.; Li, X.; Jing, M.; Huo, C.; Zong, W.; Liu, R. Molecular insights into the binding model and response mechanisms of triclosan with lysozyme. *J. Mol. Liq.* **2022**, *357*, 119080. [CrossRef]

31. Livak, K.J.; Schmittgen, T.D. Analysis of relative gene expression data using realtime quantitative PCR and the $2^{-\Delta\Delta CT}$ method. *Methods* **2001**, *25*, 402–408. [CrossRef]

32. Bast, A.; Haenen, G.R. Lipoic acid: A multifunctional antioxidant. *Biofactors* **2003**, *17*, 207–213. [CrossRef]

33. Siagian, D.R.; Jusadi, D.; Ekasari, J.; Setiawati, M. Dietary α-lipoic acid supplementation to improve growth, blood chemistry, and liver antioxidant status of African catfish *Clarias gariepinus*. *Aquacult. Int.* **2021**, *29*, 1935–1947. [CrossRef]

34. Cremer, D.R.; Rabeler, R.; Roberts, A.; Lynch, B. Long-term safety of α-lipoic acid (LA) consumption: A 2-year study. *Regul. Toxicol. Pharmacol.* **2006**, *46*, 193–201. [CrossRef] [PubMed]

35. Kim, M.S.; Park, J.Y.; Namkoong, C.; Jang, P.G.; Ryu, J.W.; Song, H.S.; Yun, J.Y.; Namgoong, I.S.; Ha, J.; Park, I.S.; et al. Anti-obesity effects of α-lipoic acid mediated by suppression of hypothalamic AMP-activated protein kinase. *Nat. Med.* **2004**, *10*, 727–733. [CrossRef] [PubMed]

36. Saltiel, A.R.; Kahn, C.R. Insulin signalling and the regulation of glucose and lipid metabolism. *Nature* **2001**, *414*, 799–806. [CrossRef]

37. Avruch, J. Insulin signal transduction through protein kinase cascades. *Mol. Cell Biochem.* **1998**, *182*, 31–48. [CrossRef]

38. Taniguchi, C.M.; Emanuelli, B.; Kahn, C.R. Critical nodes in signalling pathways: Insights into insulin action. *Nat. Rev. Mol. Cell Biol.* **2006**, *7*, 85–96. [CrossRef]

39. Li, S.; Li, Z.; Sang, C.; Zhang, J.; Chen, N. The variation of serum glucose, hepatic glycogen content and expression of glucose metabolism-related genes in hybrid grouper (female *Epinephelus fuscoguttatus* × male *Epinephelus lanceolatus*) in response to intraperitoneal insulin infusion. *Fish. Sci.* **2018**, *84*, 641–647. [CrossRef]

40. Li, S.; Sang, C.; Turchini, G.M.; Wang, A.; Zhang, J.; Chen, N. Starch in aquafeeds: The benefits of a high amylose to amylopectin ratio and resistant starch content in diets for the carnivorous fish, largemouth bass (*Micropterus salmoides*). *Br. J. Nutr.* **2020**, *124*, 1145–1155. [CrossRef]

41. Nordlie, R.C.; Foster, J.D.; Lange, A.J. Regulation of glucose production by the liver. *Annu. Rev. Nutr.* **1999**, *19*, 379–406. [CrossRef]

42. Michael, M.D.; Kulkarni, R.N.; Postic, C.; Previs, S.F.; Shulman, G.I.; Magnuson, M.A.; Kahn, C.R. Loss of insulin signaling in hepatocytes leads to severe insulin resistance and progressive hepatic dysfunction. *Mol. Cell* **2000**, *6*, 87–97. [CrossRef]

43. Ding, Z.; Xiong, Y.; Zheng, J.; Zhou, D.; Kong, Y.; Qi, C.; Liu, Y.; Ye, J.; Limbu, S.M. Modulation of growth, antioxidant status, hepatopancreas morphology, and carbohydrate metabolism mediated by alpha-lipoic acid in juvenile freshwater prawns Macrobrachium nipponense under two dietary carbohydrate levels. *Aquaculture* **2022**, *546*, 737314. [CrossRef]

44. Huang, C.; Sun, J.; Ji, H.; Kaneko, G.; Xie, X.; Chang, Z.; Deng, W. Systemic effect of dietary lipid levels and α-lipoic acid supplementation on nutritional metabolism in zebrafish (*Danio rerio*): Focusing on the transcriptional level. *Fish Physiol. Biochem.* **2020**, *46*, 1631–1644. [CrossRef]

45. Zhou, C.; Huang, Z.; Lin, H.; Ma, Z.; Wang, J.; Wang, Y.; Yu, W. Rhizoma curcumae Longae ameliorates high dietary carbohydrate-induced hepatic oxidative stress, inflammation in golden pompano Trachinotus ovatus. *Fish Shellfish Immunol.* **2022**, *130*, 31–42. [CrossRef]

46. Zhao, H.; Li, P.; Peng, K.; Chen, B.; Hu, J.; Huang, W.; Cao, J.; Sun, Y. The positive effects of dietary arginine on juvenile hybrid snakehead (*Channa maculata*♀ × *Channa argus*♂) fed high-carbohydrate diets: Liver inflammation antioxidant response, and glucose metabolism. *Aquacult. Rep.* **2023**, *30*, 101602. [CrossRef]

47. Zhao, L.; Liang, J.; Chen, F.; Tang, X.; Liao, L.; Liu, Q.; Luo, J.; Du, Z.; Li, Z.; Luo, W.; et al. High carbohydrate diet induced endoplasmic reticulum stress and oxidative stress, promoted inflammation and apoptosis, impaired intestinal barrier of juvenile largemouth bass (*Micropterus salmoides*). *Fish Shellfish Immunol.* **2021**, *119*, 308–317. [CrossRef]

48. Lu, D.; Limbu, S.M.; Lv, H.; Ma, Q.; Chen, L.; Zhang, M.; Du, Z. The comparisons in protective mechanisms and efficiencies among dietary α-lipoic acid, β-glucan and l-carnitine on Nile tilapia infected by *Aeromonas hydrophila*. *Fish Shellfish Immunol.* **2019**, *86*, 785–793. [CrossRef]

49. Packer, L.; Kraemer, K.; Rimbach, G. Molecular aspects of lipoic acid in the prevention of diabetes complications. *Nutrition* **2001**, *17*, 888–895. [CrossRef]

50. Zhang, C.N.; Zhang, J.L.; Ren, H.T.; Zhou, B.H.; Wu, Q.J.; Sun, P. Effect of tributyltin on antioxidant ability and immune responses of zebrafish (*Danio rerio*). *Ecotoxicol. Environ. Saf.* **2017**, *138*, 1–8. [CrossRef]

51. Opal, S.M.; De-Palo, V.A. Anti-inflammatory cytokines. *Chest* **2000**, *117*, 1162–1172. [CrossRef]

52. Liu, H.X.; Zhou, X.Q.; Jiang, W.D.; Wu, P.; Liu, Y.; Zeng, Y.Y.; Jiang, J.; Kuang, S.Y.; Tang, L.; Feng, L. Optimal α-lipoic acid strengthens immunity of young grass carp (*Ctenopharyngodon idella*) by enhancing immune function of head kidney, spleen and skin. *Fish Shellfish Immunol.* **2018**, *80*, 600–617. [CrossRef]

53. Yu, L.L.; Yu, H.H.; Liang, X.F.; Li, N.; Wang, X.; Li, F.H.; Wu, X.F.; Zheng, Y.H.; Xue, M.; Liang, X.F. Dietary butylated hydroxytoluene improves lipid metabolism, antioxidant and anti-apoptotic response of largemouth bass (*Micropterus salmoides*). *Fish Shellfish Immunol.* **2018**, *72*, 220–229. [CrossRef]

 *fishes*

Article

Effects of Soybean Isoflavones on the Growth Performance and Lipid Metabolism of the Juvenile Chinese Mitten Crab *Eriocheir sinensis*

Mengyu Shi [1], Yisong He [1], Jiajun Zheng [1], Yang Xu [1], Yue Tan [1], Li Jia [1], Liqiao Chen [2], Jinyun Ye [1,*] and Changle Qi [1,*]

[1] National-Local Joint Engineering Laboratory of Aquatic Animal Genetic Breeding and Nutrition, Zhejiang Provincial Key Laboratory of Aquatic Resources Conservation and Development, College of Life Science, Huzhou University, Huzhou 313000, China; znytsmy@126.com (M.S.); 19816907180@163.com (Y.H.); 18987437460@163.com (J.Z.); x15237127582@163.com (Y.X.); tan18853548898@163.com (Y.T.); jiali890411@163.com (L.J.)

[2] Laboratory of Aquaculture Nutrition and Environmental Health, School of Life Sciences, East China Normal University, Shanghai 200062, China; lqchen@bio.ecnu.edu.cn

* Correspondence: yjy@zjhu.edu.cn (J.Y.); 02862@zjhu.edu.cn (C.Q.); Tel.: +86-18868215672 (C.Q.)

Abstract: In order to study the effects of soybean isoflavones on the growth performance and lipid metabolism of juvenile Chinese mitten crabs, six experimental diets were formulated by gradient supplementation with 0%, 0.004% and 0.008% soybean isoflavones at different dietary lipid levels (10% and 15%). The groups were named as follows: NF-0 group (10% fat and 0% SIFs), NF-0.004 group (10% fat and 0.004% SIFs), NF-0.008 group (10% fat and 0.008% SIFs), HF-0 group (15% fat and 0% SIFs), HF-0.004 group (15% fat and 0.004% SIFs) and HF-0.008 group (15% fat and 0.008% SIFs). All crabs with an initial weight of 0.4 ± 0.03 g were fed for 8 weeks. The results showed that dietary supplementation with 0.004% or 0.008% SIFs significantly increased the weight gain and specific growth rate of crabs. Diets supplemented with 0.004% or 0.008% SIFs significantly reduced the content of non-esterified free fatty acids and triglycerides in the hepatopancreas of crabs at the 10% dietary lipid level. Dietary SIFs significantly decreased the relative mRNA expressions of elongase of very-long-chain fatty acids 6 (*elovl6*), triglyceride lipase (*tgl*), sterol regulatory element-binding protein 1 (*srebp-1*), carnitine palmitoyltransferase-1a (*cpt-1a*), fatty acid transporter protein 4 (*fatp4*), carnitine palmitoyltransferase-2 (*cpt-2*), Δ9 fatty acyl desaturase (*Δ9 fad*), carnitine palmitoyltransferase-1b (*cpt-1b*), fatty acid-binding protein 10 (*fabp10*) and microsomal triglyceride transfer protein (*mttp*) in the hepatopancreas of crabs. At the 15% dietary lipid level, 0.008% SIFs significantly increased the relative mRNA expressions of fatty acid-binding protein 3 (*fabp3*), carnitine acetyltransferase (*caat*), *fatp4*, *fabp10*, *tgl*, *cpt-1a*, *cpt-1b* and *cpt-2* and significantly down-regulated the relative mRNA expressions of Δ9 *fad* and *srebp-1*. In conclusion, SIFs can improve the growth and utilization of a high-fat diet by inhibiting genes related to lipid synthesis and promoting lipid decomposition in juvenile Chinese mitten crabs.

Keywords: soybean isoflavones; Chinese mitten crab; growth performance; lipid metabolism

Key Contribution: Soybean isoflavones improved the growth and utilization of a high-fat diet by inhibiting genes related to lipid synthesis and promoting lipid decomposition in juvenile Chinese mitten crabs.

Citation: Shi, M.; He, Y.; Zheng, J.; Xu, Y.; Tan, Y.; Jia, L.; Chen, L.; Ye, J.; Qi, C. Effects of Soybean Isoflavones on the Growth Performance and Lipid Metabolism of the Juvenile Chinese Mitten Crab *Eriocheir sinensis*. *Fishes* **2024**, *9*, 335. https://doi.org/10.3390/fishes9090335

Academic Editor: Laura Susana López-Greco

Received: 12 May 2024
Revised: 16 August 2024
Accepted: 21 August 2024
Published: 26 August 2024

1. Introduction

Soybean isoflavones (SIFs) are natural phytoestrogens that mainly exists in soybean and other leguminous plants. It has been widely reported that SIFs can enhance non-specific immunity and alleviate human diseases [1,2]. Moreover, they can improve lipid metabolism and associate with the gonadal development of animal. Therefore, SIFs could be a potential feed additive for improving animal growth and health.

Soy isoflavones play an important role in improving immunity and lipid metabolism. It has been reported that diets supplemented with SIFs improved the growth performance of pigs [3] and chickens [4] fed normal-fat diets. The supplementation of high-fat diets with SIFs promoted insulin resistance and growth in rats [5]. However, certain studies have reported contradictory results [6,7].

In mammals, a large number of studies have shown that soy isoflavones are an effective additive for regulating lipid metabolism. Dietary supplementation with SIFs can alleviate lipid accumulation by reducing triglyceride (TG) and low-density lipoprotein cholesterol (LDL-C) levels in the liver of obese rats [8,9]. Further studies suggested that SIFs improved lipocatabolic metabolism by up-regulating the relative expression of peroxisome proliferate-activated receptor α (PPARα) [10] and sterol regulatory element-binding protein 1 (srebp-1) [11]. Moreover, some studies reported that SIFs can stimulate liver X receptor α or liver X receptor β phosphorylation and trigger lipid catabolism [12]. In addition to promoting lipolysis, SIFs can also alleviate lipid accumulation through reducing fat synthesis. Some previous studies reported that dietary SIFs inhibited the fat synthesis of obese Zucker rats by down-regulating the expression of peroxisom-proliferator activated receptor $\gamma2$ (PPAR$\gamma2$) and adipose-specific protein 27 (FSP27) [13]. In aquatic animals, diets supplemented with 100 mg/kg and 500 mg/kg genistein significantly down-regulated the genes associated with lipid synthesis of *Cyprinus carpio* [14]. Some similar results were widely reported in *Oncorhynchus mykiss* and *Paralichthys olivaceus* [15,16]. On the contrary, some studies reported that SIFs can increase lipid synthesis. For example, dietary SIFs significantly increased TG and TC content in the serum of *Allogynogenetic crucian carp* and *Paralichthys olivaceus* [17,18]. Unfortunately, it is difficult to explain functional variations between species. As such, more studies are needed to understand the function of isoflavones in aquatic animals.

The Chinese mitten crab (*Eriocheir sinensis*) is an economic crustacean that has been widely farmed. Different from mammals, shrimp and crabs do not have specific adipose tissue, and the lipid metabolism pattern is also different from that of mammals. Whether soy isoflavones can regulate the fat metabolism of crabs is still poorly understood. Therefore, we speculated that soybean isoflavones can improve the utilization of dietary lipid, so soybean isoflavones were used to supplement high-fat diets and normal-fat diets, respectively, to examine the effects of soy isoflavones on the growth performance and lipid metabolism of the Chinese mitten crab.

2. Material and Methods

2.1. Experimental Diets

Six experimental diets were formulated by gradient supplementation with 0%, 0.004% and 0.008% soybean isoflavones at different dietary lipid levels (10% and 15%). The groups were named NF-0, NF-0.004, NF-0.008, HF-0, HF-0.004 and HF-0.008. The formulation and proximate composition of the experimental diets are shown in Table 1.

The ingredients were finely ground and sieved through a 0.25 mm mesh strainer (Pulverizer, 2500Y, Anhui Hualing Xichu Equipment Co., Ltd., Maanshan, China; Strainer, Huafeng Hardware Instrument Co., Ltd., Shaoxing, China). The ingredients were weighed according to the formulation (Table 1) and mixed using an electric mixer (B10K, Foshan Shunhengji Kitchenware Co., Ltd., Foshan, China). The oil and distilled water were subsequently added to make a dough. Finally, the dough was pelleted using a screw-press pelletizer (F-26, South China University of Technology, Guangzhou, China). The pallets were air dried to reach a moisture content of less than 10%. After drying, diet foods were stored at $-20\,^{\circ}$C.

Table 1. Formulation and proximate composition of the experimental diets (dry matter, %).

	Normal Fat			High Fat		
Ingredients	**0% SIFs** **NF-0**	**0.004% SIFs** **NF-0.004**	**0.008% SIFs** **NF-0.008**	**0% SIFs** **HF-0**	**0.004% SIFs** **HF-0.004**	**0.008% SIFs** **HF-0.008**
Fish meal	20	20	20	20	20	20
Casein	21	21	21	21	21	21
Gelatin	7	7	7	7	7	7
Corn starch	23	23	23	23	23	23
Soybean lecithin	2	2	2	2	2	2
Cholesterol	0.5	0.5	0.5	0.5	0.5	0.5
Choline chloride [a]	0.5	0.5	0.5	0.5	0.5	0.5
Fish oil	3	3	3	7	7	7
Soybean oil	3	3	3	7	7	7
Arginine	1.8	1.8	1.8	1.8	1.8	1.8
Methionine	0.5	0.5	0.5	0.5	0.5	0.5
Lysine	0.5	0.5	0.5	0.5	0.5	0.5
Vitamin premix [a]	1.5	1.5	1.5	1.5	1.5	1.5
Mineral premix [b]	1.5	1.5	1.5	1.5	1.5	1.5
Sodium carboxymethyl cellulose	2	2	2	2	2	2
Attractant	3	3	3	3	3	3
Butylated hydroxytoluene	0.1	0.1	0.1	0.1	0.1	0.1
SIFs	0	0.004	0.008	0	0.004	0.008
Microcrystalline Cellulose	9.1	9.096	9.092	1.1	1.096	1.092
Proximate analysis (%)						
Crude protein	45.18	45.52	44.63	43.68	43.97	45.71
Crude lipid	9.76	9.28	9.57	15.80	16.15	16.55
Moisture	9.13	10.32	10.80	9.63	9.97	10.25
Ash	7.56	7.59	7.59	7.55	7.48	7.49

[a] Vitamin premix (per 100 g premix): Ca pantothenate, 0.3 g; para-aminobenzoic acid, 0.1 g; cholecalciferol, 0.0075 g; riboflavin, 0.0625 g; menadione, 0.05 g; ascorbic acid, 0.5 g; biotin, 0.005 g; retinol acetate, 0.043 g; folic acid, 0.025 g; pyridoxine hydrochloride, 0.225 g; thiamin hydrochloride, 0.15 g; niacin, 0.3 g; α-tocopherol acetate, 0.5 g; The remaining part will be used α-cellulose to 100 g. [b] Mineral premix (per 100 g premix): KI, 0.023 g; $CuCl_2 \cdot 2H_2O$, 0.015 g; $Ca(H_2PO_4)_2$, 26.5 g; $MnSO_4 \cdot 6H_2O$, 0.143 g; $AlCl_3 \cdot 6H_2O$, 0.024 g; KH_2PO_4, 21.5 g; NaH_2PO_4, 10.0 g; $CoCl_2 \cdot 6H_2O$, 0.14 g; KCl, 2.8 g; $ZnSO_4 \cdot 7H_2O$, 0.476 g; Calcium lactate, 16.50 g; $CaCO_3$, 10.5 g; $MgSO_4 \cdot 7H_2O$, 10.0 g; Fe-citrate, 1 g; The remaining part will be used α-cellulose to 100 g.

2.2. Feeding Trial, Sampling and Growth Evaluation

Adult crabs were obtained from a farm in Huzhou, China (East longitude 119 degrees 14′ to 120 degrees 29′, north latitude 30 degrees 22′ to 31 degrees 11′). Crabs were acclimatized to the experimental conditions in 300 L tanks (100 × 80 × 60 cm) before the feeding trial. A total of 1050 crabs (0.4 ± 0.03 g, mean ± SEM) were weighed and put into 30 tanks (100 × 80 × 60 cm), with each tank containing 35 crabs. Four parallel tanks were randomly allotted to one of the experimental diets. Three plastic nets were placed in each tank as shelters to reduce attacking behavior. Diets with a daily ration of 4% body weight were hand-fed to crabs twice daily (6:00 and 18:00). Feces was removed in the morning (09:00), and water 30% of the tank volume was exchanged daily. Dead crabs were immediately removed from the tank, weighed and recorded. The feed intake of each tank was recorded throughout the trial period. During the experimental period, the experimental water temperature in the tanks varied from 18 °C to 24 °C, the dissolved oxygen concentration was >7 mg/L, and the ammonia nitrogen value was <0.05 mg/L.

At the end, crabs were euthanized, after which the hepatopancreases were frozen in liquid nitrogen and kept in an ultra-low temperature freezer (-80 °C) for enzyme activity, gene expression and nutrient composition analyses.

Weight gain, specific growth rate, survival and the hepatopancreas index were calculated using the formulas below:

Weight gain (WG, %) = (final crab weight − initial crab weight)/initial crab weight × 100.

$$\text{Specific growth rate (SGR, \%)} = (\text{LN final weight} - \text{LN initial weight})/\text{days} \times 100.$$

$$\text{Survival (\%)} = \text{final number/initial number} \times 100.$$

$$\text{Hepatopancreas index (\%)} = \text{hepatopancreas weight of crab/whole crab weight} \times 100.$$

2.3. Chemical Composition Analysis

The chemical compositions of the experimental diets and crabs were measured according to the standard procedures for proximate composition analysis [19]. Four duplicate samples were measured in each treatment ($n = 4$). The moisture content was measured after the samples were oven-dried at 105 °C. Crude protein was quantified using the Kjeltec™ 8200 (Foss, Hoganas, Sweden). Crude lipid was extracted using a 1000 mL Soxhlet extraction tube (Fujian minbo toughened glass Co., Ltd., Fuzhou, China). Ash was analyzed using a muffle furnace (SX2-8-10, Shanghai Yiheng Technology Co., Ltd., Shanghai, China) at 550 °C for 6 h. Four duplicate samples were analyzed for each treatment ($n = 4$).

2.4. Analysis of Biochemical Parameters in the Hepatopancreas

The biochemical parameters in the hepatopancreas were measured using commercial assay kits (Nanjing Jiancheng Bioengineering Institute, Nanjing, China) in accordance with the instructions of the manufacturer. The source and information of each kit used in this study were as follows: non-esterified fatty acid (NEFA; Cat. No. A042-1-1), lipase (LPS; Cat. No. A054-1-1), total cholesterol (TC; Cat. No. A111-2-1) and triglyceride (TG; Cat. No. A110-2-1). Four duplicate samples were analyzed for each treatment ($n = 4$).

2.5. Analysis of Gene Expression

Total RNA was extracted from the hepatopancreas using an RNAiso Plus (CAT # 9109, Takara, Kusatsu, Japan). The total RNA concentration and quality were estimated using the Nano Drop 2000 spectrophotometer (Thermo, Waltham, MA, USA). If the ratio of A260/A280 was between 1.8 and 2.0, the sample was used for reverse transcription using a PrimeScript™ RT master mix reagent kit (Perfect Real Time, Takara, Japan). The specific primers for the genes of *E. sinensis* were designed based on the transcriptome sequencing results and NCBI database using NCBI Primer BLAST (Supplementary Table S1). The RT-PCR amplification reactions were performed in a volume of 10 µL containing 5 µL 2×SYBR Premix Ex TaqTM, 0.25 µL of 10 mM forward primer, 0.25 µL of 10 mM reverse primer and 4.5 µL of diluted cDNA, using a CFX96 Real-Time PCR system (Bio-rad, Richmond, CA, USA). PCR conditions were as follows: 94 °C for 3 min and following 40 cycles at 94 °C for 15 s, 60 °C for 50 s, and 72 °C for 20 s. Samples were run in quintuplicate and normalized with the control gene β-actin and glyceraldehyde-phosphate dehydrogenase (GAPDH). The gene expression levels were calculated by the geometric averaging of multiple internal control genes [20]. Four duplicate samples were analyzed for each treatment ($n = 4$).

2.6. Statistical Analysis

Statistical analysis was performed using the SPSS 26.0 for Windows (SPSS, Michigan Avenue, Chicago, IL, USA). All data were subject to normality tests and homogeneity of variance by using Shapiro–Wilk and Levene's equal variance tests, respectively. Data were analyzed by two-way analysis of variance (ANOVA) to determine if there was any interaction between the dietary lipid level and SIFs level. Under the same lipid condition, one-way analysis of variance (ANOVA) was used to analyze the significant differences among crabs fed the diets with different SIFs levels after the normality and homogeneity of variance tests. When the means of each treatment were significantly different, Duncan's multiple range test was used to compare means among these treatments. At the same SIFs level, an independent-samples T test was used to determine significant differences between crabs cultivated at different lipid levels. Significance was set at $p < 0.05$. The data were represented as the mean $\pm$ standard error of mean (S.E.M.).

3. Results

3.1. Growth Performance

Both with normal-fat diets and high-fat diets, dietary SIFs significantly increased the weight gain (WG), specific growth rate (SGR) and hepatopancreas index (HSI) ($p < 0.05$, Figure 1a,c,d). At the normal-fat level, the WG, SGR and HSI of crabs fed the diets supplemented with 0.004% and 0.008% SIFs were significantly higher than crabs fed 0% SIFs ($p < 0.05$, Figure 1a,c,d). At the high-fat level, the WG, SGR and HSI of crabs fed the diets supplemented with 0.008% SIFs were significantly higher than crabs fed 0% SIFs ($p < 0.05$, Figure 1a,c,d). In the absence of SIFs, the WG and SGR of the crabs in the NF-0 group was significantly higher than that in the HF-0 group ($p < 0.01$, Figure 1c,d). The dietary lipid level and SIFs did not significantly affect the survival of crabs ($p > 0.05$, Figure 1b).

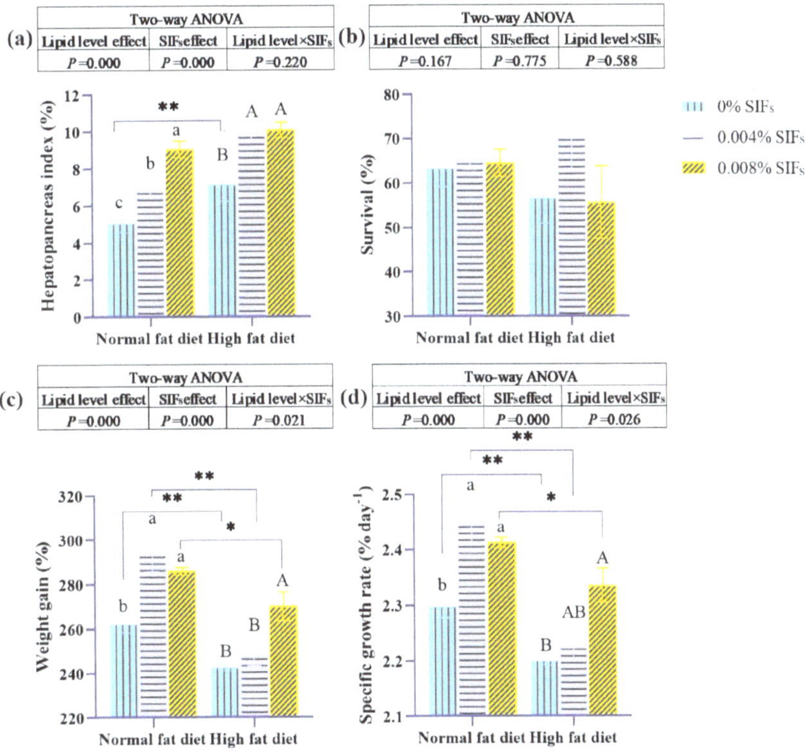

Figure 1. Effects of soybean isoflavones on the growth performance of juvenile Chinese mitten crab. Note: (**a**) hepatopancreas index, (**b**) survival, (**c**) weight gain, (**d**) specific growth rate. The different superscripts on the columns represent significant differences ($p < 0.05$, one-way ANOVA and Duncan multiple comparisons). The lines connecting the columns represent significant differences ($p < 0.05$ (marked *) or $p < 0.01$ (marked **), independence *t*-test). The table above the column represents the results of the two-factor analysis of variance.

3.2. Biochemical Parameters in the Hepatopancreas

At the normal-fat level, diets supplemented with 0.004% and 0.008% SIFs significantly decreased the NEFA and TG contents in the hepatopancreas ($p < 0.05$, Table 2). In the absence of SIFs, the NEFA and TG contents in the hepatopancreas of crabs in the NF-0 group was significantly higher than that in the HF-0 group ($p < 0.05$, Table 2). The dietary lipid level and SIFs did not significantly affect the LPS activities of crabs ($p > 0.05$, Table 2).

Table 2. Effects of soybean isoflavones on the biochemical parameters in the hepatopancreas of Chinese mitten crab.

Diets	Parameters			
	NEFA (μmol/gprot)	**LPS** (U/gprot)	**TC** (mmol/gprot)	**TG** (mmol/gprot)
NF-0	36.56 ± 16.49 [a]	1.49 ± 0.74	3.98 ± 0.42 *	34.13 ± 2.26 [a]
NF-0.004	20.75 ± 7.75 [b]	1.83 ± 0.38	4.01 ± 0.29	30.29 ± 2.13 [b] *
NF-0.008	17.47 ± 9.50 [b]	2.13 ± 0.35	4.01 ± 0.74	26.80 ± 3.53 [c] *
HF-0	26.42 ± 6.00 [A]	2.44 ± 0.65	3.15 ± 0.36 *	33.61 ± 4.66
HF-0.004	22.65 ± 3.48 [A]	1.70 ± 0.59	3.46 ± 0.55	38.03 ± 3.75 *
HF-0.008	21.16 ± 9.12 [B]	2.73 ± 0.88	3.70 ± 0.54	33.31 ± 2.61 *
Two-way ANOVA (*p* value)				
Lipid level	<0.01	NS	<0.01	<0.01
SIFs	<0.01	NS	NS	<0.01
Lipid level × SIFs	<0.01	NS	NS	<0.01

Note: NEFA, non-esterified fatty acid; LPS, lipase; TC, total cholesterol; TG, triacylglycerol; lipid level; SIFs, soybean isoflavones; NS, no significant differences ($p > 0.05$). The experimental results are expressed as mean ± SEM. Different superscripts within the first or last three rows of the same column indicate significant differences ($p < 0.05$). The * marked in the same column represents statistically significant differences ($p < 0.05$) between the corresponding NF and HF.

3.3. The mRNA Expressions of Genes Related to Lipid Synthesis in the Hepatopancreas

In the normal-fat level, diets supplemented with 0.004% SIFs significantly down-regulated the mRNA expression of *fabp 3* in the hepatopancreas ($p < 0.05$, Figure 2a). However, at the high-fat level, diets supplemented with 0.004% and 0.008% SIFs significantly up-regulated the mRNA expression of *fabp 3* in the hepatopancreas ($p < 0.05$, Figure 2a). The mRNA expression of *fabp 3* of crabs fed the high-fat diets were significantly higher than crabs fed the normal-fat diets ($p < 0.01$, Figure 2a). At the normal-fat level, dietary SIFs significantly down-regulated the mRNA expression of *fabp 4*, *fabp 10*, *srebp-1*, *elovl6* and *Δ9 fad* in the hepatopancreas ($p < 0.05$, Figure 2b–f). At the high-fat level, diets supplemented with 0.008% SIFs significantly up-regulated the mRNA expression of *fabp 4* and *elovl6* in the hepatopancreas ($p < 0.05$, Figure 2b,e). Diets supplemented with 0.004% and 0.008% SIFs significantly up-regulated the mRNA expression of *fabp 10* in the hepatopancreas of crabs under high-fat conditions ($p < 0.05$, Figure 2c). Diets supplemented with 0.004% SIFs significantly up-regulated the mRNA expression of *srebp-1* in the hepatopancreas of crabs under high-fat conditions ($p < 0.05$, Figure 2d). At the high-fat level, diets supplemented with 0.004% SIFs significantly up-regulated the mRNA expression of *srebp-1* in the hepatopancreas ($p < 0.05$, Figure 2d). On the contrary, diets supplemented with 0.004% and 0.008% SIFs significantly down-regulated the mRNA expression of *Δ9 fad* in the hepatopancreas of crabs under high-fat conditions ($p < 0.05$, Figure 2f).

3.4. The mRNA Expressions of Lipolysis-Related Genes in the Hepatopancreas

In the normal-fat level, diets supplemented with 0.004% SIFs significantly down-regulated the mRNA expression of *cpt-1a*, *cpt-2*, *mttp* and *tg1* in the hepatopancreas ($p < 0.05$, Figure 3a,c,d,f). However, at the high-fat level, diets supplemented with 0.004% SIFs significantly up-regulated the mRNA expression of *cpt-1a* and *cpt-1b* in the hepatopancreas ($p < 0.05$, Figure 3a,b). Diets supplemented with 0.004% and 0.008% SIFs significantly up-regulated the mRNA expression of *caat* and *tg1* in the hepatopancreas of crabs under high-fat conditions ($p < 0.05$, Figure 3e,f). On the contrary, diets supplemented with 0.004% and 0.008% SIFs significantly down-regulated the mRNA expression of *mttp* in the hepatopancreas of crabs under high-fat conditions ($p < 0.05$, Figure 3d). Diets supplemented with 0.008% SIFs significantly down-regulated the mRNA expression of *caat* in the hepatopancreas of crabs under high-fat conditions ($p < 0.05$, Figure 3e). In the absence of SIFs, the mRNA expression of *cpt-1a* and *cpt-1b* in the hepatopancreas of the crabs in the NF-0 group was significantly higher than that in the HF-0 group ($p < 0.01$, Figure 3a,b). However, dietary SIFs significantly up-regulated the mRNA expression of *cpt-1a* and *cpt-1b* of crabs fed the high-fat diets ($p < 0.05$, Figure 3a,b). Similar results were observed in *cpt-2*, *caat* and *tg1* ($p < 0.05$, Figure 3c,f).

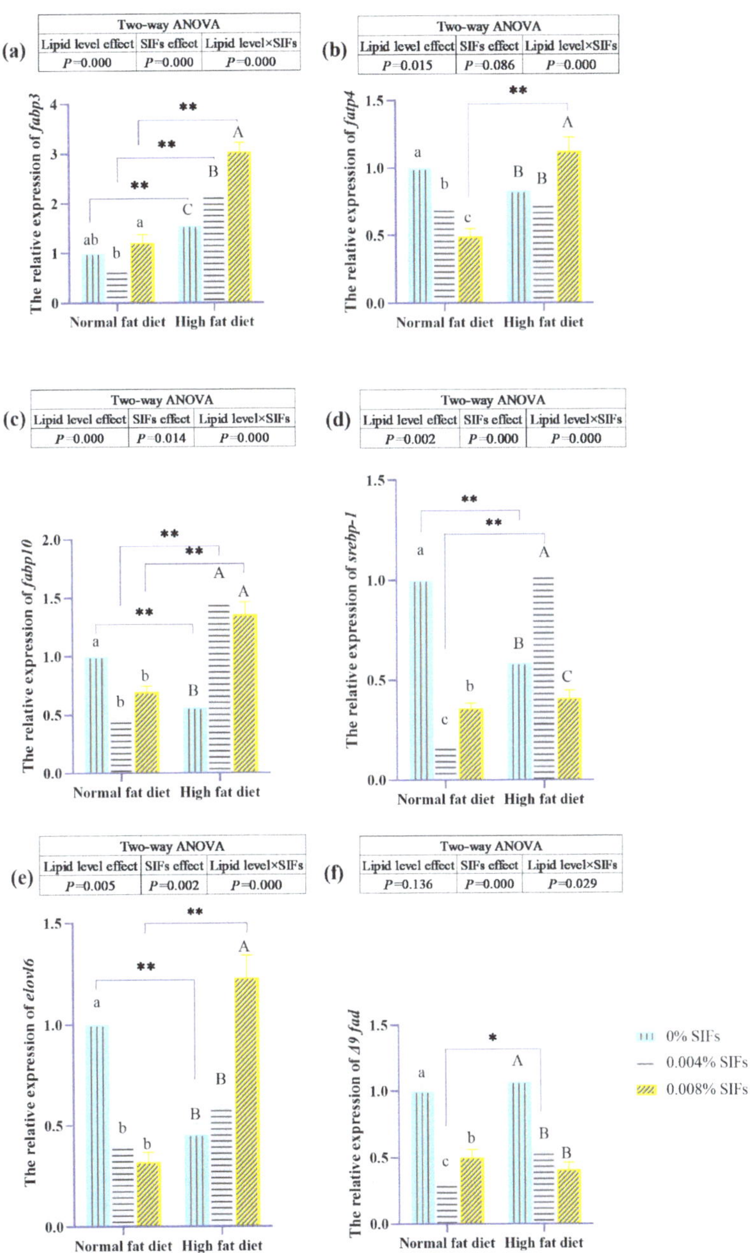

Figure 2. Effects of soybean isoflavones on the mRNA expressions of genes related to lipid synthesis in the hepatopancreas of juvenile Chinese mitten crab. Note: (**a**) fatty acid-binding protein 3; (**b**) fatty acid transport protein 4; (**c**) fatty acid-binding protein 10; (**d**) sterol regulatory element-binding protein 1; (**e**) elongase of very-long-chain fatty acids 6, (**f**) Δ9 fatty acyl desaturase. The different superscripts on the columns represent significant differences ($p < 0.05$, one-way ANOVA and Duncan multiple comparisons). The lines connecting the columns represent significant differences ($p < 0.05$ (marked *) or $p < 0.01$ (marked **), independence t-test). The table above the column represents the results of the two-factor analysis of variance.

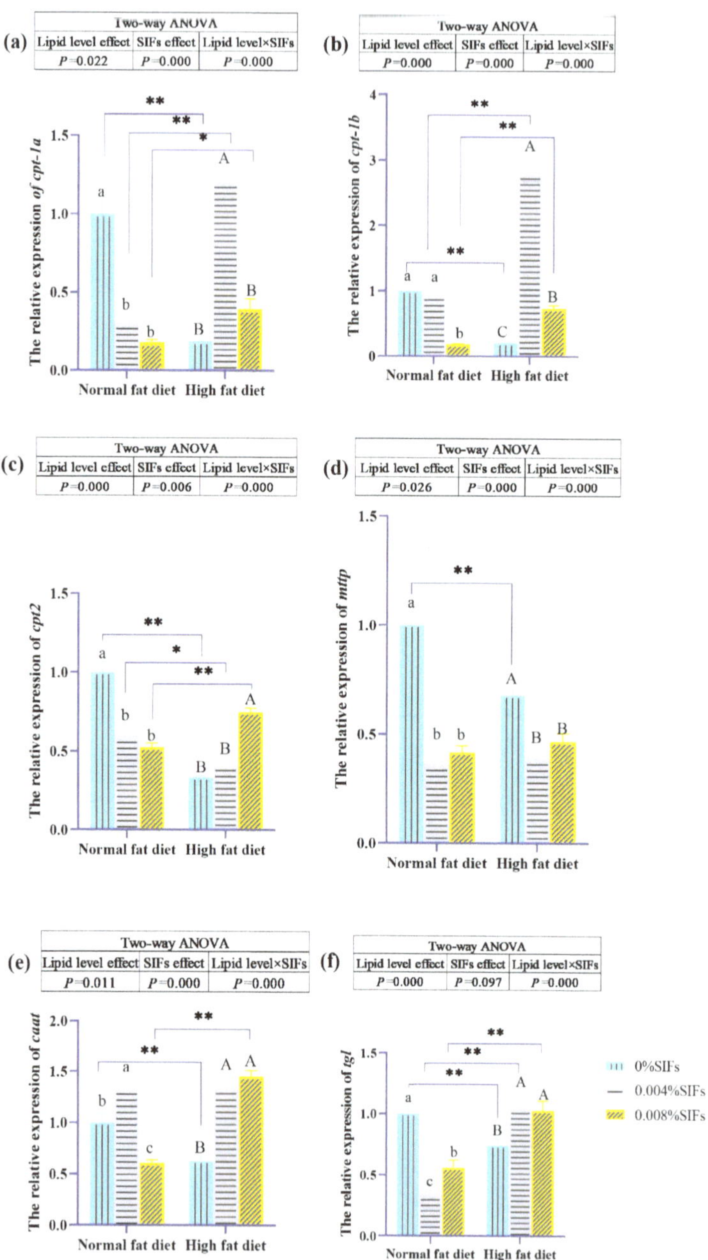

Figure 3. Effects of soybean isoflavones on the mRNA expressions of lipolysis-related genes in the hepatopancreas of juvenile Chinese mitten crab. Note: (**a**) carnitine palmitoyl transterase 1a, (**b**) carnitine palmitoyl transterase 1b, (**c**) carnitine palmitoyl transterase 2, (**d**) microsomal triglyceride transfer protein, (**e**) carnitine acetyltransferase, (**f**) triacylglycerol lipase. The different superscripts on the columns represent significant differences ($p < 0.05$, one-way ANOVA and Duncan multiple comparisons). The lines connecting the columns represent significant differences ($p < 0.05$ (marked *) or $p < 0.01$ (marked **), independence *t*-test). The table above the column represents the results of the two-factor analysis of variance.

4. Discussion

Lipids are important nutrients for aquatic animals, and an optimal dietary lipid level can ensure the growth and development of aquatic animals. In the present study, 15% lipid diets significantly reduced the growth performance of juvenile Chinese mitten crabs, and similar results were reported in *Cyprinus carpio* [21]. These results showed that an excessive dietary lipid level has a negative impact on aquatic animals. In the present study, the high-fat diet decreased the WG and SGR of juvenile crabs, but dietary SIFs increased the growth performance of juvenile crabs. The possible reason for these results was that dietary SIFs promoted the utilization of the high-fat diet by optimizing the lipid metabolism process, thereby improving the growth performance of juvenile Chinese mitten crabs.

Previous studies have reported that SIFs have lipid catabolism effects in humans and rats [22,23]. Studies have reported that dietary SIFs can reduce blood glucose, serum TG and LDL-C levels in a mouse model of type 2 diabetes induced by a high-fat and high-sugar diet [24]. In the present study, at the 10% lipid level, dietary SIFs significantly reduced the contents of NEFA and TG in the hepatopancreas of the crabs, which indicates that SIFs may be involved in lipolysis and provide energy for the growth of juvenile Chinese mitten crabs. However, the 15% lipid diet decreased the TG and NEFA content in the hepatopancreas when the diet was not supplemented with SIFs, but the dietary supplementation of SIFs increased the TC content, which indicates that SIFs may improve dietary fat utilization in juvenile crabs.

To date, the regulatory mechanisms of SIFs on lipid metabolism in crustaceans are still unknown. Previous studies have shown that SIFs can affect the activities of enzymes and β-oxidation involved in fat metabolism [25–28]. In the present study, the results showed that SIFs supplementation significantly up-regulated the relative expression of genes related to the fat synthesis and lipolysis of crabs fed the high-fat diets, which indicated that SIFs could improve lipid synthesis and oxidation in the Chinese mitten crabs. A similar result was reported in rainbow trout, which reported that SIFs enhanced the expression of lipid synthesis genes and lipid-binding protein genes [16]. The function of FABPs is the targeted transportation of hepatic fat to catabolic and anabolic sites, which is an important indicator for fat anabolism [29]. In the present study, SIFs supplementation significantly increased the expression of FABPs and elovl6 of crabs fed 15% fat diets. This result indicated that SIFs can improve lipid synthesis in Chinese mitten crabs fed high-fat diets. In contrast, some other studies found that dietary SIFs significantly inhibited the expression of genes related to fat synthesis in rats [30]. The reason for these divergent results may be related to species differences.

Lipid oxidation is one of the key steps in lipid metabolism in mammals, fish and crabs, and β-oxidation is the main pathway of fatty acid oxidation [31,32]. CPT1 and CPT2 form the mitochondrial carnitine palmitoyltransferase system, which plays an important role in the transfer of long-chain fatty acids from cytosolic compartments to the cytoplasm [33]. In the present study, the results showed that SIFs increased the relative expression of genes related to lipolysis (*cpt-1a*, *cpt-1b*, *cpt-2* and *caat*) in crabs fed high-fat diets. These results suggested that SIFs supplemented alongside a high-fat diet may accelerate lipid oxidation and improve the utilization of dietary lipids in Chinese mitten crabs.

5. Conclusions

Diets supplemented with 0.004% SIFs or 0.008% SIFs could inhibit the expression of the fatty acid synthesis-related genes *srebp-1* and *Δ9 fad* and promote the expression of the lipolysis-related genes *caat*, *tgl*, *cpt-1a*, *cpt-1b* and *cpt2*, thereby improving growth performance and lipid utilization in juvenile Chinese mitten crabs.

Supplementary Materials: The following supporting information can be downloaded at: https://www.mdpi.com/article/10.3390/fishes9090335/s1, Table S1: Sequences of primers.

Author Contributions: Methodology, M.S.; Software, Y.H.; Investigation, M.S.; Resources, L.C. and J.Y.; Data curation, J.Z.; Writing—review and editing, Y.T. and L.J.; Visualization, Y.X.; Supervision, C.Q. All authors have read and agreed to the published version of the manuscript.

Funding: This research was supported by grants from the Zhejiang Provincial Natural Science Foundation of China under Grant No. LTGN23C190003, the Huzhou Natural Science Foundation (2021YZ14), the National Natural Science Foundation of China (No. 32072986) and Zhejiang Province R&D Plan (2022C02058).

Institutional Review Board Statement: The animal study was reviewed and approved by the Committee on the Ethics of Animal Experiments in Huzhou University (approval code 20220701, approved on 1 August 2022).

Informed Consent Statement: Not applicable.

Data Availability Statement: Data are contained within the article or Supplementary Materials.

Conflicts of Interest: The authors declare no conflicts of interest.

References

1. Sacks, F.M.; Lichtenstein, A.; Horn, D.; Kris-Etherton, P.; Winston, M. Soy protein isoflavones and cardiovascular health. *Circulation* **2006**, *113*, 1689–1692. [CrossRef] [PubMed]
2. Anthony, M.S.; Clarkson, T.B.; Koudy, W.J. Effectsof soy isoflavones on atherosclerosis: Potential mechanisms. *Am. J. Clin. Nutr.* **1998**, *68*, 1390–1393. [CrossRef]
3. Li, Y.; Jiang, X.; Wei, Z.; Cai, L.; Yin, J.; Li, X. Effects of soybean isoflavones on the growth performance, intestinal morphology and antioxidative properties in pigs. *Animal* **2020**, *14*, 2262–2270. [CrossRef]
4. Jiang, Z.; Jiang, S.; Lin, Y.; Xi, P.; Yu, D.; Wu, T. Effects of soybean isoflavone on growth performance, meat quality, and antioxidation in male broilers. *Poult. Sci.* **2007**, *86*, 1356–1362. [CrossRef] [PubMed]
5. Zhang, H.M.; Chen, S.W.; Zhang, L.S.; Feng, X.F. The effects of soy isoflavone on insulin sensitivity and adipocytokines in insulin resistant rats administered with high-fat diet. *Nat. Prod. Res.* **2008**, *22*, 1637–1649. [CrossRef] [PubMed]
6. Payne, R.; Bidner, T.; Southern, L.; Geaghan, J. Effects of dietary soy isoflavones on growth, carcass traits, and meat quality in growing-finishing pigs. *J. Anim. Sci.* **2001**, *79*, 1230–1239. [CrossRef]
7. Mai, K.; Zhang, Y.; Chen, W.; Xu, W.; Ai, Q.; Zhang, W. Effects of dietary soy isoflavones on feed intake, growth performance and digestibility in juvenile Japanese flounder (*Paralichthys olivaceus*). *J. Ocean Univer. Chin.* **2012**, *11*, 511–516. [CrossRef]
8. Lee, S.O.; Renouf, M.; Ye, Z.; Murphy, P.A.; Hendrich, S. Isoflavone glycitein diminished plasma cholesterol in female golden Syrian hamsters. *J. Agric. Food Chem.* **2007**, *55*, 11063–11067. [CrossRef]
9. Ali, A.A.; Velasquez, M.T.; Hansen, C.T.; Mohamed, A.I.; Bhathena, S.J. Effects of soybean isoflavones, probiotics, and their interactions on lipid metabolism and endocrine system in an animal model of obesity and diabetes. *J. Nutr. Biochem.* **2004**, *15*, 583–590. [CrossRef]
10. Yao, J.T. *Effects of Soy Isoflavones on Serum FFA and Hepatic PPARα mRNA Expression in Rats with Feeding-Induced Metabolic Syndrome*; Heilongjiang University of Chinese Medicine: Harbin, China, 2011.
11. Kalaiselvan, V.; Kalaivani, M.; Vijayakumar, A.; Sureshkumar, K.; Venkateskumar, K. Current knowledge and future direction of research on soy isoflavones as a therapeutic agents. *Phcog. Rev.* **2010**, *4*, 111. [CrossRef]
12. González-Granillo, M.; Steffensen, K.; Granados, O.; Torres, N.; Korach-André, M.; Ortíz, V.; Aguilar-Salinas, C.; Jakobsson, T.; Díaz-Villaseñor, A.; Loza-Valdes, A.J.D. Soy protein isoflavones differentially regulate liver X receptor isoforms to modulate lipid metabolism and cholesterol transport in the liver and intestine in mice. *Diabetologia* **2012**, *55*, 2469–2478. [CrossRef]
13. Mezei, O.; Banz, W.J.; Steger, R.W.; Peluso, M.R.; Winters, T.A.; Shay, N. Soy isoflavones exert antidiabetic and hypolipidemic effects through the PPAR pathways in obese Zucker rats and murine RAW 264.7 cells. *J. Nutr.* **2003**, *133*, 1238–1243. [CrossRef]
14. Yang, L.; Zhang, W.; Zhi, S.; Zhao, M.; Liu, M.; Qin, C.; Feng, J.; Yan, X.; Nie, G. Evaluation of dietary genistein on the antioxidant capacity, non-specific immune status, and fatty acid composition of common carp (*Cyprinus carpio L.*). *Aquaculture* **2022**, *550*, 737822. [CrossRef]
15. Grgic, D.; Varga, E.; Novak, B.; Müller, A.; Marko, D. Isoflavones in animals: Metabolism and effects in livestock and occurrence in feed. *Toxins* **2021**, *13*, 836. [CrossRef] [PubMed]
16. Cleveland, B.M.; Manor, M.L. Effects of phytoestrogens on growth-related and lipogenic genes in rainbow trout (*Oncorhynchus mykiss*). *Comp. Biochem.* **2015**, *170*, 28–37. [CrossRef] [PubMed]
17. Deng, J.; Mai, K.; Ai, Q.; Zhang, W.; Wang, X.; Xu, W.; Liufu, Z.; Cai, Y.; Chen, W. Effects of antinutritional factors on plasma lipoprotein levels in Japanese flounder *Paralichthys olivaceus*. *J. Fish Biol.* **2012**, *80*, 286–300. [CrossRef] [PubMed]
18. Zhang, W. *Effects of Soybean Saponins and Soybean Isoflavones on Growth, Physiology and Intestinal Health of Carassius auratus*; Suzhou University: Suzhou, China, 2010.
19. AOAC. *Official Methods of Analysis of AOAC International*; Association of Official Analytical Chemists: Washington, DC, USA, 2005.

20. Vandesompele, J.; Preter, K.D.; Pattyn, F.; Poppe, B.; Roy, N.V.; Paepe, A.D.; Speleman, F. Accurate normalization of real-time quantitative RT-PCR data by geometric averaging of multiple internal control genes. *Genome Biol.* **2002**, *3*, research0034.1. [CrossRef] [PubMed]
21. Wang, J.L.; Wan, T.T.; Liu, X.X.; Jia, J.L.; Qin, C.B.; Nie, G.X. Regulative effects of Rehmanniae rehmanniae or Yam on growth, serum biochemical indexes and lipid metabolism of common Carp in high fat diet. *Chin. Acad. Fish. Sci.* **2023**, *30*, 48–59.
22. Nestel, P.J.; Yamashita, T.; Sasahara, T.; Pomeroy, S.; Dart, A.; Komesaroff, P.; Owen, A.; Abbey, M. Soy isoflavones improve systemic arterial compliance but not plasma lipids in menopausal and perimenopausal women. *Arterioscler. Thromb. Vasc. Bio.* **1997**, *17*, 3392–3398. [CrossRef]
23. Ørgaard, A.; Jensen, L. The effects of soy isoflavones on obesity. *Exp. Bio. Med.* **2008**, *233*, 1066–1080. [CrossRef]
24. Wan, H.M. *Effects of Soy Isoflavones on Lipid Metabolism and Inflammatory Factors in Type 2 Diabetic Mice*; Shanxi Medical University: Taiyuan, China, 2021.
25. Park, S.A.; Choi, M.S.; Cho, S.Y.; Seo, J.S.; Jung, U.J.; Kim, M.J.; Sung, M.K.; Park, Y.B.; Lee, M.K. Genistein and daidzein modulate hepatic glucose and lipid regulating enzyme activities in C57BL/KsJ-db/db mice. *Life Sci.* **2006**, *79*, 1207–1213. [CrossRef] [PubMed]
26. Kawakami, Y.; Tsurugasaki, W.; Nakamura, S.; Osada, K. Comparison of regulative functions between dietary soy isoflavones aglycone and glucoside on lipid metabolism in rats fed cholesterol. *J. Nutr. Biochem.* **2005**, *16*, 205–212. [CrossRef] [PubMed]
27. Mezei, O.; Li, Y.; Mullen, E.; Ross-Viola, J.S.; Shay, N.F. Dietary isoflavone supplementation modulates lipid metabolism via PPARα-dependent and -independent mechanisms. *Physiol. Genom.* **2006**, *26*, 8–14. [CrossRef]
28. Lee, Y.M.; Choi, J.S.; Kim, M.H.; Jung, M.H.; Lee, Y.S.; Song, J. Effects of dietary genistein on hepatic lipid metabolism and mitochondrial function in mice fed high-fat diets. *Nutrition* **2006**, *22*, 956–964. [CrossRef]
29. Storch, J.; Thumser, A.E. Tissue-specific functions in the fatty acid-binding protein family. *J. Biol. Chem.* **2010**, *285*, 32679–32683. [CrossRef] [PubMed]
30. Xiao, C.W.; Wood, C.M.; Weber, D.; Aziz, S.A.; Mehta, R.; Griffin, P.; Cockell, K.A. Dietary supplementation with soy isoflavones or replacement with soy proteins prevents hepatic lipid droplet accumulation and alters expression of genes involved in lipid metabolism in rats. *Genes Nutr.* **2014**, *9*, 373. [CrossRef]
31. Nguyen, P.; Leray, V.; Diez, M.; Serisier, S.; Bloc'h, J.L.; Siliart, B.; Dumon, H. Liver lipid metabolism. *J. Anim. Physiol. Anim. Nutr.* **2008**, *92*, 272–283. [CrossRef]
32. Greene, D.H.; Selivonchick, D.P. Lipid metabolism in fish. *Prog. Lipid Res.* **1987**, *26*, 53–85. [CrossRef]
33. Liu, L.; Long, X.; Deng, D.; Cheng, Y.; Wu, X.; Loor, J.J. Molecular characterization and tissue distribution of carnitine palmitoyl-transferases in Chinese mitten crab (*Eriocheir sinensis*) and the effect of dietary fish oil replacement on their expression in the hepatopancreas. *PLoS ONE* **2018**, *13*, e0201324. [CrossRef]

Article

Thyroid-Active Agents Triiodothyronine, Thyroxine and Propylthiouracil Differentially Affect Growth, Intestinal Short Chain Fatty Acids and Microbiota in Little Yellow Croaker *Larimichthys polyactis*

Xiao Liang, Yu Zhang, Ting Ye, Feng Liu and Bao Lou *

State Key Laboratory for Managing Biotic and Chemical Threats to the Quality and Safety of Agro-Products, Institute of Hydrobiology, Zhejiang Academy of Agricultural Sciences, Desheng Middle Road 298, Hangzhou 310021, China; liangxiao1225@yeah.net (X.L.); zhangy@zaas.ac.cn (Y.Z.); 15tye@stu.edu.cn (T.Y.); lengfeng0210@126.com (F.L.)
* Correspondence: loubao6577@163.com

Abstract: Thyroid dysfunction may affect the intestinal microbiota through short-chain fatty acids (SCFAs) in marine fish. This study investigated the effects of triiodothyronine (T3, 20 ng/g) and thyroxine (T4, 20 ng/g), and propylthiouracil (PTU, 5000 ng/g) on growth performance, intestinal SCFA profiles, and microbiota composition in little yellow croakers *Larimichthys polyactis*. The results showed that dietary thyroid-active agent supplementation significantly decreased weight gain, and specific growth ratio. Moreover, dietary T3, T4, and PTU induced the states of hyperthyroidism, hyperthyroidism, and hypothyroidism, respectively, leading to differential alterations in intestinal SCFA profiles. Specifically, only dietary T4 supplementation significantly increased the diversity of intestinal microbiota. Our findings suggest that the genera *Vibrio* and *Sediminibacterium* play key roles in multiple metabolic pathways within the host intestine. Correlation analyses further indicated that intestinal acetic acid and isobutyric acid were characteristic metabolites involved in the alteration of the genus *Vibrio* abundance. These results provide a foundation for further investigation into the effects of thyroid-disrupting activities on growth, intestinal SCFA profiles, and microbiota composition in marine fish.

Keywords: thyroid; short-chain fatty acids; intestine; microbiota; little yellow croaker

Key Contribution: This study was conducted to investigate the effects of various dietary thyroid-active agents on growth performance, intestinal short chain fatty acids, and microbiota in little yellow croaker and to explore the possible correlations.

Academic Editor: Marina Paolucci

Received: 30 December 2024
Revised: 4 February 2025
Accepted: 5 February 2025
Published: 7 February 2025

Citation: Liang, X.; Zhang, Y.; Ye, T.; Liu, F.; Lou, B. Thyroid-Active Agents Triiodothyronine, Thyroxine and Propylthiouracil Differentially Affect Growth, Intestinal Short Chain Fatty Acids and Microbiota in Little Yellow Croaker *Larimichthys polyactis*. *Fishes* **2025**, *10*, 69. https://doi.org/10.3390/fishes10020069

1. Introduction

Thyroid hormones (THs) play a crucial role in regulating the growth, development, osmoregulation, and metabolic pathways of multiple organs in fish [1]. Previous studies have demonstrated that the thyroid status in fish can be modulated through the exogenous administration of drugs such as T3, T4, and PTU, which are used to investigate the effects of different thyroid functions on physiological processes. However, the underlying regulatory mechanisms of these effects in fish remain unclear [2,3]. Exogenous administration of T3 and PTU can induce hyperthyroidism and hypothyroidism in GH-transgenic coho salmon (*Oncorhynchus kitsutch*), respectively, leading to inverse effects on growth and skeletal abnormalities [4]. Both T3 and T4 have been shown to enhance the activity of a-GPD and increase mitochondrial protein content in the liver and muscle of adult toads (*Bufo melanostictus*) [5]. Furthermore, T3 and PTU exert differential effects on thyroidal status in vivo, which in turn modulates various immune responses in the head kidney, spleen, and peripheral blood of rainbow trout (*Oncorhynchus mykiss*) [2,6]. Recent studies also indicate that THs regulate gastrointestinal physiological processes, such as intestinal development, nutrient absorption, and metabolism in mammals [7,8]. However, the role of THs in regulating gastrointestinal physiology in fish remains underexplored, and there is limited information regarding the potential immunomodulatory effects of THs in fish. Physiological processes in the gut are known to depend on the interactions between the gut microbiota and host metabolism [9]. The gut microbiota has been shown to perform functional organ-like roles, primarily residing on the mucosal surfaces of the host [10]. Studies have revealed that the composition of gut microbiota can affect energy absorption and the onset of diseases in the host through their metabolites [11–13]. The vagus nerve plays an essential role in the microbiota–gut–brain axis; specific bacteria can synthesize and release neuroactive substances to regulate the host behavior [14,15]. Recent research emphasizes that the microbiota–gut–brain axis functions as a bidirectional regulatory pathway [16–18]. These processes are mediated by immune and endocrine systems, including humoral and cellular regulation, hypothalamic–pituitary–adrenal (HPA) and hypothalamic–pituitary–thyroid (HPT) axes [19,20]. Many autoimmune disorders, like hypothyroidism and hyperthyroidism, have been linked to excessive microbial growth or alterations in microbiota composition [21,22], indicating that thyroid function may also interact with gut microbiota. Given the critical role of the thyroid in animal physiology, understanding the stability of thyroid function and its impact on gut microbiota is an important area of research. However, to date, studies on the effects of thyroid homeostasis on gut microbiota, and their potential crosstalk, particularly in fish, remain limited.

Microbial metabolites, particularly short-chain fatty acids (SCFAs), such as propionate, butyrate, and acetate, serve as important anions in the colonic lumen and can affect both the morphology and physiological functions of the gut [23]. SCFAs can enhance T3 metabolism and stimulate the secretion of prolactin [24]. Together, SCFAs and T3 work synergistically to maintain the development and homeostasis of the intestinal epithelium [25]. Furthermore, SCFAs may indirectly adjust thyroid function by affecting gut microbiota and immune function [26]. Recent studies have highlighted that SCFAs are involved in the crosstalk between thyroid function and gut microbial metabolism, contributing to the regulation of disease resistance in aquatic species [27–29]. However, few studies were focused on the effects of thyroid functions on SCFA metabolism and gut microbiota [29].

Little yellow croaker (*Larimichthys polyactis*), a widely cultured species in the coastal areas of southeastern China, is an economically significant marine fish in China, Korea, and Japan. Environmental factors in aquaculture can easily induce metabolic disorders in the digestive system of little yellow croaker through the endocrine system [30]. Investigating the interactions within the thyroid–gut microbiota axis may provide valuable insights into the regulation of gastrointestinal physiological homeostasis in fish. In the present study, we used thyroid-active agents, including T3, T4, and PTU, to induce varying states of thyroid dysfunction in little yellow croaker. The aim of this study was to explore the effects of thyroid dysfunction on the composition of the intestinal microbiota and SCFA metabolism. The results of this study will lay the foundation for a better understanding of the interplay between thyroid function and the intestinal microbiota in fish.

2. Materials and Methods

2.1. Animal Ethics

The fish experiments were conducted following the regulations of "Guidelines for Experimental Animals" of the Ministry of Science and Technology (Beijing, China), which was approved by the Committee of Laboratory Animal Experimentation at Zhejiang Academy of Agricultural Sciences (Permit Number: 2023ZAASLA29). The scientists in charge of the experiments received training and personal authorization.

2.2. Diets

The experimental diets were labeled as Control (Con), T3, T4, and PTU. The commercial diets (45% crude protein, 12% crude lipid) were purchased from the Fuzhou Haima Feed Co., Ltd., (Fuzhou, China), and commercial diets were supplemented with these thyroid-active agents to prepare four experimental diets following standard protocols. Based on previous studies [2,3], the concentrations of T3, T4, and PTU were selected, and selected concentrations of T3, T4, and PTU were 20 ng/g, 20 ng/g, and 5 mg/g, respectively. Referring to the methods in a previous study [4], T3, T4, and PTU were prepared as a 99% alkaline ethanol solution (20 mL 99% ethanol + 0.4 mL 0.1 N NaOH) and uniformly applied to feed pellets via spray-coating, with subsequent oven-drying at 60 °C for solvent removal. The control group underwent identical processing steps with ethanol application but lacked the pharmacological additives.

2.3. Fish Feeding Experiment

Little yellow croakers were obtained from the Xiangshan harbor aquatic seedling Co., Ltd. (Xiangshan County, Ningbo, China). Prior to the feeding trial, these fish were acclimated to the experimental diet and conditions by being stocked in indoor cage tanks for 14 days. The healthy fish (initial weight 23.14 ± 1.08 g) were randomly assigned to 12 laboratory tanks (200 L), with 10 individuals in each tank. Each diet was assigned to triplicate tanks. Fish were hand-fed twice daily to apparent satiation (05:00 and 17:00), and the amount of diet consumption in each net cage was recorded daily. The feeding trial lasted for 28 days, during which the water temperature ranged from 28.0 to 30.5 °C, salinity ranged from 24.2‰ to 25.7‰, and dissolved oxygen was not less than 8.0 mg/L.

2.4. Sample Collection and Zootechnical Parameters

At the end of the feeding experiment, all fish in each tank were anesthetized with MS-222 (Sigma Diagnostics INS, St. Louis, MO, USA) at a concentration of 100 mg/L and weighed. A total of six fish from each group (two fish per tank) were randomly selected for serum collection. Blood was collected from these fish, and serum was separated via centrifugation ($1500 \times g$, 4 °C, 30 min), with the individual serum samples pooled into a single tube. The pooled serum was stored at −80 °C until further analysis. Subsequently, the posterior intestinal content from the 6 fish per group were ground in liquid N_2 and pooled into a tube. These samples were stored at −80 °C for further determination of SCFA profiles and intestinal microbiota.

Zootechnical parameters such as weight gain (WG), and specific growth rate (SGR) were calculated using the formulae described below.

$$\text{Weight gain (WG, \%)} = 100 \times (\text{final body weight (g)} - \text{initial body weight (g)})/\text{initial body weight (g)}.$$

$$\text{Specific growth ratio (SGR, \% day}^{-1}) = 100 \times (\text{Ln (final body weight)} - \text{Ln (initial body weight)})/\text{t (culture period)}.$$

2.5. Serum Biochemical Analysis

Serum T3 and T4 levels were measured via radioimmunoassay (RIA) using commercial kits (Beijing North Institute of Biotechnology, China). The detailed method for measuring was described in our previous study [31]. T3/T4 ratios were calculated for each individual.

2.6. Intestinal SCFA Quantitative Analysis

The intestinal SCFA content was determined by GC-MS (Gas Chromatography–Mass Spectrometer), and the protocol was described in a previous study [32].

2.7. Analysis of Intestinal Microbiota

The E.Z.N.A.® Soil DNA kit (Omega Biotek, Norcross, GA, USA) was used to extract microbial DNA. The extracted bacterial DNA was sent for sequencing of the V1–V9 regions of the 16S rRNA gene on dedicated PacBio Sequel cells at Shanghai Biozeron Biotechnology Co., Ltd. (Shanghai, China). After sequencing, raw reads were processed through the SMRT Portal, and OTU clustering and species annotation were performed using UPARSE (version 7.1 http://drive5.com/uparse/, accessed on 15 July 2020). Then, analyses of alpha and beta diversity, and Tac4Fun and BugBase analyses were conducted. Related manufacturers' protocols for the above were described in a previous study [33].

2.8. Statistical Analyses

Statistical analysis was performed with SPSS 24.0 (IBM, New York, NY, USA) using one-way ANOVA. For biochemical analyses in serum, intestinal SCFA content, and microbial profiles, statistical comparisons on the measured variables among the dietary treatments were investigated by Tukey's post hoc test. The level of $p < 0.05$ was considered significant and all values expressed as mean $\pm$ standard error (SD). The principal coordinates analysis (PCoA), redundancy analysis (RDA), and correlations among the variables assessed with Spearman's correlation and other graphs used the OmicShare tools, a free online platform for data analysis and charting (http://www.omicshare.com/tools, accessed on 30 April 2024).

3. Results

3.1. Growth Performance

The growth performance data are presented in Table 1. Compared with the control (Con) group, the indexes of final weight, weight gain (WG), and specific growth rate (SGR) were significantly lower in all treatment groups ($p < 0.05$). Additionally, the lowest final weight, WG, and SGR were observed in the T4 group ($p < 0.05$). The feed consumption in the Con and T3 groups was significantly higher than those in the T4 and PTU groups ($p < 0.05$).

Table 1. Growth performance, feed utilization, and morphologic index of little yellow croakers fed thyroid-active agents.

Index	Groups			
	Con	T3	T4	PTU
Initial weight	23.14 ± 1.08	23.14 ± 1.08	23.14 ± 1.08	23.14 ± 1.08
Final weight	34.21 ± 0.82 [a]	30.32 ± 0.19 [b]	28.02 ± 0.20 [c]	30.55 ± 0.14 [b]
WG (%)	47.84 ± 3.53 [a]	31.04 ± 0.82 [b]	21.10 ± 0.85 [c]	32.01 ± 0.62 [b]
SGR (% day^{-1})	1.40 ± 0.09 [a]	0.97 ± 0.02 [b]	0.68 ± 0.03 [c]	0.99 ± 0.02 [b]
Feed consumption	270.62 ± 4.74 [a]	274.92 ± 2.96 [a]	241.88 ± 3.53 [b]	249.77 ± 9.97 [b]

Note: Values with different superscripts in the same line are significantly different ($p < 0.05$).

3.2. Serum Thyroid Hormone Content

The levels of serum TH are presented in Figure 1. Compared with the Con group, both T3 content and the T3/T4 ratio were significantly higher in the T3 group, while T3 content and the T3/T4 ratio were significantly lower in the PTU group, followed by the T4 group ($p < 0.05$). Moreover, the levels of T4 in the Con group were significantly higher than those in the T3 group, but significantly lower than those in the T4 and PTU groups ($p < 0.05$).

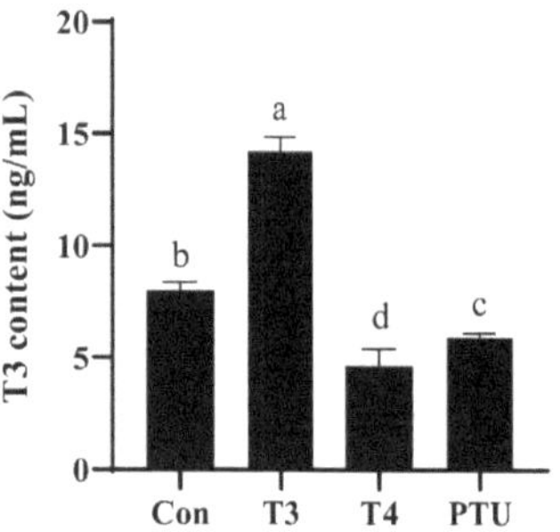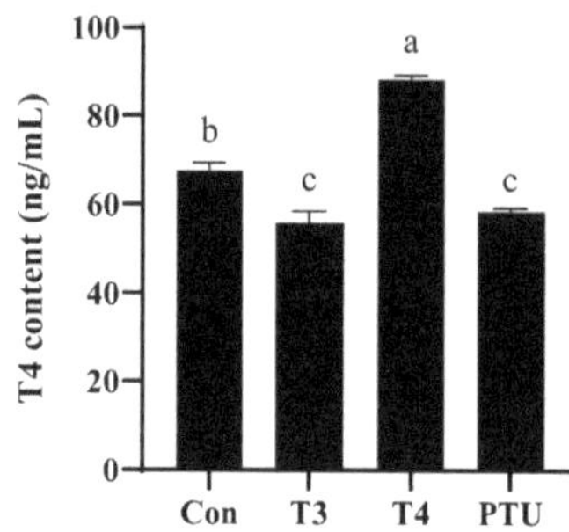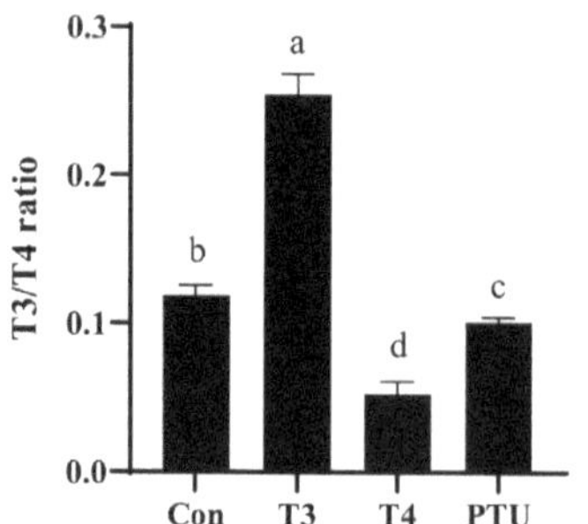

Figure 1. Serum concentrations (ng/mL) of circulating total (T3, A), total T4 (T4, B), and the ratio of total T3 to T4 (T3/T4, C) of little yellow croakers fed various thyroid-active agents. The values are expressed as mean ± SD (N = 6). Different little letters indicate significant differences among all the groups ($p < 0.05$).

3.3. Intestinal SCFA Profiles

The UHPLC-MS/MS data revealed the detection of seven SCFAs in all treatment groups (Figure 2A–G). Acetic acid (AA) was the predominant compound among BAs found in the posterior intestine of the little yellow croaker. Compared to the Con group, significantly higher levels of AA were observed in the T3 and T4 groups ($p < 0.05$), and there were no significant differences in AA content between the Con group and the PTU group ($p > 0.05$) (Figure 2A). The highest levels of butyric acid (BA) were shown in the Con group, followed by the T3 group, while the T4 and PTU groups exhibited the lowest BA levels ($p < 0.05$) (Figure 2B). The caproic acid (CA) content in the Con group was significantly lower than that in the T3 group ($p < 0.05$), but there were no significant differences between the T4/PTU group and the Con group ($p > 0.05$) (Figure 2C). There were no significant differences in the isobutyric acid (IBA) between the T3/PTU group and the Con group ($p > 0.05$), but the IBA content in the T4 group was significantly down-regulated ($p < 0.05$) (Figure 2D). A similar trend of changes in valeric acid (VA) content is observed in Figure 2E. Furthermore, isovaleric acid (IVA) content was significantly higher in both the Con and T3 groups compared to the T4 and PTU groups ($p < 0.05$) (Figure 2F). Finally, different thyroid-active agents did not affect the propionic acid (PA) content ($p > 0.05$) (Figure 2G).

3.4. Composition and Diversity of Intestinal Microbiota

Community richness and diversity were assessed using the Chao1, ACE, Shannon, and Simpson indices across different treatments (Figure 3A). The Shannon and Simpson indices in the T4 group were significantly higher in the PBO groups ($p < 0.05$), whereas no significant differences were observed in the ACE and Chao1 indices between the groups ($p < 0.05$). The up-regulated Shannon and Simpson indices in the PBO groups indicated that dietary T4 supplementation improved the diversity of the microbiota. PCoA analysis based on the Bray–Curtis distances revealed a distinct separation in microbiota composition between the Con and T4 groups, while the Con, T3, and PTU groups exhibited similar microbial profiles (Figure 3B).

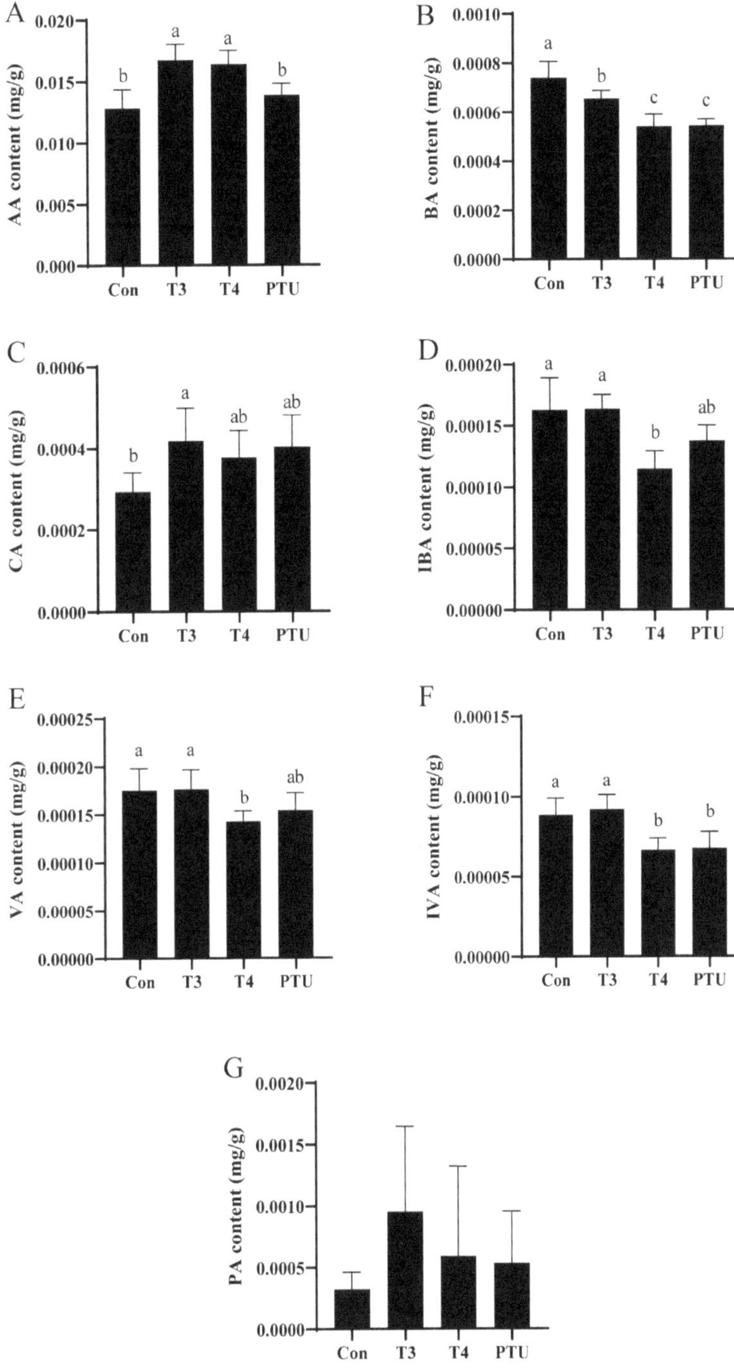

Figure 2. Significantly different content of various SCFAs (ng/g) in the intestine of little yellow croakers fed various thyroid-active agents. (**A**) acetic acid content; (**B**) butyric acid content; (**C**) caproic acid content; (**D**) isobutyric acid content; (**E**) valeric acid content; (**F**) isovaleric acid content; (**G**) propionic acid content. The values are expressed as mean ± SD (N = 6). Different letters indicate significant differences among all the groups ($p < 0.05$).

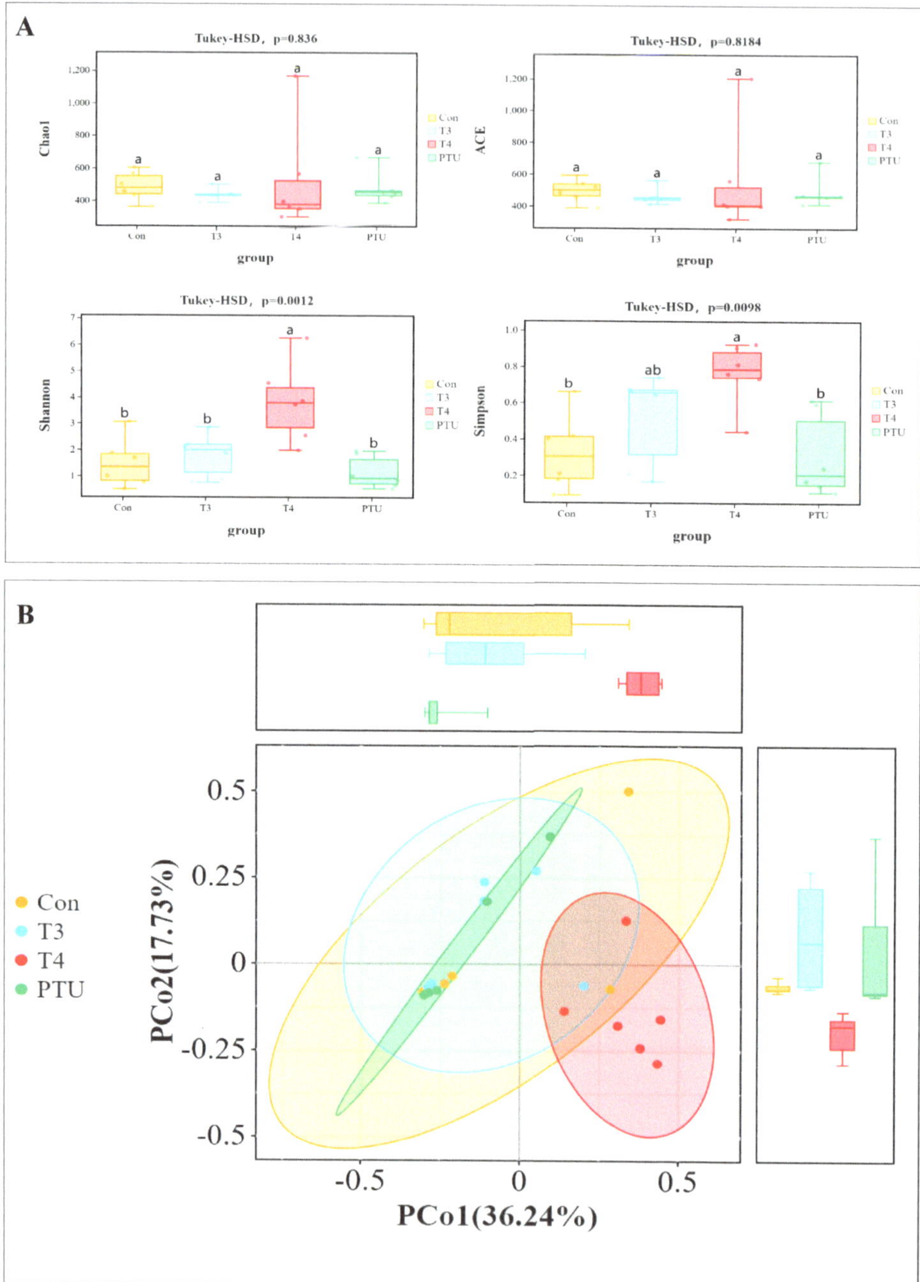

Figure 3. The α diversity index (**A**) and β diversity index (**B**) of gut microbiota. (**A**) The α diversity index contains Chao1, ACE, Shannon, and Simpson indices. (**B**) The β diversity index is revealed by principal coordinates analysis (PCoA) of gut microbiota. Different letters indicate significant differences among all the groups ($p < 0.05$).

The Tukey HSD test was applied to analyze the relative abundance at the phylum and genus levels. At the phylum level (Figure 4A), Proteobacteria and Firmicutes were identified as the dominant phyla in the gut of little yellow croakers. At the genus level (Figure 4B), *Vibrio* and *Acinetobacter* were the predominant genera. Further Lefse analysis identified *Vibrio* and *Sediminibacterium* as biomarkers for distinguishing the T4 group from the other treatments (LDA = 3.5) (Figure 4C).

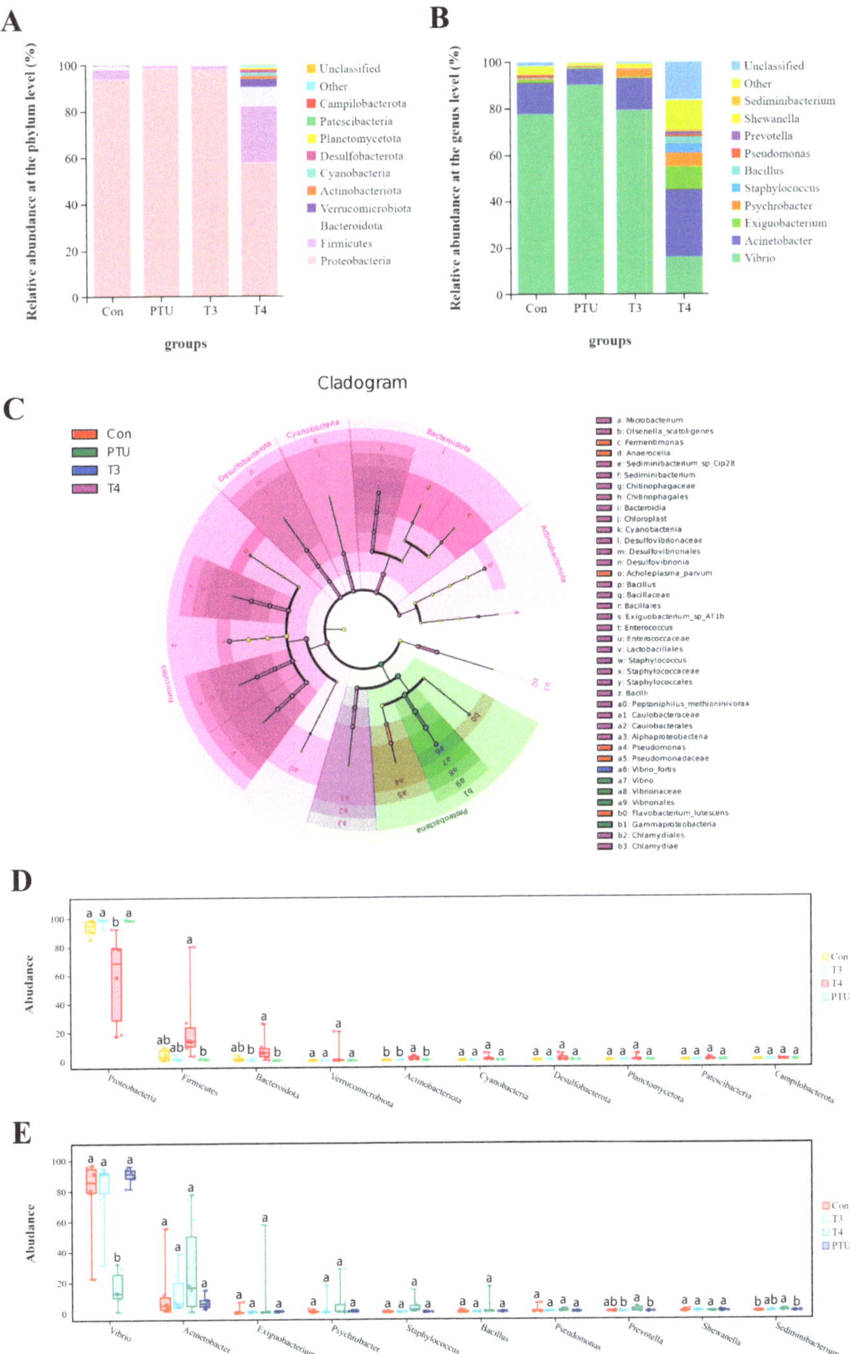

Figure 4. The relative abundance of the intestinal microbial communities in little yellow croakers fed various thyroid-active agents. (**A**) Column chart of the top 10 species in relative abundance at the phylum level. (**B**) Column chart of the top 10 species in relative abundance at the genus level. (**C**) Analysis of

bacterial taxa was identified via LEfSe, and the species with an LDA SCORE > 3.5 were defined as statistically different biomarkers. Intergroup variation in the relative abundance of the intestinal microbial communities at the phylum (**D**) and species (**E**) levels, and different letters indicate significant differences among all the groups ($p < 0.05$).

The top 10 relative abundance of bacterial phyla in little yellow croakers from the four experimental groups are shown in Figure 4D. Compared with the Con group, only the dietary T4 supplement significantly reduced the abundance of Protcobacteria, while it notably increased the abundance of Actinobacteriota ($p < 0.05$). The abundance of Firmicutes in the T4 group was significantly higher than that in the PTU group ($p < 0.05$), and the abundance of Bacteroidota in the T4 group was significantly higher than that in the T3 and PTU groups ($p < 0.05$). The top 10 relative abundance of bacterial genus in little yellow croakers from the four groups are shown in Figure 4E. In comparison to the Con group, only the dietary T4 supplement significantly decreased the abundance of *Vibrio*, while significantly increasing the abundance of *Sediminibacterium* ($p < 0.05$). Furthermore, the abundance of *Prevotella* in the T4 group was significantly higher than that in the T3 and PTU groups ($p < 0.05$).

3.5. Redundancy Analysis and Correlation Coefficients Among the Parameters

To elucidate the relationship between various parameters, the RDA was conducted using THs as explanatory variables and intestinal microbiota abundance at the genus level as response variables (Figure 5A). The first two axes cumulatively explained 100% of the variation in the microbial community structure. Combined with the arrow lines of T3 and T4, these results clearly showed that the T4 was positively correlated with the microbial profile in the T4 group, whereas T3 was positively correlated with the microbial profile in the other groups. Furthermore, the Spearman correlation analyses were used to further investigate the significant correlation between TH profiles and intestinal microbiota abundance (Figure 5B). The genera of *Staphylococcus*, *Bacillus*, *Sediminibacterium*, and *Enterococcus* were clustered together, with their abundance showing a significantly negative correlation with T3 content and a significantly positive correlation with T4 content ($p < 0.05$). In contrast, the abundances of *Vibrio* and *Cobetia* were significantly positively correlated with T3 content, but the abundance of the *Vibrio* genus was significantly negatively correlated with T4 content ($p < 0.05$).

Another RDA was used to further investigate the relationship between intestinal SCFA content and intestinal microbiota abundance at the genus level (Figure 5C). The first two axes cumulatively explained 94.72% of the variation in the microbial community structure. Combined with the length of the arrow lines, the AA was positively correlated with the microbial profile in the T4 group, while other SCFAs were positively correlated with the microbial profile in the other groups. Moreover, correlation analyses were conducted to explore the significant correlation between SCFA profiles and intestinal microbiota abundance (Figure 5D). It was investigated that the abundance of the *Vibrio* genus was significantly negatively correlated with AA content, but significantly positively correlated with IBA content ($p < 0.05$). The abundances of *Staphylococcus*, *Bacillus*, *Sediminibacterium*, and *Enterococcus* were negatively correlated with BA, IVA, VA, and IBA content. The abundance of *Sediminibacterium* was significantly negatively correlated with IBA content ($p < 0.05$).

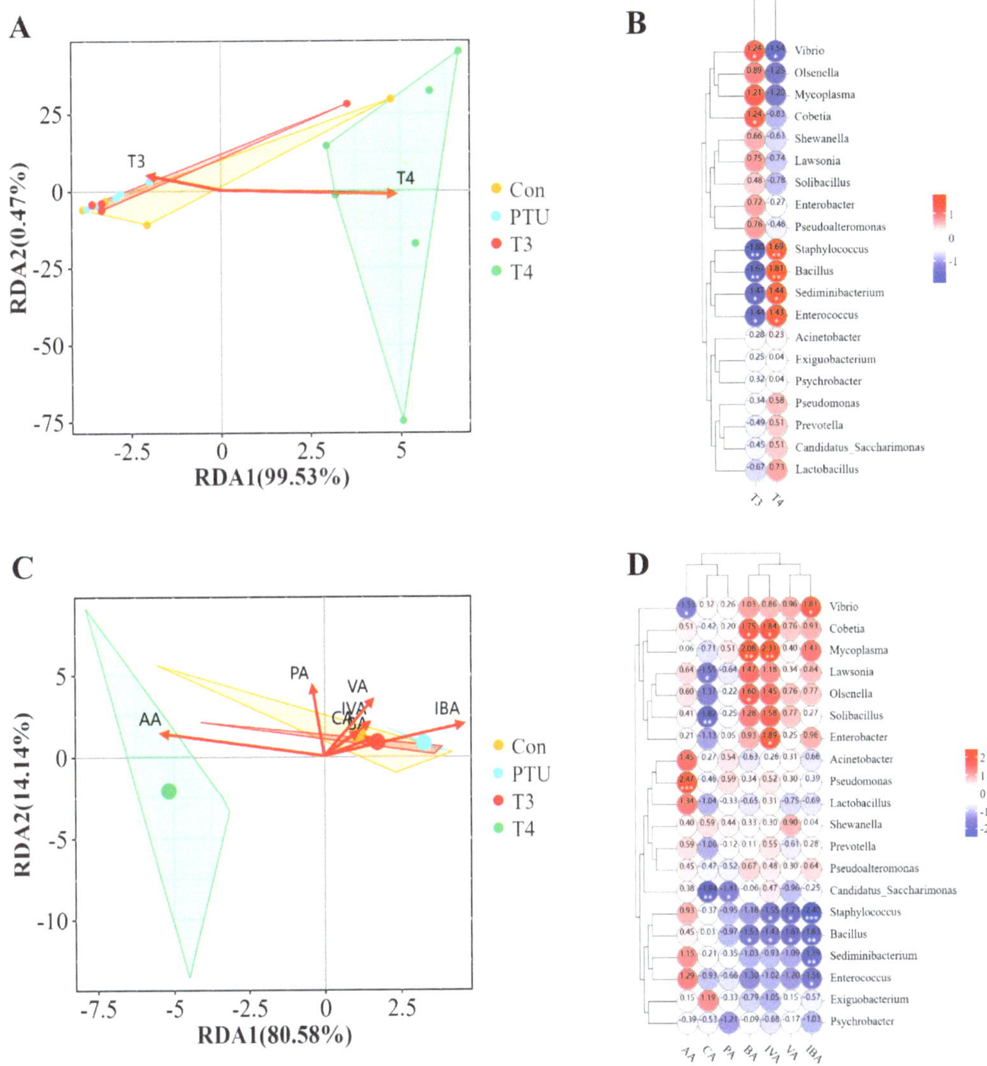

Figure 5. RDA plots showing factors that significantly explain the variance in the microbiota composition. (**A**) RDA of TH levels and relative abundance of gut microbiota at the genus level. (**B**) RDA of SCFA levels and relative abundance of gut microbiota at the genus level. (**C**) Heatmap representation of the Spearman correlation coefficient between TH levels and relative abundance of the top 20 predominant genera. (**D**) Heatmap representation of the Spearman correlation coefficient between SCFA levels and relative abundance of the top 20 predominant genera. Red represents a positive correlation, and blue represents a negative correlation.

3.6. Functional Prediction

The microbial function pathways predicted with the Tax4Fun tool in the T4 group were significantly different from those observed in the other groups (Figure 6A). Compared with the Con group, the T4 group exhibited lower abundances of bacteria related to membrane transport and signal transduction, but higher abundances of bacteria associated with amino acid metabolism, energy metabolism, replication and repair, translation, and lipid metabolism ($p < 0.05$) (Figure 6C).

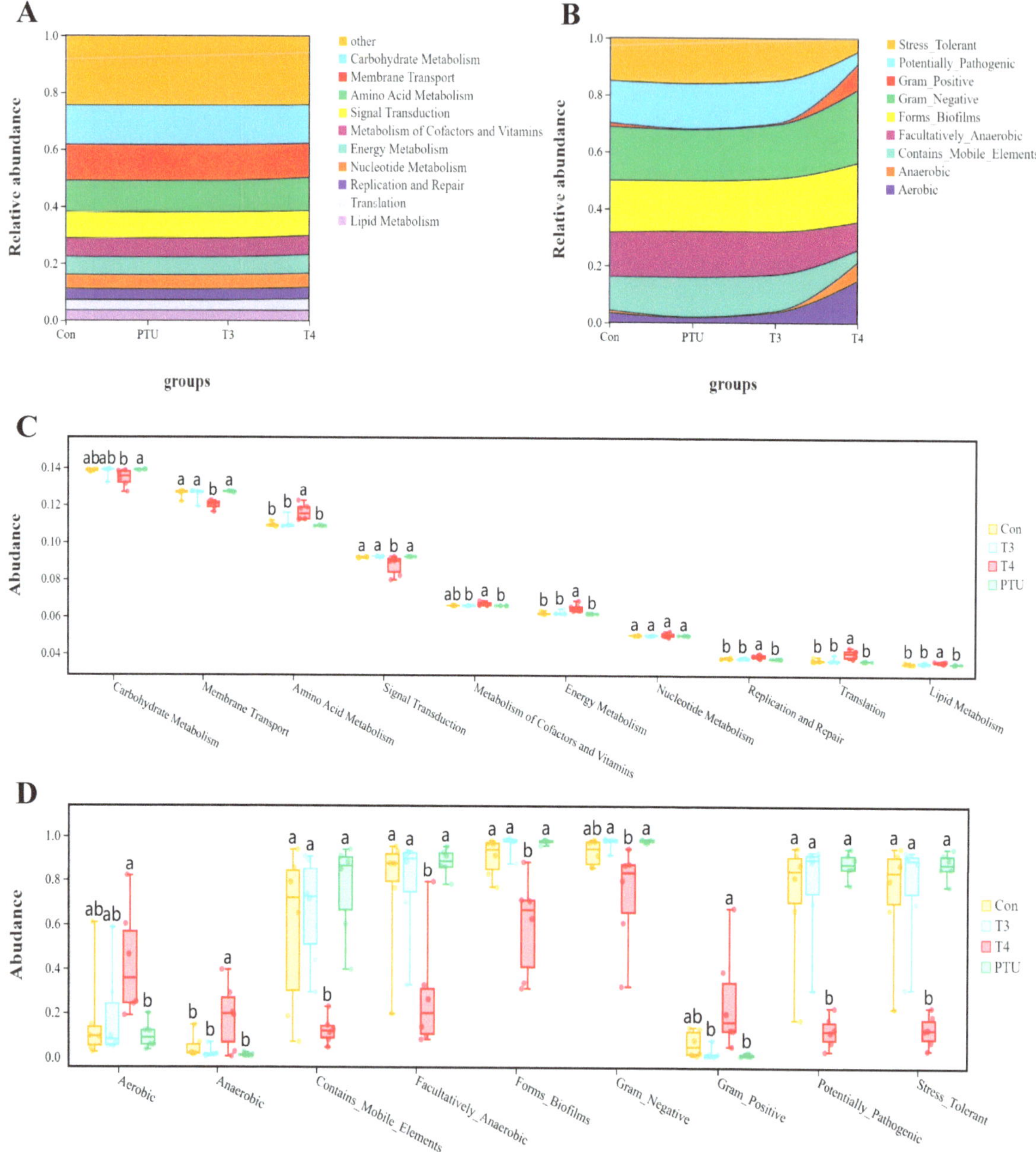

Figure 6. (**A**,**C**) Tax4Fun analysis results of predicted functional pathways in gut microbiota of little yellow croakers fed various thyroid-active agents. (**B**,**D**) BugBase analysis results of predicted bacterial phenotypic classifications of gut microbiota in little yellow croakers fed various thyroid-active agents. Different letters indicate significant differences among all the groups ($p < 0.05$).

The bacterial phenotypic classifications predicted with the BugBase tool in the T4 group also differed markedly from those in the other groups (Figure 6B). Compared with the Con group, the T4 group showed higher abundances of anaerobic bacteria, and lower abundances of bacteria related to the values of contains_mobile_elements, facultatively_anaerobic, forms_biofilms, potentially_pathogenic, and stress tolerant ($p < 0.05$) (Figure 6D).

4. Discussion

Administration of THs or other thyroactive preparations to fish can alter body growth [34]. In coho salmon (*O. kisutch*), dietary T3 (12 µg/g) was found to increase the growth index, whereas dietary PTU (6 mg/g) resulted in a decrease [34]. Moreover, exposure to T4 (10 nM) for 30 days did not affect WG, SGR, and FER in red tilapia (*Oreochromis mossambicus* × *O. urolepis hornorum*), but a 60-day exposure led to significant increases in WG, SGR, and FBW [35]. Our study revealed that administration of T3, T4, and PTU significantly decreased WGand SGR in the little yellow croakers. Despite these findings, relatively few studies have focused on the effects of thyroid-active agents on growth performance in fish.

Based on previous studies in humans, studies have employed THs or anti-thyroid drugs to induce the status of hyper- or hypothyroidism through disruption of the thyroid axis [36,37]. PTU, a common anti-thyroid drug, is employed to induce hypothyroidism [3,38]. In tilapia (*O. mossambicus*), plasma T3 and T4 levels were significantly down-regulated following PTU feeding (20 µg/g) for 15 days [38]. Similarly, a higher dose of PTU (1.5 or 6 mg/g) administered to coho salmon (*O. kisutch*) for 15 days also led to significant reductions in plasma T3 and T4 concentrations [34]. In line with these findings, our study demonstrated that feeding PTU (5 mg/g) to little yellow croakers for 28 days significantly decreased serum T3 and T4 levels, which could indicate hypothyroidism. On the other hand, both plasma T3 and T4 levels in Nile tilapia (*Oreochromis niloticus*) showed no significant changes after feeding T3 (48 µg/g) or T4 (48 µg/g) for 28 days [3]. In contrast to previous studies, feeding a dose of T3 or T4 (20 µg/g) for 28 days resulted in a significant increase in serum T3 or T4 levels in little yellow croakers, which could serve as indicators of different types of hyperthyroidism. Based on the existing data [3,38], we propose that effective regulation of hyperthyroid or hypothyroid states in fish is relevant to several factors, including drug dosage, treatment duration, and the species' sensitivity to the medication. Consequently, further fundamental research is needed to explore these variables and their interactions.

SCFAs, products of anaerobic fermentation by the intestinal microbiota, play an essential role in primarily anti-inflammatory mechanisms as well as the indirect modulation of energy metabolism [39]. The primary SCFAs in the human intestine are AA, PA, and BA, which are typically present in a ratio of 60:25:15, respectively [26]. Similarly, our results also found that AA, PA, and BA were the predominant SCFAs in the intestines of little yellow croakers. SCFAs have been shown to act synergistically with thyroid hormones, particularly T3, to regulate enterocyte differentiation [40]. This suggests that SCFAs may be linked to thyroid function. Furthermore, impaired thyroid function has been associated with decreased levels of AA, PA, and BA in rats [41]. Further evidence suggests that patients with autoimmune thyroid diseases exhibit significantly lower levels of propionate and butyrate, alongside elevated levels of isovalerate, with propionate being negatively correlated with free T3 and free T4 levels [42]. Currently, research on SCFA metabolism modulated by hyperthyroid or hypothyroid states in fish remains limited. Our results demonstrated that T3-induced hyperthyroidism significantly increased the levels of AA and CA, while decreasing BA levels. T4-induced hyperthyroidism was associated with reduced levels of BA, IBA, VA, and IVA, while elevating AA levels. Differently, PTU-induced hypothyroidism resulted in significantly lower levels of BA and IVA. Therefore, evidence suggests that the compositions of intestinal SCFA are related to thyroid status in little yellow croakers.

Intestinal microbiota of various fish species is dominated by several phyla, including Proteobacteria, Firmicutes, Bacteroidota, Actinobacteria, and Fusobacteria [43]. The phyla Firmicutes and Bacteroidota are commonly recognized for their beneficial role in maintaining intestinal health, but the up-regulated abundance of Proteobacteria is closely correlated with pathogenicity [44]. In this study, various dietary thyroid-active agents changed the diversity of dominant phyla in little yellow croakers, such as Proteobacteria and Firmicutes. The imbalance of intestinal microbiota has been linked to thyroidal functional status. For instance, patients with Hashimoto's thyroiditis, characterized by elevated free thyroxine (FT4) content (as markers of thyroid function), exhibit a higher abundance of Proteobacteria and Actinobacteria [45]. In contrast, patients with Graves' disease, who have elevated levels of T3, T4, free triiodothyronine (FT3), and FT4, tend to show a lower abundance of Proteobacteria [46]. However, some patients with higher levels of FT3 and FT4 suggest a higher abundance of Proteobacteria and Actinobacteria [45], suggesting that the underlying mechanisms of these differing patterns require further investigation. In the present study, elevated serum T4 levels significantly reduced the abundance of Proteobacteria while increasing the abundance of Actinobacteria. Proteobacteria in little yellow croakers was primarily represented by the genus *Vibrio*, a major group of bacterial pathogens commonly found in marine fish cultures. Research has shown that serum T4 gradually decreases during the natural progression of vibriosis of silver sea bream (*Sparus sarba*) [47]. Additionally, studies indicate that the TH synthesis pathway in some fish species plays a crucial role in regulating infections caused by *Vibrio* spp. [48,49]. Furthermore, we found that Bacteroidota in little yellow croakers was primarily represented by the genera *Sediminibacterium* and *Prevotella*, and elevated serum T4 significantly increased the abundance of Bacteroidota. These findings align with previous research on thyroid dysfunction in rats [50]. Correlation analyses in our study revealed that serum T3 content was significantly associated with the abundances of genera *Vibrio* and *Sediminibacterium*, but serum T4 content showed a reverse association with these two genera. Inhibition of pathogenic bacteria can effectively improve the disease resistance in fish species [48,49]. Thus, a thorough understanding of how T3 and T4 influence these pathogens is crucial for enhancing disease management in aquaculture.

SCFAs play a critical role in maintaining the microbial ecosystem, with their composition significantly influencing interactions within the thyroid–gut microbiota axis [26]. A previous study demonstrated that after *Vibrio splendidus* infection, the content of AA gradually decreased, while IBA content progressively increased in the mid-intestine of *Apostichopus japonicus* [51]. Our study also found a significant correlation between the abundance of the *Vibrio* genus and the levels of AA and IBA. On the other hand, various genera within the phyla Firmicutes and Bacteroidota are known to efficiently produce SCFAs, which in turn modulate the intestinal barrier and energy metabolism [44]. The genus *Sediminibacterium*, a member of the phylum Bacteroidota, has been shown to exhibit altered abundance following infections. For instance, *Vibrio cholerae* infection results in a decrease in the relative abundance of *Sediminibacterium* sp. and an increase in various *Vibrio* ASVs in the intestine of zebrafish (*Danio rerio*), with IBA acting as a key metabolite involved in these interactions [52]. Here, our result revealed that the abundance of the genus *Sediminibacterium* was significantly negatively correlated with IBA content. Both *Vibrio* and *Sediminibacterium* genera are commonly regarded as disease biomarkers [53]. We hypothesize that intestinal AA and IBA might exert a crucial function in modulating the abundance of *Vibrio* under the hyperthyroidism status in little yellow croakers. However, the precise regulatory mechanisms by which AA and IBA influence the microbiota remain to be explored.

The gut microbiota profile can provide insights into the metabolic processes and health status of fish [26]. This study found that the intestinal microbiota primarily contributes to metabolic processes, particularly those related to amino acid and energy metabolism. Moreover, genera *Vibrio* and *Sediminibacterium* can modulate the process of amino acid and energy metabolism [54–56]. Notably, in the T4-treated group, the abundance of *Vibrio* and *Sediminibacterium* was significantly different from that in the other groups, suggesting that these genera are involved in the modulation of amino acid and energy metabolism under hyperthyroid conditions induced by T4. In addition, BugBase analysis predicted a higher abundance of aerobic bacteria and a lower abundance of bacteria with pathogenic and stress-tolerant characteristics in the T4 group. However, the underlying mechanisms driving the changes in the intestinal microbiota following dietary T4 supplementation remain unclear and warrant further investigation.

5. Conclusions

In summary, dietary thyroid-active agent (T3, T4, and PTU) supplementation can inhibit growth and regulate the thyroid status in little yellow croakers. These types of thyroid dysfunction result in differential intestinal SCFA profiles. Specifically, dietary supplementation with T3 or PTU had no significant effect on the richness or diversity of the intestinal microbiota in little yellow croakers, whereas dietary T4 supplementation notably increased microbial diversity. Our findings suggested that the genera *Vibrio* and *Sediminibacterium* played key roles in multiple metabolic pathways within the host intestine. Correlation analyses further indicated that intestinal AA and IBA were characteristic metabolites involved in the alteration of the genus *Vibrio*. These results provide a foundation for investigating the effects of thyroid-disrupting activities on the intestinal SCFA profiles and microbiota composition in marine fish. Additionally, the administration dosage of thyroid-active agents may have different impacts on the thyroid–gut microbiota axis, warranting further experimental investigation.

Author Contributions: Methodology, Y.Z.; software, T.Y.; investigation, X.L.; resources, X.L. and T.Y.; data curation, Y.Z.; writing—review and editing, X.L. and F.L.; project administration, X.L.; supervision, B.L. All authors have read and agreed to the published version of the manuscript.

Funding: This research was funded by the Natural Science Foundation of Zhejiang Province, China (LQ22C190007), and the National Natural Science Foundation of China (32303031).

Institutional Review Board Statement: The study was approved by the Committee of Laboratory Animal Experimentation at Zhejiang Academy of Agricultural Sciences (Permit Number: 2023ZA-ASLA29, approved on 22 February 2023).

Informed Consent Statement: Not applicable.

Data Availability Statement: The authors confirm that the data supporting the findings of this study are available within the manuscript, tables, and figures.

Conflicts of Interest: The authors declare no conflicts of interest.

References

1. Deal, C.K.; Volkoff, H. The role of the thyroid axis in fish. *Front. Endo-Crinol.* **2020**, *11*, 596585. [CrossRef]
2. Quesada-García, A.; Encinas, P.; Valdehita, A.; Baumann, L.; Segner, H.; Coll, J.M.; Navas, J.M. Thyroid active agents T3 and PTU differentially affect immune gene transcripts in the head kidney of rainbow trout (*Oncorynchus mykiss*). *Aquat. Toxicol.* **2016**, *174*, 159–168. [CrossRef]
3. Van der Geyten, S.; Byamungu, N.; Reyns, G.E.; Kühn, E.R.; Darras, V.M. Iodothyronine deiodinases and the control of plasma and tissue thyroid hormone levels in hyperthyroid tilapia (*Oreochromis niloticus*). *J. Endocrinol.* **2005**, *184*, 467–479. [CrossRef]
4. Kang, D.Y.; Devlin, R.H. Effects of 3, 5, 3′-triiodo-L-thyronine (T3) and 6-n-propyl-2-thiouracil (PTU) on growth of GH-transgenic coho salmon, *Oncorhynchus kitsutch*. *Fish Physiol. Biochem.* **2003**, *29*, 77–85. [CrossRef]

5. Dey, S.S.; Medda, A.K. Effect of Triiodothyronine and Thyroxine on Mitochondrial α-Glycerophosphate Dehydrogenase Activity and Mitochondrial Protein Content of Liver, Muscle and Brain of Toad, *Bufo melanostictus*. *Horm. Metab. Res.* **1990**, *22*, 418–422. [CrossRef]

6. Quesada-García, A.; Valdehita, A.; Kropf, C.; Casanova-Nakayama, A.; Segner, H.; Navas, J.M. Thyroid signaling in immune organs and cells of the teleost fish rainbow trout (*Oncorhynchus mykiss*). *Fish Shellfish Immunol.* **2014**, *38*, 166–174. [CrossRef]

7. Hodin, R.A.; Shei, A.; Morin, M.; Meng, S. Thyroid hormone and the gut: Selective transcriptional activation of a villus-enterocyte marker. *Surgery* **1996**, *120*, 138–143. [CrossRef]

8. Vítek, L.; Haluzík, M. The role of bile acids in metabolic regulation. *J. Endocrinol.* **2016**, *228*, 85. [CrossRef]

9. Tremaroli, V.; Backhed, F. Functional interactions between the gut microbiota and host metabolism. *Nature* **2012**, *71*, 242–249. [CrossRef]

10. Nicholson, J.K.; Holmes, E.; Kinross, J.; Burcelin, R.; Gibson, G.; Jia, W.; Pettersson, S. Host-gut microbiota metabolic interactions. *Science* **2012**, *336*, 1262–1267. [CrossRef]

11. Lin, L.; Zhang, J. Role of intestinal microbiota and metabolites on gut homeostasis and human diseases. *BMC Immunol.* **2017**, *18*, 2. [CrossRef] [PubMed]

12. Schroeder, B.O.; Bäckhed, F. Signals from the gut microbiota to distant organs in physiology and disease. *Nat. Med.* **2016**, *22*, 1079. [CrossRef] [PubMed]

13. Zhang, Z.; Li, D.; Refaey, M.M.; Xu, W. High spatial and temporal variations of microbial community along the southern catfish gastrointestinal tract: Insights into dynamic food digestion. *Front. Microbiol.* **2017**, *8*, 1531. [CrossRef]

14. Lyte, M.; Miller, V. Microbial endocrinology in the microbiome-gut-brain axis: How bacterial production and utilization of neurochemicals influence behavior. *PLoS Pathog.* **2013**, *9*, e1003726. [CrossRef]

15. Lyte, M. Microbial endocrinology: Host-microbiota neuroendocrine interactions influencing brain and behavior. *Gut Microbes* **2014**, *5*, 381–389. [CrossRef]

16. Foster, J.A.; Lyte, M.; Meyer, E.; Cryan, J.F. Gut microbiota and brain function: An evolving field in neuroscience. *Int. J. Neuropsychopharmacol.* **2016**, *19*, 114. [CrossRef]

17. Kiran, N.S.; Yashaswini, C.; Chatterjee, A. Zebrafish: A trending model for gut-brain axis investigation. *Aquat. Toxicol.* **2024**, *270*, 106902. [CrossRef]

18. Sang, H.R.; Pothoulakis, C.; Mayer, E.A. Principles and clinical implications of the brain-gut-enteric microbiota axis. *Nat. Rev. Gastroenterol. Hepatol.* **2009**, *6*, 306–314.

19. O'Mahony, S.M.; Clarke, G.; Dinan, T.G.; Cryan, J.F. Early-life adversity and brain development: Is the microbiome a missing piece of the puzzle? *Neuroscience* **2017**, *342*, 37–54. [CrossRef]

20. Wang, Y.; Kasper, L.H. The role of microbiome in central nervous system disorders. *Brain Behav. Immun.* **2014**, *38*, 1–12. [CrossRef]

21. Lauritano, E.C.; Bilotta, A.L.; Gabrielli, M.; Scarpellini, E.; Lupascu, A.; Laginestra, A.; Novi, M.; Sottili, S.; Serricchio, M.; Cammarota, G.; et al. Association between hypothyroidism and small intestinal bacterial overgrowth. *J. Clin. Endocrinol. Metab.* **2007**, *92*, 4180–4184. [CrossRef] [PubMed]

22. Zhou, L.; Li, X.; Ahmed, A.; Wu, D.C.; Liu, L.; Qiu, J.J.; Yan, Y.; Jin, M.L.; Xin, Y. Gut microbe analysis between hyperthyroid and healthy individuals. *Curr. Microbiol.* **2014**, *69*, 675–680. [CrossRef] [PubMed]

23. Sa'ad, H.; Peppelenbosch, M.P.; Roelofsen, H.; Vonk, R.J.; Venema, K. Biological effects of propionic acid in humans; metabolism, potential applications and underlying mechanisms. *Biochim. Biophys. Acta Mol. Cell Biol. Lipids* **2010**, *1801*, 1175–1183.

24. Stanley, F.; Samuels, H.H. n-Butyrate effects thyroid hormone stimulation of prolactin production and mRNA levels in GH1 cells. *J. Biol. Chem.* **1984**, *259*, 9768–9775. [CrossRef]

25. Malo, M.S.; Zhang, W.; Alkhoury, F.; Pushpakaran, P.; Abedrapo, M.A.; Mozumder, M.; Fleming, E.; Siddique, A.; Henderson, J.W.; Hodin, R.A. Thyroid hormone positively regulates the enterocyte differentiation marker intestinal alkaline phosphatase gene via an atypical response element. *Mol. Endocrinol.* **2004**, *18*, 1941–1962. [CrossRef]

26. Mendoza-León, M.J.; Mangalam, A.K.; Regaldiz, A.; González-Madrid, E.; Rangel-Ramírez, M.A.; Álvarez-Mardonez, O.; Vallejos, O.P.; Méndez, C.; Bueno, S.M.; Melo-González, F.; et al. Gut microbiota short-chain fatty acids and their impact on the host thyroid function and diseases. *Front. Endocrinol.* **2023**, *14*, 1192216. [CrossRef]

27. Hoseinifar, S.H.; Sun, Y.Z.; Caipang, C.M. Short-chain fatty acids as feed supplements for sustainable aquaculture: An updated view. *Aquac. Res.* **2017**, *48*, 1380–1391. [CrossRef]

28. Li, S.; Heng, X.; Guo, L.; Lessing, D.J.; Chu, W. SCFAs improve disease resistance via modulate gut microbiota, enhance immune response and increase antioxidative capacity in the host. *Fish Shellfish Immunol.* **2022**, *120*, 560–568. [CrossRef]

29. Zhu, B.; Shao, C.; Xu, W.; Dai, J.; Fu, G.; Hu, Y. Effects of thyroid powder on tadpole (*Lithobates catesbeiana*) metamorphosis and growth: The role of lipid metabolism and gut microbiota. *Animals* **2024**, *14*, 208. [CrossRef]

30. Liu, F.; Zhang, T.L.; He, Y.; Zhan, W.; Xie, Q.P.; Lou, B. Integration of transcriptome and proteome analyses reveals the regulation mechanisms of *Larimichthys polyactis* liver exposed to heat stress. *Fish Shellfish Immunol.* **2023**, *135*, 108704. [CrossRef]

31. Liang, X.; Wang, Y.; Liu, L.; Zhang, X.; Li, L.; Tang, R.; Li, D. Acute nitrite exposure interferes with intestinal thyroid hormone homeostasis in grass carp (*Ctenopharyngodon idellus*). *Ecotoxicol. Environ. Saf.* **2022**, *237*, 113510. [CrossRef] [PubMed]

32. Ebrahimi, M.; Daeman, N.H.; Chong, C.M.; Karami, A.; Kumar, V.; Hoseinifar, S.H.; Romano, N. Comparing the effects of different dietary organic acids on the growth, intestinal short-chain fatty acids, and liver histopathology of red hybrid tilapia (*Oreochromis* sp.) and potential use of these as preservatives. *Fish Physiol. Biochem.* **2017**, *43*, 1195–1207. [CrossRef] [PubMed]

33. Yuan, X.Y.; Zhang, X.T.; Xia, Y.T.; Zhang, Y.Q.; Wang, B.; Ye, W.W.; Ye, Z.F.; Qian, S.C.; Huang, M.M.; Yang, S.; et al. Transcriptome and 16S rRNA analyses revealed differences in the responses of largemouth bass (*Micropterus salmoides*) to early *Aeromonas hydrophila* infection and immunization. *Aquaculture* **2021**, *541*, 736759. [CrossRef]

34. Sullivan, C.V.; Darling, D.S.; Dickhoff, W.W. Effects of triiodothyronine and propylthiouracil on thyroid function and smoltification of coho salmon (*Oncorhynchus kisutch*). *Fish Physiol. Biochem.* **1987**, *4*, 121–135. [CrossRef]

35. Ismail, R.F.; Assem, S.S.; Sharaf, H.E.; Zeitoun, A.A.; Srour, T.M. The effect of thyroxine (T4) and goitrogen on growth, liver, thyroid, and gonadal development of red tilapia (*O. mossambicus× O. urolepis hornorum*). *Aquac. Int.* **2024**, *32*, 1151–1168. [CrossRef]

36. Campinho, M.A.; Morgado, I.; Pinto, P.I.; Silva, N.; Power, D.M. The goitrogenic efficiency of thioamides in a marine teleost, sea bream (*Sparus auratus*). *Gen. Comp. Endocrinol.* **2012**, *179*, 369–375. [CrossRef]

37. Corpas, E.; Sánchez-Franco, F.; Correa, R.; Vinales, K.; Larrad-Jiménez, Á.; Blackman, M.R. Physiology and diseases of the thyroid gland in the elderly: Physiological changes, hypothyroidism, and hyperthyroidism. In *Endocrinology of Aging, Corpas, E*; Elsevier: Amsterdam, The Netherlands, 2021; pp. 171–224.

38. Peter, M.C.; Peter, V.S. Action of thyroid inhibitor propyl thiouracil on thyroid and interrenal axes in the freshwater tilapia *Oreochromis mossambicus* Peters. *J. Endocrinol. Reprod.* **2009**, *1*, 37–44.

39. Kumar, J.; Rani, K.; Datt, C. Molecular link between dietary fibre, gut microbiota and health. *Mol. Biol. Rep.* **2020**, *47*, 6229–6237. [CrossRef]

40. Meng, S.; Wu, J.T.; Archer, S.Y.; Hodin, R.A. Short-chain fatty acids and thyroid hormone interact in regulating enterocyte gene transcription. *Surgery* **1999**, *126*, 293–298. [CrossRef]

41. Dobrowolska-Iwanek, J.; Zagrodzki, P.; Prochownik, E.; Jarkiewicz, A.; Paśko, P. Influence of brassica sprouts on short chain fatty acids concentration in stools of rats with thyroid dysfunction. *Acta Pol. Pharm. Drug Res.* **2019**, *76*, 1005–1014. [CrossRef]

42. Yan, K.; Sun, X.; Fan, C.; Wang, X.; Yu, H. Unveiling the role of gut microbiota and metabolites in autoimmune thyroid diseases: Emerging perspectives. *Int. J. Mol. Sci.* **2024**, *25*, 10918. [CrossRef] [PubMed]

43. Butt, R.L.; Volkoff, H. Gut microbiota and energy homeostasis in fish. *Front. Endocrinol.* **2019**, *10*, 9. [CrossRef] [PubMed]

44. Zheng, X.; Xu, X.; Liu, M.; Yang, J.; Yuan, M.; Sun, C.; Zhou, Q.; Chen, J.; Liu, B. Bile acid and short chain fatty acid metabolism of gut microbiota mediate high-fat diet induced intestinal barrier damage in *Macrobrachium rosenbergii*. *Fish Shellfish Immunol.* **2024**, *146*, 109376. [CrossRef] [PubMed]

45. Zhao, H.; Yuan, L.; Zhu, D.; Sun, B.; Du, J.; Wang, J. Alterations and mechanism of gut microbiota in Graves' disease and Hashimoto's thyroiditis. *Pol. J. Microbiol.* **2022**, *71*, 173–189. [CrossRef] [PubMed]

46. Chen, J.; Wang, W.; Guo, Z.; Huang, S.; Lei, H.; Zang, P.; Lu, B.; Shao, J.; Gu, P. Associations between gut microbiota and thyroidal function status in Chinese patients with Graves' disease. *J. Endocrinol. Investig.* **2021**, *44*, 1913–1926. [CrossRef]

47. Deane, E.E.; Li, J.; Woo, N.Y. Hormonal status and phagocytic activity in sea bream infected with vibriosis. *Comp. Biochem. Physiol. B Biochem. Mol. Biol.* **2001**, *129*, 687–693. [CrossRef]

48. Guo, L.; Wang, Z.; Shi, W.; Wang, Y.; Li, Q. Transcriptome analysis reveals roles of polian vesicle in sea cucumber *Apostichopus japonicus* response to *Vibrio splendidus* infection. *Comp. Biochem. Physiol. D Genom. Proteom.* **2021**, *40*, 100877. [CrossRef]

49. Zhang, X.; Hao, X.; Ma, W.; Zhu, T.; Zhang, Z.; Wang, Q.; Liu, K.; Shao, C.; Wang, H.Y. Transcriptome analysis indicates immune responses against *Vibrio harveyi* in Chinese tongue sole (*Cynoglossus semilaevis*). *Animals* **2022**, *12*, 1144. [CrossRef]

50. Dong, X.; Yao, S.; Deng, L.; Li, H.; Zhang, F.; Xu, J.; Li, Z.; Zhang, L.; Jiang, J.; Wu, W. Alterations in the gut microbiota and its metabolic profile of PM2.5 exposure-induced thyroid dysfunction rats. *Sci. Total Environ.* **2022**, *838*, 156402. [CrossRef]

51. Song, M.; Zhang, Z.; Li, Y.; Xiang, Y.X.; Li, C.H. Midgut microbiota affects the intestinal barrier by producing short-chain fatty acids in *Apostichopus japonicus*. *Front. Microbiol.* **2023**, *14*, 1263731. [CrossRef]

52. Breen, P.; Winters, A.D.; Theis, K.R.; Withey, J.H. The *Vibrio cholerae* type six secretion system is dispensable for colonization but affects pathogenesis and the structure of zebrafish intestinal microbiome. *Infect. Immun.* **2021**, *89*, e0015121. [CrossRef] [PubMed]

53. Ma, C.; Guo, H.; Chang, H.; Huang, S.; Jiang, S.; Huo, D.; Zhang, J.; Zhu, X. The effects of exopolysaccharides and exopolysaccharide-producing *Lactobacillus* on the intestinal microbiome of zebrafish (*Danio rerio*). *BMC Microbiol.* **2020**, *20*, 300. [CrossRef] [PubMed]

54. Yu, H.; Sun, L.; Fan, W.; Chen, H.; Guo, Y.; Liu, Y.; Gao, W.; Zhang, W.; Mai, K. Effects of dietary arginine on growth, anti-oxidative enzymes, biomarkers of immunity, amino acid metabolism and resistance to *Vibrio parahaemolyticus* challenge in abalone *Haliotis discus hannai*. *Aquaculture* **2022**, *549*, 737707. [CrossRef]

55. Cheng, Z.X.; Yang, M.J.; Peng, B.; Peng, X.X.; Lin, X.M.; Li, H. The depressed central carbon and energy metabolisms is associated to the acquisition of levofloxacin resistance in *Vibrio alginolyticus*. *J. Proteom.* **2018**, *181*, 83–91. [CrossRef]
56. Liu, P.; Wan, Y.; Zhang, Z.; Ji, Q.; Lian, J.; Yang, C.; Wang, X.; Qin, B.; Yu, J. Toxic effects of combined exposure to cadmium and nitrate on intestinal morphology, immune response, and microbiota in juvenile Japanese flounder (*Paralichthys olivaceus*). *Aquat. Toxicol.* **2023**, *264*, 106704. [CrossRef]

Article

Effects of Melatonin on the Growth and Diurnal Variation of Non-Specific Immunity, Antioxidant Capacity, Digestive Enzyme Activity, and Circadian Clock-Related Gene Expression in Crayfish (*Procambarus clarkii*)

Jinglong Chen [1], Youhai Du [1], Mengyue Zhang [1], Jiahui Wang [1], Jianhua Ming [1,*], Xianping Shao [1], Aimin Wang [2,*], Hongyan Tian [2], Wuxiao Zhang [2], Silei Xia [2], Weigen Cheng [2], Jinlan Xu [2], Xiaochuan Zheng [3] and Bo Liu [3]

[1] Zhejiang Provincial Key Laboratory of Aquatic Resources Conservation and Development, College of Life Sciences, Huzhou University, Huzhou 313000, China; jinglongchen2025@163.com (J.C.)

[2] College of Marine and Biology Engineering, Yancheng Institute of Technology, Yancheng 224051, China

[3] Freshwater Fisheries Research Center, Chinese Academy of Fishery Sciences, Wuxi 214081, China

* Correspondence: mingjh686@zjhu.edu.cn (J.M.); blueseawam@ycit.edu.cn (A.W.)

Academic Editor: Dariel Tovar-Ramírez

Received: 28 January 2025
Revised: 24 February 2025
Accepted: 3 March 2025
Published: 5 March 2025

Citation: Chen, J.; Du, Y.; Zhang, M.; Wang, J.; Ming, J.; Shao, X.; Wang, A.; Tian, H.; Zhang, W.; Xia, S.; et al. Effects of Melatonin on the Growth and Diurnal Variation of Non-Specific Immunity, Antioxidant Capacity, Digestive Enzyme Activity, and Circadian Clock-Related Gene Expression in Crayfish (*Procambarus clarkii*). *Fishes* 2025, 10, 114. https://doi.org/10.3390/fishes10030114

Abstract: This study aimed to investigate the effects of dietary melatonin supplementation on growth and diurnal non-specific immunity, antioxidant capacity, digestive enzyme activities, and circadian clock-related gene expression in crayfish (*Procambarus clarkii*). A total of 500 healthy juvenile crayfish (6.68 ± 0.31 g) were randomly distributed into five groups with four replicates each and fed five different diets supplemented with melatonin at 0, 25, 50, 75, and 100 mg/kg for 60 days. The results indicated that dietary supplementation of 50 mg/kg melatonin significantly increased the weight gain rate (WGR), specific growth rate (SGR), and survival rate (SR) of juvenile *Procambarus clarkii*. However, no significant differences were observed in the hepatosomatic index (HSI), meat yield, and condition factor ($p > 0.05$). When the dietary melatonin level was 50 mg/kg, the activities of LZM and ALP in the hemolymph of *Procambarus clarkii* were higher than the levels at both 15:00 and 03:00, while the activities of AST and ALT remained at lower levels during these two time points. It also significantly upregulated the mRNA expression levels of *Clock*, *Per1*, *Cry1*, *Tim1*, and *Tim2* in the hepatopancreas ($p < 0.05$). Furthermore, dietary melatonin at 50 mg/kg significantly reduced the activities of alanine aminotransferase (ALT) and aspartate aminotransferase (AST), as well as the malondialdehyde (MDA) content across day and night ($p < 0.05$). No significant differences were found in acid phosphatase (ACP) at 15:00, alkaline phosphatase (ALP), and amylase (AMS) activities in the hepatopancreas and intestine at 3:00 among the groups ($p > 0.05$). At 15:00, supplementation with 50 mg/kg significantly upregulated *Bmal1* mRNA expression ($p < 0.05$). Melatonin supplementation at 50–75 mg/kg resulted in significantly higher levels of TP, LZM, ALP, and CAT activities, as well as significantly higher mRNA expression of *Clock*, *Bmal1*, *Cry1*, *Per1*, *Tim1*, and *Tim2* in the hepatopancreas at 3:00 compared to 15:00 ($p < 0.05$), with the opposite trend observed for MDA content ($p < 0.05$). No significant differences were found in ACP, ALT, and AST activities between 3:00 and 15:00 among the groups ($p > 0.05$). Thus, dietary supplementation of 50 mg/kg melatonin could promote the growth of juvenile *Procambarus clarkii*, enhance their non-specific immunity and antioxidant capacity during both day and night, increase the activities of digestive enzymes in the hepatopancreas and intestine, and regulate the expression of circadian clock-related genes.

Keywords: *Procambarus clarkii*; melatonin; non-specific immunity; antioxidant capacity; digestive enzymes; circadian clock gene

Key Contribution: This study aimed to investigate the effects of dietary melatonin supplementation on the growth of *P. clarkii*, as well as on diurnal and nocturnal non-specific immunity, antioxidant capacity, digestive enzyme activities, and circadian clock-related gene expression. The findings intended to provide a reference for the application of melatonin in *P. clarkii* feed and the establishment of precise feeding strategies.

1. Introduction

Melatonin is widely present in various organisms. In vertebrates, melatonin is mainly secreted by the pineal gland, while in invertebrates, it is mainly secreted by the eyestalk ganglion [1]. As a neurotransmitter, melatonin regulates various physiological functions such as antioxidation, immunity, and anti-stress in crustaceans [2–4]. Currently, several studies have investigated the incorporation of melatonin into the feed of aquaculture animals. The addition of an appropriate amount of melatonin to the feed can promote growth and improve non-specific immunity, antioxidant capacity, and intestinal trypsin activity in juvenile black carp (*Mylopharyngodon piceus*) [5]. Yang et al. [6] investigated the effects of melatonin on blood immunity, antioxidant defense, and disease resistance in the Chinese mitten crab (*Eriocheir sinensis*). The study found that incorporating melatonin into the diet significantly enhanced the antioxidant capacity, immune response, and antibacterial properties of the Chinese mitten crab, while also increasing melatonin concentration in eyestalks. Yang et al. [7] added melatonin to the diet of crayfish (*Cherax destructor*), which improved growth performance, enhanced the antioxidant capacity of the hepatopancreas, and elevated the immune indices of the hemolymph. The optimal dosage is between 75 and 81 mg/kg. Similar findings have been observed in Pacific white shrimp (*Litopenaeus vannamei*) [8]. The above results indicated that adding a proper amount of melatonin in feed had effects on the growth, immunity, and antioxidant capacity of aquatic animals. However, the effects of melatonin on the changes of non-specific immunity, antioxidant capacity, and digestive enzyme activity in aquatic animals during the day and night have not been reported.

The biological clock is an endogenous timekeeping system in living organisms [9]. It is widely distributed and found in almost all living organisms, interacting through a complex regulatory network to control downstream genes [10]. Circadian clock genes have been reported to include circadian locomotor output cycle kaput (*Clock*), brain and muscle ARNT-like 1 (*Bmal1*), timeless (*Tim*), period circadian protein 1/2/3 (*Per1/2/3*), pigment dispersion factor (*Pdf*), cryptochrome 1/2 (*Cry1/2*), etc. [11]. The circadian rhythm of the biological clock relies on positive feedback loops involving proteins such as CLOK/BMAL1, negative feedback loops involving proteins such as CRY, and regulation by clock genes such as *Per* and *Tim*. Melatonin can influence the expression of genes associated with the biological clock [12–14]. Previous studies have demonstrated that melatonin affected the expression of genes such as *Clock*, *Bmal1*, *Cry1*, and *Per1* in mice (*Mus musculus*) [15–17]; melatonin also enhanced the expression of *Bmal1* mRNA in humans [18]. Currently, there are few studies on the effects of melatonin on the expression of circadian clock-related genes in aquatic animals.

P. clarkii is favored by aquaculture farmers and consumers due to its strong environmental adaptability, fast growth, delicious meat, and rich nutrition [19,20]. Before the experiment, the feeding rhythm of *P. clarkii* was examined, revealing diurnal variations. This suggests that physiological functions, such as nutrient absorption and digestive enzyme activity, might also exhibit diurnal fluctuations influenced by circadian rhythms. Therefore, understanding how melatonin impacts these diurnal physiological changes is essential. However, the effects of dietary melatonin on the diurnal physiological state changes in *P. clarkii* and the expression of circadian clock genes is still lacking. This study investigated the effects of dietary melatonin supplementation on the growth and diurnal variation of non-specific immunity, antioxidant capacity, digestive enzyme activity, and circadian clock-related gene expression in *P. clarkii*. These findings not only address the research gap concerning the varying effects of melatonin on *P. clarkii* during day and night but also offer a scientific foundation for the development of melatonin-based feed and the establishment of precise feeding strategies and models.

2. Materials and Methods

2.1. Feed Preparation

The composition and nutrient levels of the experimental diets are presented in Table 1. Melatonin (purity $\geq$ 99%) was obtained from Sigma-Aldrich Co., Ltd. (Sigma, St. Louis, MO, USA). The basic feed was formulated using imported fish meal, soybean meal, fermented cottonseed meal, and fermented rapeseed meal as protein sources, along with fish oil and soybean oil as fat sources, flour as a carbohydrate source, and vitamin and mineral premixes. The basal diet was supplemented with melatonin at concentrations of 0 (control group), 25, 50, 75, and 100 mg/kg, respectively. For the preparation of the experimental diets, all dry ingredients were ground through a 60-mesh sieve. The ingredients for each diet were first homogenized, after which oil and distilled water were added to the mixture and thoroughly mixed. This mixture was then pelleted into granules with a diameter of 2 mm using a small feed pelleting machine and dried for approximately 12 h in a ventilated oven at 45 °C. All dry diets were stored at -20 °C until use.

Table 1. Formulation and proximate composition of experimental diets (air-dry basis).

Ingredients	Dietary Melatonin Levels/(mg/kg)				
	0	25	50	75	100
Fish meal	12.00	12.00	12.00	12.00	12.00
Soybean meal	21.00	21.00	21.00	21.00	21.00
Fermented rapeseed meal	16.00	16.00	16.00	16.00	16.00
Fermented cottonseed meal	14.00	14.00	14.00	14.00	14.00
Corn starch	10.00	10.00	10.00	10.00	10.00
Flour	19.55	19.55	19.55	19.55	19.55
Fish oil	1.00	1.00	1.00	1.00	1.00
Soybean oil	2.50	2.50	2.50	2.50	2.50
$Ca(H_2PO_4)_2$	2.00	2.00	2.00	2.00	2.00
Zeolite powder	0.50	0.4975	0.495	0.4925	0.49
Choline chloride	0.40	0.40	0.40	0.40	0.40
Vitamin premix [1]	0.20	0.20	0.20	0.20	0.20
Mineral premix [1]	0.30	0.30	0.30	0.30	0.30
Cholesterin	0.50	0.50	0.50	0.50	0.50
Butylated Hydroxytoluene	0.05	0.05	0.05	0.05	0.05
Melatonin	0.00	0.0025	0.005	0.0075	0.01
Total:	100	100	100	100	100
Nutrient levels (%) [2]					
Moisture	5.52	5.69	5.68	5.42	5.68
Crude protein	33.81	33.79	33.83	33.84	33.86
Ether extract	5.82	5.56	6.07	5.79	6.09
Ash	8.06	8.06	8.05	8.05	8.06

Note: [1] Vitamin premix and mineral premix were provided by Beijing Dabeinong Science and Technology Group Co., Ltd. [2] The following values were measured according to the methods of AOAC.

2.2. Crayfish and Feeding Management

Juvenile red swamp crayfish were obtained from a local aquaculture farm in Huzhou (Huzhou, China). After a week of indoor acclimatization, 500 healthy juvenile *P. clarkii*, with an initial body weight of 6.68 ± 0.31 g, were randomly distributed into 20 tanks, 4 tanks per group, 25 crayfish per tank (300 L, height 78 cm). The tanks were randomly assigned to five experimental groups, each consisting of four replicates. Artificial shelters were provided as hiding places, and regular cleaning and water exchanges were performed. Crayfish were fed to apparent satiation once daily at 18:00. Continuous aeration was maintained using air stones to ensure adequate dissolved oxygen levels. The rearing trial lasted for 60 days. During the experimental period, the feeding trial was conducted under a natural light and dark cycle (12L:12D). The water depth was kept at 20–30 cm, the water temperature ranged from 26 °C to 28 °C, and the water quality parameters were as follows: pH 7.2–7.6, dissolved oxygen (DO) > 6.5 mg/L, and ammonia nitrogen (NH_3-N) < 0.01 mg/L.

2.3. Sampling and Processing

After 60 days of formal feeding, the crayfish were fasted for 24 h and sampled at 15:00 and 3:00, respectively. The crayfish were quickly placed into an ice–water mixture to induce rapid loss of consciousness. Each tank's crayfish were weighed and counted, and three crayfish with similar body weights were randomly selected from each tank. Following this, the crayfish were weighed and measured for body length. Hemolymph was extracted from the posterior margin of the carapace using a disposable medical syringe and mixed with an anticoagulant. The hemolymph samples were then allowed to stand overnight in a refrigerator at 4 °C. Subsequently, the samples were centrifuged at 3000 rpm for 10 min at 4 °C to prepare a serum, which was frozen at −80 °C for later use. After hemolymph collection, the viscera and hepatopancreas were removed and weighed. Appropriate amounts of liver and intestine were taken for routine and molecular biological analysis. After being quickly frozen in liquid nitrogen, these samples were stored at −80 °C.

2.4. Indicator Detection

2.4.1. Determination of Growth Performance

Weight gain rate (WGR), specific growth rate (SGR), feed coefficient (FC), survival rate (SR), hepatosomatic index (HSI), meat ratio (MR), and condition factor (CF) were calculated according to the following formula:

$$\text{Weight gain rate (WGR, \%)} = 100 \times (W_t - W_0)/W_0$$

$$\text{Specific growth rate (SGR, \%/d)} = 100 \times (\ln W_t - \ln W_0)/t$$

$$\text{Feed coefficient (FC)} = F_I/(W_t - W_0)$$

$$\text{Survival rate (SR, \%)} = 100 \times N_t/N_0$$

$$\text{Hepatosomatic index (HSI)} = W_h/(W_t - W_0)$$

$$\text{Meat ratio (MR, \%)} = W_j/W_t \times 100\%$$

$$\text{Condition factor (CF) (g/cm}^3) = W_t/L^3 \times 100$$

In the formula, W_0 (g) is the initial average weight of a crayfish; W_t (g) is the final average weight of a crayfish carcass; t (d) is the number of feeding days; F_I (g) is the average total feed intake per crayfish (air-dried crayfish weight); N_0 is the initial number of crayfish; number of N_t is the final number of crayfish; W_h (g) is the weight of hepatopancreas per crayfish; W_j (g) is the muscle weight of each crayfish; L (cm) is the body length of each crayfish.

2.4.2. Measurement of Biochemical Indicators

At 3:00 and 15:00, measurements were taken for the following parameters in *Procambarus clarkii*: total protein (TP) level, lysozyme (LZM), acid phosphatase (ACP), alkaline phosphatase (ALP), alanine aminotransferase (ALT), and aspartate aminotransferase (AST) activities in the hemolymph; catalase (CAT), glutathione peroxidase (GPx), glutathione reductase (GR) activities, and malondialdehyde (MDA) levels were measured in the hepatopancreas; trypsin, amylase, and lipase activities were assessed in the hepatopancreas and intestine. The above indicators were measured in accordance with the instructions of the kits from Nanjing Jiancheng Bioengineering Institute (Nanjing, China).

2.4.3. Gene Expression Analysis

The total RNA was extracted from the hepatopancreas of *P. clarkii* using the EASY spin Plus Total RNA Extraction Kit (Aidlab, Beijing, China). The purity and concentration were measured using a microvolume spectrophotometer (NanoDrop 2000, Thermo, Waltham, MA, USA) with an OD260/OD280 ratio between 2.1 to 2.2. Subsequently, the total RNA was reverse transcribed to cDNA using the PrimeScript™ RT Reagent Kit (Takara Bio Inc., Kusatsu, Shiga, Japan) and stored at -20 °C. Based on the nucleotide sequences of the *Clock*, *Bmal1*, *Cry1*, *Per1*, *Tim1*, and *Tim2* genes in the *P. clarkii* genome database JAIWQB000000000.1, specific primers for the aforementioned genes and the housekeeping gene β-actin were designed using Primer 5.0 (Table 2). The primers were synthesized by Sangon Biotech Co., Ltd. (Shanghai, China).

Table 2. The primers for real-time qPCR.

Genes	Forward Primer Sequence (5′-3′)	Reverse Primer Sequence (5′-3′)
Clock	F: GGCGGATCAAGTAGTAAACGAG	R: AGCATCAGAACACGGAGAAGG
Bmal1	F: TCCGAATGGCAGTTCAGCA	R: CAACCCAOGACAAACAAGAAAC
Cry1	F: AATGCTGGGTCCTGGATGTG	R: TTCTGGCTCTGCTTGATGTGAT
Per1	F: AATGGGAATAATACTGCCGAGAA	R: GAGCCTTGATOCTGATTGGTG
Tim1	F: AGGAACCCAAGCAATCTCAATG	R: CCAACAACTGCGTCTGTAACCA
Tim2	F: ATCTGTCCACGATCAGGTGTTG	R: CCGCATTTCCAGGAGTTCTTT
β-action	F: ATTCTCACCGAGCGTGGCT	R: AGGCGGCAGTGGTCATTTC

Note: F, forward; R, reverse; *Clock*: circadian locomotor output cycles kaput; *Bmal*1: brain and muscle ARNT-like protein-1; *Cry*1: cryptochrome circadian regulator 1; *Per*1: period circadian regulator 1; *Tim*1: timeless circadian regulator 1; *Tim*2: timeless circadian regulator 2.

Real-time quantitative PCR was performed using the TB Green® Premix Ex Taq™ II (Takara Bio Inc., Kusatsu, Shiga, Japan) kit on a CFX96 Real-Time PCR Detection System (Bio-Rad, Hercules, CA, USA). The PCR reaction system consisted of a total volume of 20 μL, including 10 μL of 2 × TB Green Premix Ex Tag II, 10 μmol/L forward and reverse primers of 0.8 μL each, 1 μL of cDNA template, and 7.4 μL of ddH$_2$O. The amplification conditions were as follows: pre-denaturation at 95 °C for 30 s; denaturation at 95 °C for 5 s; annealing at the optimal temperature for each gene for 30 s for a total of 40 cycles. Under conditions where the amplification efficiencies of each gene and β-actin were approximately equal, the mRNA expression levels of the related genes were compared and analyzed using the $2^{-\Delta\Delta Ct}$ method [21], with the mRNA expression level of the control group as the baseline.

2.5. Data Processing and Analysis

Experimental data were statistically analyzed using SPSS 25. Data from different melatonin supplementation groups at two sampling time points were analyzed using one-way ANOVA and Tukey's multiple comparison test, while *t*-tests were performed for the same melatonin supplementation group at two time points. Results are expressed as means ± standard deviation (Means ± SD), with $p < 0.05$ indicating significant differences.

3. Results

3.1. Effects of Melatonin on Growth Performance of Procambarus clarkii

As shown in Table 3, with the increase of melatonin supplementation in the feed, the weight gain rate, specific growth rate, and survival rate of juvenile *P. clarkii* exhibited a trend of initially rising and then declining, while the feed conversion ratio showed a trend of initially decreasing and then increasing. When the melatonin supplementation level was 50 mg/kg, the weight gain rate, specific growth rate, and survival rate reached relatively high levels, while the feed conversion ratio decreased to a relatively low level ($p < 0.05$). There were no significant differences in the hepatosomatic index, meat yield, and condition factor among the different experimental groups of *P. clarkii* ($p > 0.05$).

Table 3. Effects of melatonin on growth and body indexes of *P. clarkii*

Items	Dietary Melatonin Levels (mg/kg)				
	0	25	50	75	100
IBW [1] (g)	6.90 ± 0.16	6.77 ± 0.39	6.39 ± 0.29	6.55 ± 0.10	6.80 ± 0.37
FBW [1] (g)	19.36 ± 0.34 [bc]	19.99 ± 0.49 [ab]	20.49 ± 0.44 [a]	19.50 ± 0.24 [bc]	19.14 ± 0.31 [c]
WGR [1] (%)	180.57 ± 5.11 [b]	196.13 ± 21.25 [ab]	220.97 ± 8.09 [a]	197.65 ± 2.17 [ab]	182.09 ± 17.68 [b]
SGR [1] (%)	1.71 ± 0.01 [b]	1.84 ± 0.07 [ab]	1.96 ± 0.08 [a]	1.82 ± 0.01 [b]	1.74 ± 0.09 [b]
FCR [1]	0.88 ± 0.05 [ab]	0.83 ± 0.14 [ab]	0.73 ± 0.01 [b]	0.80 ± 0.01 [ab]	0.91 ± 0.09 [a]
SR [1] (%)	79.17 ± 5.32 [ab]	84.72 ± 2.78 [ab]	87.50 ± 5.32 [a]	79.17 ± 2.78 [ab]	77.78 ± 4.54 [b]
HIS [2] (%)	8.01 ± 1.23	7.23 ± 1.11	7.50 ± 0.92	7.56 ± 0.73	7.89 ± 1.57
MR [2] (%)	13.79 ± 2.39	12.98 ± 2.39	13.77 ± 2.13	13.53 ± 2.36	13.61 ± 2.40
CF [2]	2.76 ± 0.28	3.01 ± 0.40	2.81 ± 0.33	2.95 ± 0.31	2.77 ± 0.30

Note: [1]. Values are the mean ± SD (standard deviation) of 4 replicates ($n = 4$). [2]. Values are the mean ± SD of 9 samples. Experimental data were analyzed using SPSS 25 software with one-way ANOVA and Tukey's multiple comparison test. Different superscript letters within the same row indicate significant differences ($p < 0.05$). Identical letters or no superscript indicate no significant difference ($p > 0.05$).

3.2. Effects of Melatonin on Diurnal Variation in Non-Specific Immune Indices and AST and ALT Activities in Procambarus clarkii Hemolymph

As shown in Figure 1A, the total protein (TP) content in the hemolymph of *P. clarkii* increased following the addition of melatonin to the diet. Specifically, TP levels measured at 3:00 and 15:00 were significantly higher than those in the control group ($p < 0.05$). However, no significant differences in TP levels were observed among other dietary groups ($p > 0.05$). Furthermore, when the dietary melatonin concentration ranged from 50 to 100 mg/kg, the TP levels at 3:00 were significantly higher than those measured at 15:00 ($p < 0.05$).

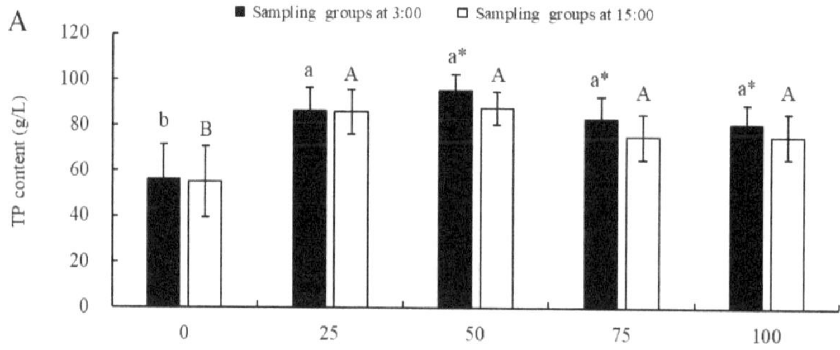

Figure 1. *Cont.*

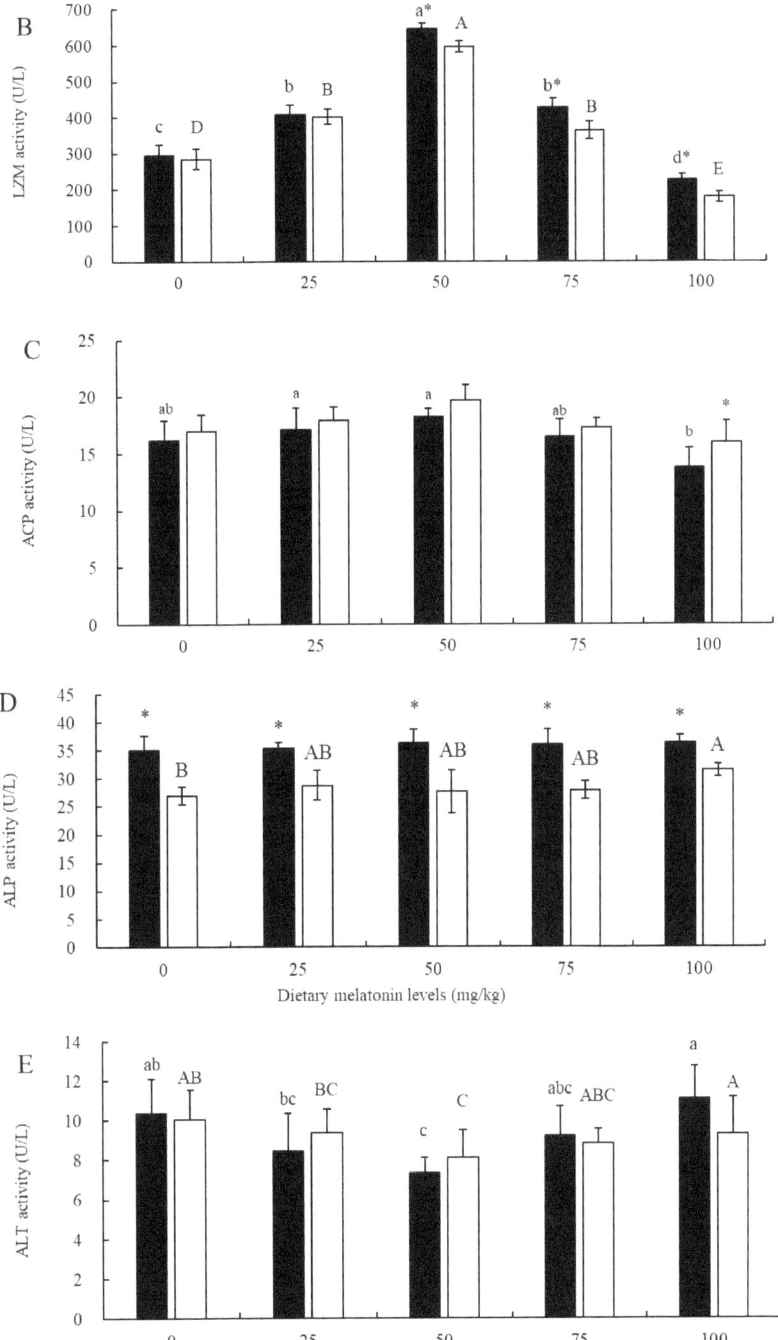

Figure 1. *Cont.*

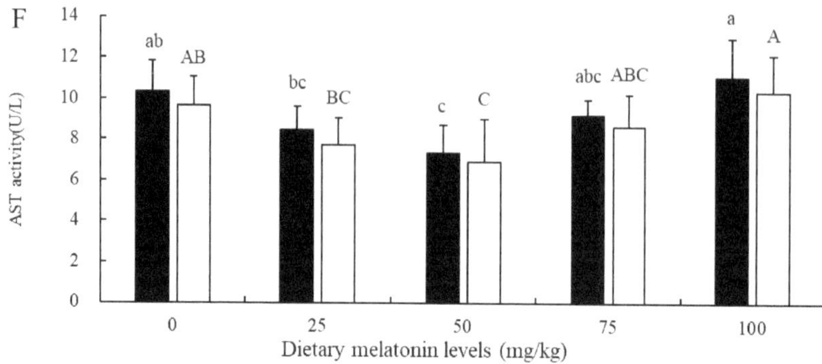

Figure 1. Effects of dietary melatonin on hemolymph levels of TP (**A**), ACP (**B**), ALP (**C**), LZM (**D**), ALT (**E**), and AST (**F**) in *P. clarkii*. Bars represent the means $\pm$ SD ($n = 4$). Asterisks above the bars indicate significant differences between the 3:00 and 15:00 sampling groups within the same melatonin dose group, according to t-test ($p < 0.05$). Different lowercase and uppercase letters above the bars indicate significant differences among different melatonin dose groups in the 3 h and 15 h sampling groups, respectively, according to one-way ANOVA and Tukey's multiple range test ($p < 0.05$). TP, total protein; ACP, acid phosphatase; ALP, alkaline phosphatase; LZM, lysozyme; ALT, alanine aminotransferase; AST, aspartate aminotransferase.

Figure 1B indicated that lysozyme (LZM) activity in the hemolymph increased initially and then decreased with increasing dietary melatonin levels at 3:00 and 15:00. When the melatonin level was 50 mg/kg, LZM activity was significantly higher than in other groups ($p < 0.05$). At dietary melatonin levels of 50–100 mg/kg, LZM activity at 3:00 was significantly higher than at 15:00 ($p < 0.05$).

As shown in Figure 1C, the acid phosphatase (ACP) activity in the hemolymph at 3:00 initially increased and then decreased with rising dietary melatonin levels. At levels of 25–50 mg/kg, ACP activity was higher but showed no significant difference compared to the control group ($p > 0.05$). No significant differences in ACP activity were observed among the groups at 15:00 ($p > 0.05$), nor between 3:00 and 15:00 within any group ($p > 0.05$).

Figure 1D shown that with increasing dietary melatonin levels, no significant differences in alkaline phosphatase (ALP) activity were observed among the groups at 3:00 ($p > 0.05$). However, at 15:00, ALP activity exhibited an increasing trend, with levels significantly higher than the control group when melatonin was added at 100 mg/kg ($p < 0.05$), though no significant differences were noted among the other groups ($p > 0.05$). ALP levels at 3:00 were significantly higher than at 15:00 across all groups ($p < 0.05$).

Figure 1E,F illustrated that the activities of alanine aminotransferase (ALT) and aspartate aminotransferase (AST) in the hemolymph showed a decreasing and then increasing trend with the addition of dietary melatonin. When melatonin was at 50 mg/kg, both ALT and AST activities were at relatively low levels ($p < 0.05$). No significant differences were observed in ALT and AST activities between 3:00 and 15:00 within any group ($p > 0.05$).

3.3. Effects of Melatonin on Diurnal Antioxidant Capacity in Procambarus clarkii Hepatopancreas

As shown in Figure 2A, increasing melatonin supplementation in the feed, catalase (CAT) activity within the hepatopancreas of *P. clarkii*, showing an initial increase followed by a subsequent decrease at both 3:00 and 15:00. Notably, at a melatonin supplementation level of 50 mg/kg, CAT activities were significantly elevated compared to the other groups ($p < 0.05$). Furthermore, at melatonin supplementation levels ranging from 50 to 100 mg/kg, CAT activity at 3:00 was significantly greater than that observed at 15:00 ($p < 0.05$).

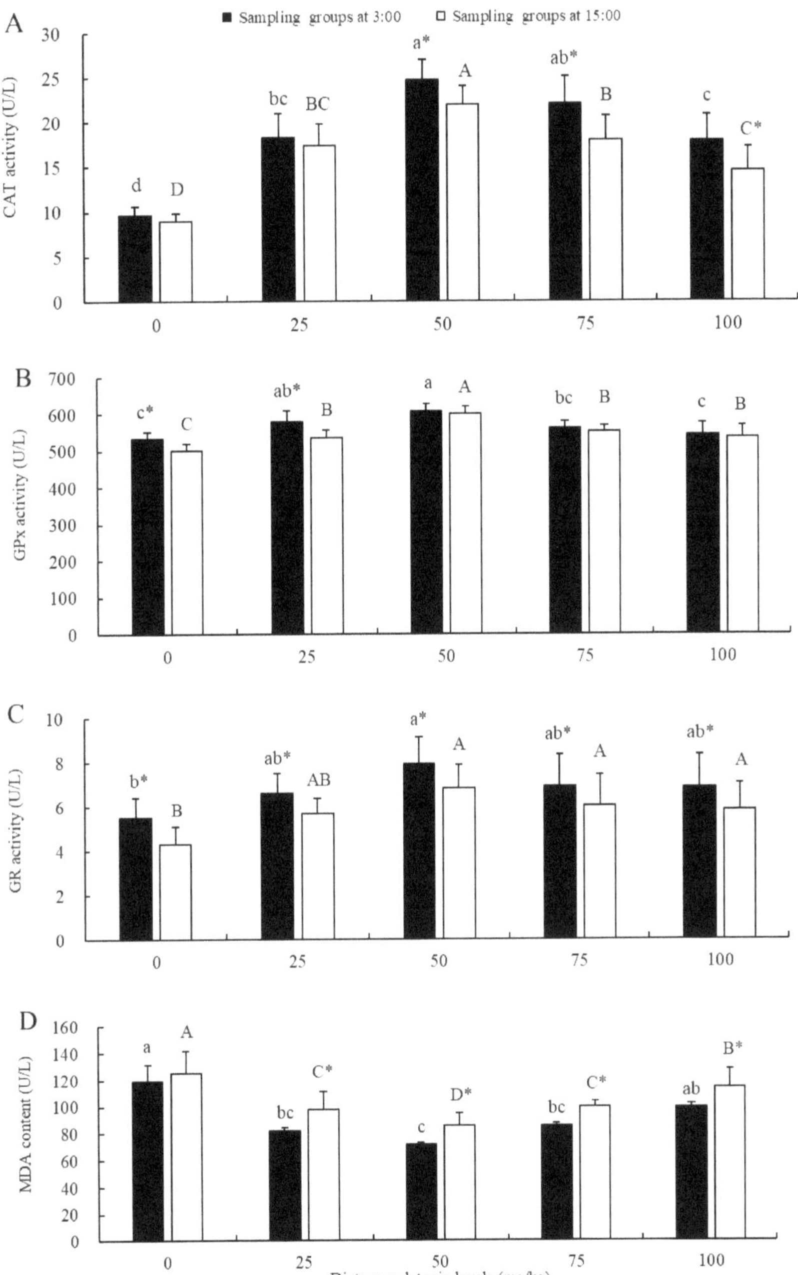

Figure 2. Effects of dietary melatonin on hepatopancreatic levels of CAT (**A**), GPx (**B**), GR (**C**), and MDA (**D**) of *P. clarkii*. Bars represent the means $\pm$ SD (n = 4). Asterisks above the bars indicate significant differences between the 3 h and 15 h sampling groups within the same melatonin dose group, according to t-test ($p < 0.05$). Different lowercase and uppercase letters above the bars indicate significant differences among different melatonin dose groups in the 3 h and 15 h sampling groups, respectively, according to one-way ANOVA and Tukey's multiple range test ($p < 0.05$). CAT, catalase; GPx, glutathione peroxidase; GR, glutathione reductase; MDA, malondialdehyde.

Figure 2B,C indicated that with increasing melatonin supplementation, the activities of glutathione peroxidase (GPx) and glutathione reductase (GR) in the hepatopancreas of *P. clarkii* at 3:00 and 15:00 exhibit an initial increase followed by a decrease. At a melatonin level of 50 mg/kg, both GPx and GR activities reached relatively high levels ($p < 0.05$). At melatonin supplementation levels of 0–25 mg/kg, GPx and GR activities at 3:00 were significantly higher than those at 15:00 ($p < 0.05$).

As illustrated in Figure 2D, increasing levels of melatonin supplementation resulted in a decrease in malondialdehyde (MDA) content within the hepatopancreas of *P. clarkii* at both 3:00 and 15:00, followed by an increase. At a melatonin supplementation level of 50 mg/kg, MDA content reached a relatively low level ($p < 0.05$). Furthermore, at melatonin supplementation levels ranging from 25 to 100 mg/kg, MDA content at 3:00 was significantly lower than that at 15:00 ($p < 0.05$).

3.4. Effects of Melatonin on Diurnal Digestive Enzyme Activities in Procambarus clarkii Hepatopancreas and Intestines

As shown in Figure 3A,B, increasing the level of melatonin supplementation in the feed initially enhances the activities of trypsin (TPS) and lipase (LPS) in the hepatopancreas of *P. clarkii*, followed by a subsequent decline. When the melatonin supplementation level ranged from 50 to 100 mg/kg, TPS and LPS activities reached significantly elevated levels ($p < 0.05$). At melatonin supplementation levels of 0 to 25 mg/kg, TPS activity at 3:00 was significantly higher than at 15:00 ($p < 0.05$). For LPS, at melatonin levels of 0 to 50 mg/kg, activity at 3:00 was also significantly higher than at 15:00 ($p < 0.05$).

Figure 3C demonstrated that with increasing melatonin supplementation, there were no significant differences in amylase (AMS) activity among the groups at 3:00 ($p > 0.05$). At 15:00, AMS activity in the hepatopancreas of *P. clarkii* initially increases and then decreases, reaching a relatively high level at a melatonin supplementation of 50 mg/kg ($p < 0.05$). At melatonin levels of 25–50 mg/kg, AMS content at 3:00 is significantly higher than at 15:00 ($p < 0.05$).

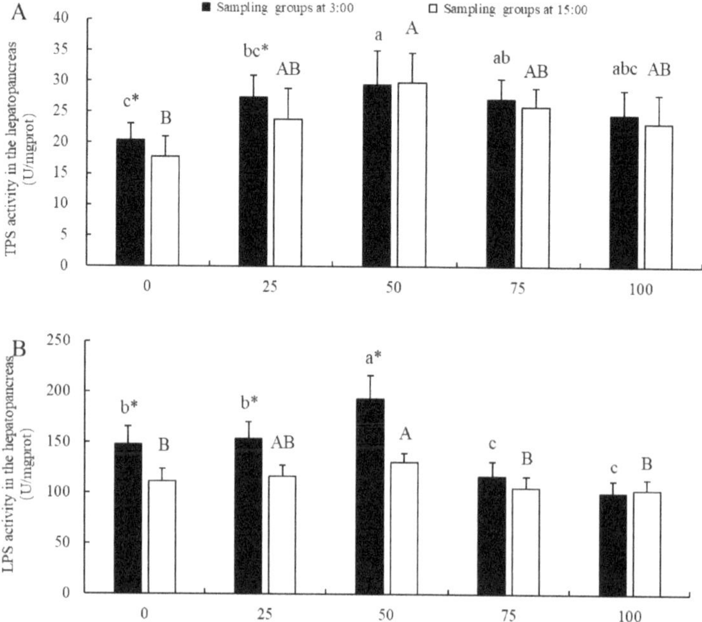

Figure 3. *Cont.*

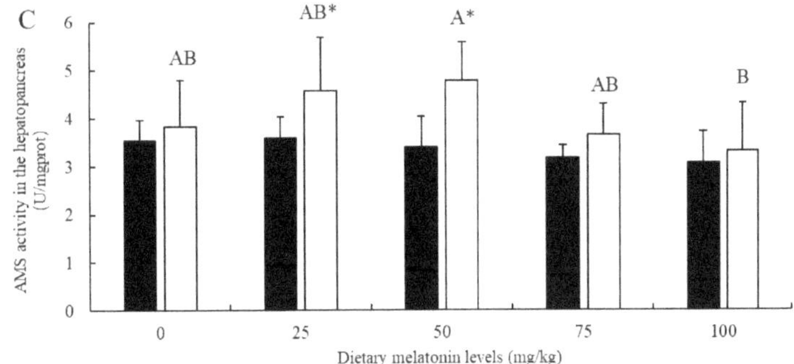

Figure 3. Effects of melatonin on the activities of TPS (**A**), LPS (**B**), and ALS (**C**) in the hepatopancreas of *P. clarkii*. Bars represent the means $\pm$ SD (n = 4). Asterisks above the bars indicate significant differences between the 3 h and 15 h sampling groups within the same melatonin dose group, according to *t*-test ($p < 0.05$). Different lowercase and uppercase letters above the bars indicate significant differences among different melatonin dose groups in the 3 h and 15 h sampling groups, respectively, according to one-way ANOVA and Tukey's multiple range test ($p < 0.05$) TPS, trypsin; LPS, lipase; AMS, amylase.

Figure 4A,B demonstrated that with increasing melatonin supplementation in the feed, the activities of trypsin (TPS) and lipase (LPS) in the hepatopancreas of *P. clarkii* initially increased and then decreased. When the melatonin supplementation level was 50–100 mg/kg, TPS and LPS activities reached relatively high levels ($p < 0.05$). At melatonin supplementation levels of 0–25 mg/kg, TPS activity at 3:00 was significantly higher than at 15:00 ($p < 0.05$). For LPS, at melatonin levels of 0–50 mg/kg, activity at 3:00 was also significantly higher than at 15:00 ($p < 0.05$).

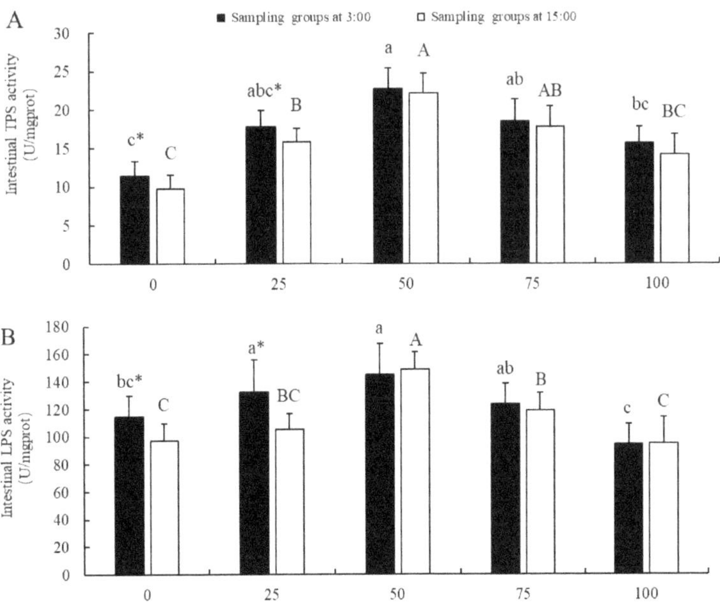

Figure 4. *Cont.*

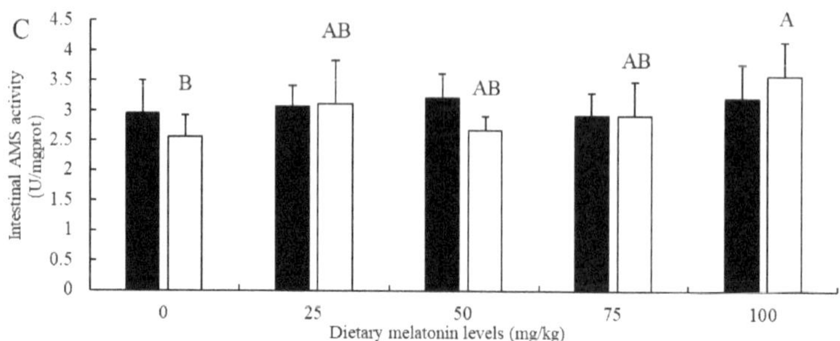

Figure 4. Effects of melatonin on intestinal TPS (**A**), LPS (**B**) and ALS (**C**) activities of *P. clarkii*. Bars represent the means ± SD (*n* = 4). Asterisks above the bars indicate significant differences between the 3 h and 15 h sampling groups within the same melatonin dose group, according to *t*-test ($p < 0.05$). Different lowercase and uppercase letters above the bars indicate significant differences among different melatonin dose groups in the 3h and 15 h sampling groups, respectively, according to one-way ANOVA and Tukey's multiple range test ($p < 0.05$). TPS, trypsin; LPS, lipase; AMS, amylase.

As illustrated in Figure 4C, there were no significant differences in amylase (AMS) activity among the groups at 3:00 with increasing melatonin supplementation ($p > 0.05$). However, at 15:00, AMS activity in the hepatopancreas of *P. clarkii* initially increased and subsequently decreased, reaching a relatively high level at a melatonin supplementation of 50 mg/kg ($p < 0.05$). Furthermore, at melatonin levels of 25–50 mg/kg, AMS content at 3:00 was significantly higher than that at 15:00 ($p < 0.05$).

3.5. Effects of Melatonin on Expression of Circadian Clock-Related Genes in Procambarus clarkii

As illustrated in Figure 5A, increasing levels of melatonin supplementation in the feed correspond to a rising trend in the mRNA expression of *Clock* in *P. clarkii* at 3:00. At melatonin supplementation levels of 50–100 mg/kg, the *Clock* mRNA expression reached significantly elevated levels ($p < 0.05$). At 15:00, the *Clock* mRNA expression initially increased before subsequently decreasing, with the expression at 50 mg/kg being significantly higher than that in the other groups ($p < 0.05$). Furthermore, at melatonin levels of 25–75 mg/kg, the Clock mRNA expression at 3:00 was significantly lower than that observed at 15:00 ($p < 0.05$).

Figure 5B,C illustrated that with increasing melatonin supplementation, there were no significant differences in the mRNA expression of *Bmal1* and *Cry1* among the groups at 3:00 ($p > 0.05$). At 15:00, the mRNA expression of *Bmal1* and *Cry1* initially increased before subsequently decreasing. The mRNA expression of *Bmal1* was significantly higher at 50 mg/kg compared to the other groups ($p < 0.05$), while the mRNA expression of *Cry1* reached relatively elevated levels at dosages of 50–75 mg/kg ($p < 0.05$). At melatonin dosages of 50–100 mg/kg, the mRNA expression of *Bmal1* at 3:00 was significantly lower than that observed at 15:00 ($p < 0.05$). Furthermore, the mRNA expression of *Cry1* at 3:00 was significantly lower than at 15:00 for melatonin dosages of 0 and 50–100 mg/kg ($p < 0.05$).

As illustrated in Figure 5D, increasing melatonin supplementation resulted in an initial rise in *Per1* mRNA expression in *P. clarkii*, followed by a subsequent decline. At 3:00, *Per1* mRNA expression reached relatively high levels with melatonin dosages of 50–100 mg/kg ($p < 0.05$). By 15:00, *Per1* mRNA expression was significantly higher at 50 mg/kg compared to the other groups ($p < 0.05$). Furthermore, at melatonin levels of 50–100 mg/kg, *Per1* mRNA expression at 3:00 was significantly lower than at 15:00 ($p < 0.05$).

Figure 5E illustrated that with increasing melatonin supplementation, the expression of *Tim1* mRNA in *P. clarkii* initially rose and then declined. At 3:00, the *Tim1* mRNA expression was significantly higher at a dosage of 25 mg/kg compared to the other groups ($p < 0.05$). At 15:00, the *Tim1* mRNA expression was significantly elevated at a dosage of 75 mg/kg relative to the other groups ($p < 0.05$). Furthermore, at melatonin levels ranging from 50 to 100 mg/kg, the *Tim1* mRNA expression at 3:00 was significantly lower than that observed at 15:00 ($p < 0.05$).

Figure 5F illustrated that with increasing melatonin supplementation, the expression of *Tim2* mRNA in *P. clarkii* initially increased, then declined. At 3:00, the *Tim2* mRNA expression at a dosage of 50 mg/kg was significantly higher than that in the other groups ($p < 0.05$). Conversely, at 15:00, the *Tim2* mRNA expression was significantly elevated at a dosage of 25 mg/kg compared to the other groups ($p < 0.05$). Furthermore, at melatonin levels ranging from 25 to 100 mg/kg, the *Tim2* mRNA expression at 3:00 was significantly lower than that at 15:00 ($p < 0.05$).

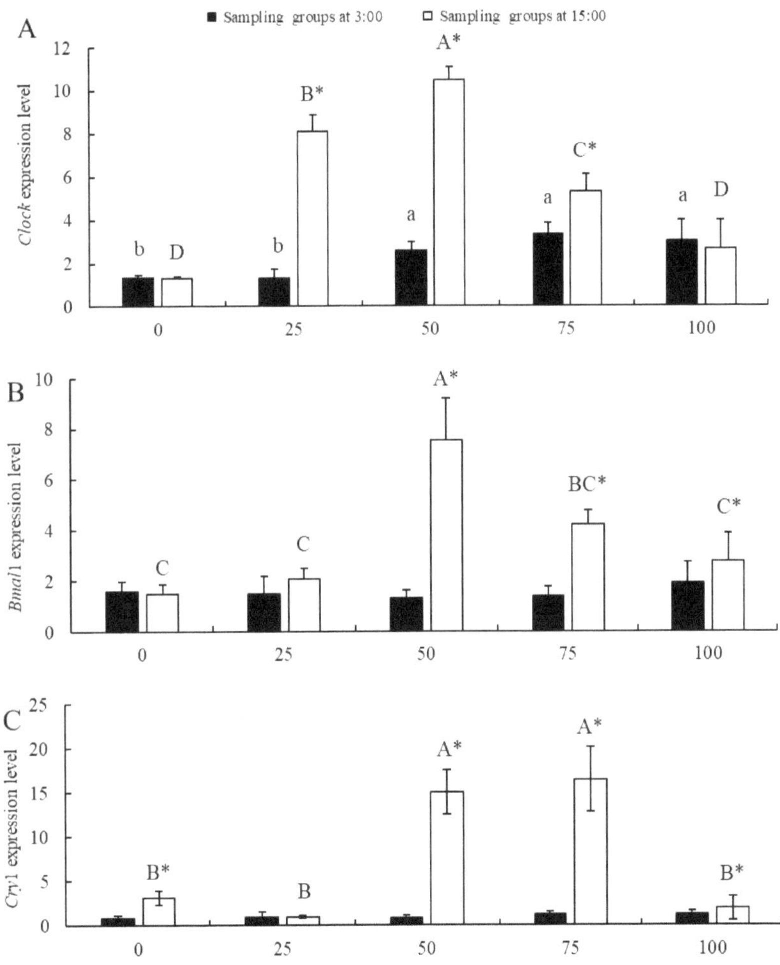

Figure 5. *Cont.*

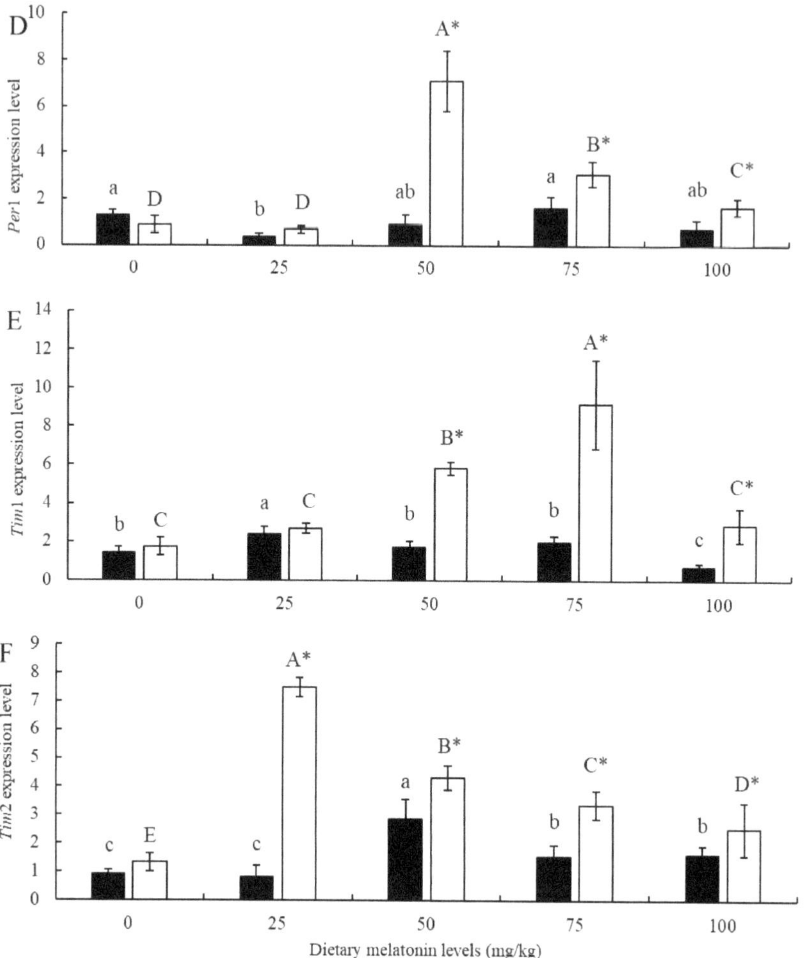

Figure 5. Effects of melatonin on the expression of circadian clock-related genes *Clock* (**A**), *Bmal1* (**B**), *Cry1* (**C**), *Per1* (**D**), *Tim1* (**E**), and *Tim2* (**F**) of crayfish (*P. clarkii*). Bars represent the means ± SD ($n = 4$). Asterisks above the bars indicate significant differences between the 3 h and 15 h sampling groups within the same melatonin dose group, according to t-test ($p < 0.05$). Different lowercase and uppercase letters above the bars indicate significant differences among different melatonin dose groups in the 3 h and 15 h sampling groups, respectively, according to one-way ANOVA and Tukey's multiple range test ($p < 0.05$). *Clock*: circadian locomotor output cycles kaput; *Bmal1*: brain and muscle ARNT-like protein-1; *Cry1*: cryptochrome circadian regulator 1; *Per1*: period circadian regulator 1; *Tim1*: timeless circadian regulator 1; *Tim2*: timeless circadian regulator 2.

4. Discussion

4.1. Effects of Melatonin on Growth Performance of Procambarus clarkii

The study of exogenous melatonin in mammals is relatively extensive, demonstrating significant improvements in milk production in dairy cows [5], as well as in weight gain and feed efficiency in domestic pigs (*Sus scrofa*) [22,23] and chickens (*Gallus gallus domesticus*) [24,25]. However, research on melatonin in aquaculture species is comparatively limited. Studies have shown that melatonin injections can enhance the growth performance of goldfish (*Carassius auratus*) and turbot (*Scophthalmus maximus*) [26,27]. Dietary supplementation with 9.28 mg/kg of melatonin has been found to promote growth in juvenile

M. piceus [5], while a dosage of 82.7 mg/kg of melatonin in the diet significantly increased the survival rate, weight gain, and specific growth rate in *L. vannamei* [28]. In Asian swamp eel (*Monopterus albus*), a diet containing 120 mg/kg of melatonin significantly improved weight gain and specific growth rates while reducing the feed conversion ratio [29]. *C. destructor* fed with melatonin at 75–81 mg/kg exhibited enhanced growth performance, as reported by Yang et al. [7]. In the current study, the addition of 50 mg/kg of melatonin to the diet of juvenile *P. clarkii* significantly enhanced their weight gain, specific growth rate, and survival rate, while simultaneously reducing the feed conversion ratio. The results indicated that dietary melatonin at 50 mg/kg could improve the growth performance of *P. clarkii*. Various researchers have proposed different optimal levels of melatonin supplementation, which may be influenced by factors such as species, growth stages, and the aquacultural environment of the studied organisms. Further research is necessary to elucidate the specific reasons and underlying mechanisms.

4.2. Effects of Melatonin on Diurnal Variation in Non-Specific Immune Indices and AST, ALT Activities in Procambarus clarkii Hemolymph

Dietary supplementation with melatonin not only influences the growth of aquatic animals but also significantly impacts non-specific immunity. The total protein (TP) content in hemolymph serves as an indicator of protein absorption and utilization in aquatic species [30]. Additionally, lysozyme (LZM), acid phosphatase (ACP), and alkaline phosphatase (ALP) are closely associated with the humoral immunity of crustaceans [31]. LZM can exert bactericidal effects indirectly by hydrolyzing the cell walls of pathogens and facilitating phagocytosis [32,33]. ACP acts as a marker enzyme for macrophage lysosomes [34], while ALP modifies the surface structure of pathogenic bacteria, thereby enhancing the organism's ability to recognize and phagocytize them [35]. Variations in the activities of alanine aminotransferase (ALT) and aspartate aminotransferase (AST) can indicate damage to the hepatopancreas [36]. Li et al. [37] found that dietary supplementation with melatonin at 41.2 mg/kg significantly increased LZM activity in the hemolymph of *L. vannamei*. Yang et al. [6] reported that adding 50.80 mg/kg of melatonin to the diet could enhance ALP activity in the hemolymph of *E. sinensis* post-challenge. Furthermore, Yang et al. [38] indicated that dietary melatonin supplementation of 75–81 mg/kg significantly reduced AST and ALT activities in *C. destructor*. These studies explored the effect of dietary melatonin on various non-specific immune parameters in different aquatic animals, but the impact of melatonin on the diurnal non-specific immunity of aquatic animals has not been reported. In the current study, increased levels of dietary melatonin were associated with a trend of first increasing and then decreasing TP content and LZM activity in the hemolymph of *P. clarkii* at 03:00 and 15:00. The activity of ACP at 03:00 exhibited a similar pattern. At a dietary melatonin level of 50 mg/kg, the activities of LZM and ALP in the hemolymph of *P. clarkii* remained elevated at both 15:00 and 03:00, while the activities of AST and ALT were comparatively lower at these two time points. Consistent with these findings, dietary melatonin supplementation promoted the enhancement of diurnal non-specific immunity and supported hepatopancreatic health in *P. clarkii*. At dietary melatonin levels ranging from 50 to 100 mg/kg, TP and LZM levels at 03:00 were significantly higher than those at 15:00. This study demonstrated that melatonin supplementation enhances both diurnal and nocturnal non-specific immune responses in juvenile *P. clarkii*.

4.3. Effects of Melatonin on Diurnal Antioxidant Capacity in Procambarus clarkii Hepatopancreas

The hepatopancreas is the primary organ responsible for energy storage and digestion in crustaceans, performing essential functions such as lipid storage, digestive enzyme secretion, and detoxification [39]. Melatonin acts as a potent direct free radical scavenger and serves as an indirect antioxidant [40,41]. Catalase (CAT) plays a crucial role in protecting cells from oxidative damage by catalyzing the decomposition of hydrogen peroxide (H_2O_2). Glutathione reductase (GR) is responsible for reducing oxidized glutathione disulfide (GSSG) back to reduced glutathione (GSH), while glutathione peroxidase (GPx) is an antioxidant enzyme that decomposes peroxides, including H_2O_2 [42]. In aquatic animals, the intestinal concentration in Indian carp (*Catla catla*) is positively correlated with the activities of CAT and GPx [43]. Yang et al. [6] demonstrated that the addition of 50.80 mg/kg of melatonin to the feed significantly increased the activities of CAT and GPx in the hepatopancreas of *F. sinensis* under stress conditions. Furthermore, the inclusion of 75–81 mg/kg of melatonin in the feed enhanced GR activity in the hepatopancreas of *C. destructor* [7]. Additionally, in *E. sinensis*, the addition of 50.80 mg of melatonin significantly increased the activities of CAT and GPx in the gills after 4 h of hypoxia, while also alleviating the elevated malondialdehyde (MDA) content resulting from hypoxia [44,45]. These experimental results investigated the effects of melatonin supplementation in feed on antioxidant capacity and other indicators in the hepatopancreas of various aquatic animals. However, the effects of melatonin on the diurnal antioxidant capacity of the hepatopancreas in aquatic animals have not been reported. In this study, with melatonin supplementation in the feed increased, the activities of CAT, GPx, and GR in the hepatopancreas of *P. clarkii* exhibited a trend of initially increasing and then decreasing at 3:00 and 15:00, while MDA displayed the opposite trend. At a melatonin concentration of 50 mg/kg, the activities of CAT, GPx, and GR at both 15:00 and 3:00 were relatively higher levels, whereas MDA content was at a comparatively lower level, aligning with previous studies. Melatonin supplementation in the feed appears to enhance the diurnal and nocturnal antioxidant capacity of the hepatopancreas in *P. clarkii*. Furthermore, this study demonstrated that when melatonin supplementation ranged from 50 to 100 mg/kg, CAT activity at 3:00 was significantly higher than that at 15:00, while MDA content showed the opposite pattern. Additionally, when melatonin supplementation was between 0 and 25 mg/kg, GPx and GR activities at 3:00 were significantly greater than those at 15:00. These results indicate that melatonin administration significantly enhanced the antioxidant capacity in juvenile *P. clarkii* across circadian cycles.

4.4. Effects of Melatonin on Diurnal Digestive Enzyme Activities in Procambarus clarkii Hepatopancreas and Intestines

Trypsin (TPS), lipase (LPS), and amylase (AMS) are common digestive enzymes found in crustaceans, involved in the digestion and absorption of food [46]. In *C. catla*, endogenous melatonin levels are positively correlated with the activity of these digestive enzymes [44]. Supplementing 120 mg/kg of melatonin in the feed significantly enhanced the activities of amylase, trypsin, and lipase in the intestines of *M. albus* [29]. Additionally, injecting 20 μL of melatonin into *E. sinensis* markedly promoted the activities of trypsin and lipase in the hepatopancreas [47]. Furthermore, incorporating 50.80 mg/kg of melatonin into the feed significantly increased the activities of trypsin and amylase in the intestines of *E. sinensis* under conditions of ammonia nitrogen stress [2]. Similarly, adding 80 mg/kg of melatonin significantly boosted the activities of trypsin and amylase in the intestines of *E. sinensis* under glyphosate stress [48]. Mardones et al. [49] conducted an experiment in which silver salmon (*Oncorhynchus kisutch*) were fed with feed immersed in melatonin solutions at concentrations of 0.002%, 0.01%, and 0.05% for 10 days, resulting in an increase in intestinal TPS activity. These studies explored that fish and crustaceans exhibit similarities in the types and functions of digestive enzymes, as well as in the fundamental mechanisms of enzyme secretion and regulation. These experimental results explored the effects of exogenous melatonin on the activities of digestive enzymes in the hepatopancreas and intestines of various aquatic animals. However, studies examining the influence of melatonin on the diurnal and nocturnal activity of digestive enzymes in these organs have not been reported. In this experiment, increased melatonin supplementation in the feed resulted in the activities of TPS and LPS in the hepatopancreas and intestines of *P. clarkii* exhibiting a trend of initially increasing followed by decreasing activity. When melatonin was added at 50 mg/kg, the activities of TPS and LPS in the hepatopancreas and intestines at 15:00 and 03:00 reached relatively high levels. At 15:00, the activity of AMS in the hepatopancreas of *P. clarkii* also demonstrated a trend of first increasing and then decreasing, peaking at the same melatonin concentration of 50 mg/kg. This finding aligned with previous studies, suggesting that melatonin supplementation in the feed enhances the levels of digestive enzymes in the hepatopancreas and intestines of *P. clarkii* during both day and night, thereby improving nutrient digestion capabilities. Furthermore, this experiment revealed that when melatonin was added from 0 to 25 mg/kg, the activities of TPS and LPS in the intestines and hepatopancreas of *P. clarkii* at 03:00 were significantly higher than those at 15:00. However, when melatonin supplementation exceeded 25 mg/kg, the difference in enzyme activity levels between 03:00 and 15:00 was not significant. Initial findings indicate that ideal dietary melatonin levels could enhance the activities of digestive enzymes in *P. clarkii* during both light and dark phases.

4.5. Effects of Melatonin on the Expression of Circadian Clock-Related Genes in Procambarus clarkii

The effects of melatonin on the expression of circadian clock-related genes have been extensively studied in mammals, but there were relatively few reports on aquatic animals. Circadian clock-related genes maintain the homeostasis of circadian rhythms by regulating melatonin secretion, which, in turn, can affect the expression of these genes [50,51]. Studies have shown that melatonin influences the expression of genes such as *Clock, Bmal1, Cry1,* and *Per1* by regulating cumulus–oocyte complexes in *Mus musculus* [15,16]. Kandalepas et al. [17] directly applied 1 μL of 1 nM melatonin to *Mus musculus* suprachiasmatic nucleus (SCN) slices, finding that melatonin induced an increase in *Per1* mRNA expression. When used as a treatment for improving sleep conditions in Parkinson's patients, melatonin increased the expression of *Bmal1* mRNA [18]. These studies explored the effects of exogenous melatonin on the expression of circadian clock-related genes in different animals, but there have been no reports on the effects of melatonin on the circadian expression of circadian clock-related genes in aquatic animals during day and night. In this experiment, with increasing melatonin supplementation in the feed, the circadian mRNA expression levels of *Clock, Bmal1, Cry1,* and *Per1* in the hepatopancreas of *P. clarkii* at 15:00 and 3:00 showed a trend of first increasing and then decreasing. When the melatonin addition was 50 mg/kg, the mRNA expression levels of *Clock, Bmal1, Cry1,* and *Per1* in the hepatopancreas at 15:00 and 3:00 were significantly upregulated. At 3:00, a melatonin addition of 25 mg/kg in the feed significantly upregulated the expression level of *Tim1* in *P. clarkii*. At 15:00, a melatonin addition of 50 mg/kg significantly upregulated the expression level of *Bmal1* mRNA in *P. clarkii*. These results are similar to the aforementioned studies, indicating that melatonin supplementation in the feed upregulates the expression of circadian clock-related genes during day and night. Furthermore, this study found that with increased melatonin supplementation, the mRNA levels of *Clock, Bmal1, Cry1, Per1, Tim1,* and *Tim2* at 3:00 were significantly lower than those at 15:00. We hypothesize that melatonin-enriched feed may synchronize circadian clock gene expression in *P. clarkii*, thereby maintaining circadian homeostasis through transcriptional-translational feedback loops. The results of this study revealed that melatonin significantly influenced the expression of circadian clock-related genes in *P. clarkii*, providing new insights into the regulatory mechanisms of circadian rhythms in aquatic invertebrates. Particularly in the context of aquaculture, this finding offers a theoretical basis and practical direction for using exogenous melatonin to regulate circadian rhythms in crustaceans.

5. Conclusions

Thus, the diet supplemented with melatonin 50 mg/kg could promote the growth of *Procambarus clarkii*, enhance diurnal non-specific immunity, antioxidant capacity, hepatopancreas, and intestinal digestive enzyme activities, and regulate the expression of circadian clock-related genes to maintain its circadian rhythm homeostasis. However, further study is required to explore how melatonin can improve the sustainability of aquaculture.

Author Contributions: Conceptualization, J.M., J.C. and A.W.; methodology, J.M. and A.W.; software, J.C., J.W., M.Z. and Y.D.; validation, A.W., H.T., W.Z. and S.X.; formal analysis, Y.D., M.Z., X.Z. and B.L.; investigation, J.W., X.S. and B.L.; resources, J.M., W.C., J.X., X.Z. and B.L.; data curation, J.C. and X.S.; writing—original draft preparation, J.M. and J.C.; writing—review and editing, J.M. and J.C.; visualization, H.T., S.X., J.X., W.Z. and B.L.; supervision, W.C. and X.Z.; project administration, J.M. and A.W. All authors reviewed the manuscript. All authors have read and agreed to the published version of the manuscript.

Funding: This research was funded by the Research on Public Welfare Technology Application of Science and Technology Project of Huzhou in China (2024GZ30), the National Key Research and Development Plan project (2023YFD2402000), the National Natural Science Foundation of China (No. 32102767), the Yancheng Fishery High-Quality Development Project (2022yc003), the Jiangsu Modern Agricultural Industrial Technology System Construction Special Fund (JATS [2023] 471), the Postgraduate Research and Innovation Project of Huzhou University (2024KYCX98) and Zhejiang Provincial College Student Innovation and Entrepreneurship Training Program (S202410347060).

Institutional Review Board Statement: Animal procedures were performed in accordance with the regulations and approved by the Institutional Animal Care and Use Committee of Huzhou University (approval ID: ZJHU-DW-2024-032; approval date: 18 March 2024).

Informed Consent Statement: Not applicable.

Data Availability Statement: Research data are available upon request to the authors.

Acknowledgments: The authors thank the editors and reviewers for their valuable comments and suggestions.

Conflicts of Interest: The authors report no declarations of conflicts of interest.

References

1. Carrillo-Vico, A.; Lardone, P.J.; Álvarez-Sánchez, N.; Rodríguez-Rodríguez, A.; Guerrero, J.M. Melatonin: Buffering the immune system. *Int. J. Mol. Sci.* **2013**, *14*, 8638–8683. [CrossRef]
2. Foster, R.G. Melatonin. *Curr. Biol.* **2021**, *31*, 1456–1458. [CrossRef] [PubMed]
3. Miller, S.C.; Pandi, P.S.; Esquifino, A.I.; Cardinali, D.P.; Maestroni, G.J. The role of melatonin in immuno-enhancement: Potential application in cancer. *Int. J. Exp. Pathol.* **2006**, *87*, 81–87. [CrossRef] [PubMed]
4. Zhang, C.; Yang, X.Z.; Xu, M.J.; Huang, G.Y.; Zhang, Q.; Cheng, Y.X.; He, L.; Ren, H.Y. Melatonin promotes cheliped regeneration, digestive enzyme function, and immunity following autotomy in the Chinese mitten crab, *Eriocheir sinensis*. *Front. Physiol.* **2018**, *9*, 269. [CrossRef]
5. Gao, W.X.; Ma, H.; Yang, M.N.; Peng, S.Y.; Qu, J.C.; Qu, Y.L. Effects of rumen-protected melatonin supplementation on melatonin and its precursors and metabolites contents in milk, melatonin synthetase activity, and performance in Holstein cows. *Chin. J. Anim. Nutr.* **2022**, *34*, 4438–4451.
6. Yang, X.; Song, X.; Zhang, C.; Pang, Y.; Song, Y.; Cheng, Y.; Nie, L.; Zong, X. Effects of dietary melatonin on hematological immunity, antioxidant defense and antibacterial ability in the Chinese mitten crab, *Eriocheir sinensis*. *Aquaculture* **2020**, *529*, 735578. [CrossRef]
7. Yang, Y.; Xu, W.; Du, X.; Ye, Y.; Tian, J.; Li, Y.; Jiang, Q.; Zhao, Y. Effects of dietary melatonin on growth performance, antioxidant capacity, and nonspecific immunity in crayfish, *Cherax destructor*. *Fish Shellfish Immunol.* **2023**, *138*, 108846. [CrossRef]
8. Yıldırım, M.; Aktaş, M. The effects of melatonin's pacific white shrimp, *Litopenaeus vannamei* (Boone 1931) on moulting, growth and meat composition. *Yunus Arast. Bul.* **2016**, *16*, 193–200.
9. Evans, R.M. Regulation of Circadian Behavior and Metabolism by Clock Genes. *J. Non-Cryst. Solids* **1981**, *44*, 171–180.

10. Fuhr, L.; Abreu, M.; Pett, P.; Relógio, A. Circadian systems biology: When time matters. *Comput. Struct. Biotechnol. J.* **2015**, *13*, 417–426. [CrossRef]

11. Lin, Y.; Han, M.; Shimada, B.; Wang, L.; Gibler, T.M.; Amarakone, A.; Awad, T.A.; Stormo, G.D.; Van Gelder, R.N.; Taghert, P.H. Influence of the period-dependent circadian clock on diurnal, circadian, and aperiodic gene expression in Drosophila melanogaster. *Proc. Natl. Acad. Sci. USA* **2002**, *99*, 9562–9567. [CrossRef] [PubMed]

12. Minnetti, M.; Hasenmajer, V.; Pofi, R.; Venneri, M.A.; Alexandraki, K.I.; Isidori, A.M. Fixing the broken clock in adrenal disorders: Focus on glucocorticoids and chronotherapy. *J. Endocrinol.* **2020**, *246*, R13–R31. [CrossRef] [PubMed]

13. Hill, S.M.; Belancio, V.P.; Dauchy, R.T.; Xiang, S.; Brimer, S.; Mao, L.; Hauch, A.; Lundberg, P.W.; Summers, W.; Yuan, L.; et al. Melatonin: An inhibitor of breast cancer. *Endocr.—Relat. Cancer* **2015**, *22*, R183–R204. [CrossRef] [PubMed]

14. Yoshiuchi, I. Analysis of evolution and ethnic diversity at glucose-associated SNPs of circadian clock-related loci with cryptochrome 1, cryptochrome 2, and melatonin receptor 1B. *Biochem. Genet.* **2021**, *59*, 1173–1184. [CrossRef]

15. Coelho, L.D.; Peres, R.; Amaral, F.G.; Reiter, R.J.; Cipolla-Neto, J. Daily differential expression of melatonin-related genes and clock genes in rat cumulus–oocyte complex: Changes after pinealectomy. *J. Pineal Res.* **2015**, *58*, 490–499. [CrossRef]

16. Tenorio, F.; Simoes, M.J.; Teixeira, V.W. Effects of melatonin and prolactin in reproduction: Review of literature. *Rev. Assoc. Med. Bras.* **2015**, *61*, 269–274. [CrossRef]

17. Kandalepas, P.C.; Mitchell, J.W.; Gillette, M.U. Melatonin signal transduction pathways require E-box-mediated transcription of Per1 and Per2 to reset the SCN clock at dusk. *PLoS ONE* **2016**, *11*, e0157824. [CrossRef]

18. Delgado-Lara, D.L.; González-Enríquez, G.V.; Torres-Mendoza, B.M.; González-Usigli, H.; Cárdenas-Bedoya, J.; Macías-Islas, M.A.; de la Rosa, A.C.; Jiménez-Delgado, A.; Pacheco-Moisés, F.; Cruz-Serrano, J.A. Effect of melatonin administration on the PER1 and BMAL1 clock genes in patients with Parkinson's disease. *Biomed. Pharmacother.* **2020**, *129*, 110485. [CrossRef]

19. Guo, K.; Ruan, G.; Fan, W.; Fang, L.; Wang, Q.; Luo, M.; Yi, T. The effect of nitrite and sulfide on the antioxidant capacity and microbial composition of the intestines of red swamp crayfish. *Fish Shellfish Immunol.* **2020**, *96*, 290–296. [CrossRef]

20. Huang, P.D. Discovery and Epidemiological Investigation of New Viral Pathogens Causing "May Decay" in Procambarus Clarkii. Master's Thesis, Nanjing Agricultural University, Nanjing, China, 2019.

21. Livak, K.J.; Schmittgen, T.D. Analysis of Relative Gene Expression Data Using Real-Time Quantitative PCR and the $2^{-\Delta\Delta CT}$ Method. *Methods* **2001**, *25*, 402–408. [CrossRef]

22. Tang, C.X. Effects of Active Immunization Against Melatonin on the Growth Performance and Carcass Quality of Growing Pigs. Master's Thesis, Sichuan Agricultural University, Ya'an, China, 2005.

23. Yan, J.M. Study on Effects of Melatonin on Growth Performance, Immune Function, and Antioxidant Capacity of Weaning Piglets. Master's Thesis, Huazhong Agricultural University, Wuhan, China, 2019.

24. Osei, P.; Robbins, K.R.; Shirley, H.V. Effects of exogenous melatonin on growth and energy metabolism of chickens. *Nutr. Res.* **1989**, *9*, 69–81. [CrossRef]

25. Lu, Y.F.; Liao, Q.H. Effects of melatonin on growth performance and immune function in broilers. *Feed Ind.* **2008**, *13*, 37–39.

26. De Vlaming, V. Effects of pinealectomy and melatonin treatment on growth in the goldfish, Carassius auratus. *Gen. Comp. Endocrinol.* **1980**, *40*, 245–250. [CrossRef] [PubMed]

27. Alvariño, J.M.; Rebollar, P.G.; Olmedo, M.; Álvarez-Blázquez, B.; Ubilla, E.; Peleteiro, J.B. Effects of melatonin implants on reproduction and growth of turbot broodstock. *Aquac. Int.* **2001**, *9*, 477–487. [CrossRef]

28. Ye, Y.; Li, S.; Zhu, B.; Yang, Y.; Du, X.; Li, Y.; Zhao, Y. Effects of dietary melatonin on growth performance, nutrient composition, and lipid metabolism of Pacific white shrimp (*Penaeus vannamei*). *Aquaculture* **2024**, *578*, 740095. [CrossRef]

29. Lv, W.; Li, M.; Mao, Y.; Huang, W.; Yuan, Q.; Li, M.; Zhou, Q.; Yang, H.; Zhou, W. Effects of dietary melatonin supplementation on growth performance and intestinal health of rice field eel (*Monopterus albus*). *Comp. Biochem. Physiol. D Genom. Proteom.* **2024**, *52*, 101273. [CrossRef]

30. Wang, J.; Li, B.; Ma, J.; Wang, S.; Huang, B.; Sun, Y.; Zhang, L. Optimum dietary protein to lipid ratio for starry flounder (*Platichthys stellatus*). *Aquac. Res.* **2017**, *48*, 189–201. [CrossRef]

31. Liu, F.; Qu, Y.K.; Geng, C.; Wang, A.M.; Zhang, J.H.; Chen, K.J.; Liu, B.; Tian, H.Y.; Yang, W.P.; Yu, Y.B. Effects of hesperidin on the growth performance, antioxidant capacity, immune responses and disease resistance of red swamp crayfish (*Procambarus clarkii*). *Fish Shellfish Immunol.* **2020**, *99*, 154–166. [CrossRef]

32. Ashouri, G.; Soofiani, N.M.; Hoseinifar, S.H.; Jalali, S.A.; Morshedi, V.; Valinassab, T.; Bagheri, D.; Van Doan, H.; Mozanzadeh, M.T.; Carnevali, O. Influence of dietary sodium alginate and *Pediococcus acidilactici* on liver antioxidant status, intestinal lysozyme gene expression, histomorphology, microbiota, and digestive enzymes activity in Asian sea bass (*Lates calcarifer*) juveniles. *Aquaculture* **2020**, *518*, 734638. [CrossRef]

33. Perez, L. Fish innate immune response to viral infection—An overview of five major antiviral genes. *Viruses* **2022**, *14*, 1546. [CrossRef]

34. Zhu, L.; Wang, X.; Hou, L.; Jiang, X.; Li, C.; Zhang, J.; Pei, C.; Zhao, X.; Li, L.; Kong, X. The related immunity responses of red swamp crayfish (*Procambarus clarkii*) following infection with *Aeromonas veronii*. *Aquac. Rep.* **2021**, *21*, 100849. [CrossRef]

35. Yang, L.X.; Xu, H.Z.; Liu, C.J.; Wang, G.L.; Ai, Y.; Jiang, W.S.; Luo, Q.H.; Li, H.; Luo, L.; Xiang, X. Effects of Vitamin C on the Structure and Function of the Digestive System of *Andrias davidianus*. *J. Fish. Sci. China* **2023**, *47*, 161–172.
36. Luo, J.; Fu, W.J.; Yang, E.J.; Huang, J.S.; Xie, R.T.; Chen, G. Effects of Quercetin on Growth Performance, Antioxidant Capacity, and Gut Microbiota of Hybrid Grouper. *J. Guangdong Ocean Univ.* **2022**, *42*, 13–22.
37. Li, Y.; Yang, Y.; Li, S.; Ye, Y.; Du, X.; Liu, X.; Jiang, Q.; Che, X. Effects of dietary melatonin on antioxidant and immune function of the Pacific white shrimp (*Litopenaeus vannamei*), as determined by transcriptomic analysis. *Comp. Biochem. Physiol. D: Genom. Proteom.* **2023**, *48*, 101146. [CrossRef] [PubMed]
38. Yang, Y.; Zhu, B.; Xu, W.; Tian, J.; Du, X.; Ye, Y.; Huang, Y.; Jiang, Q.; Li, Y.; Zhao, Y. Dietary melatonin positively impacts the immune system of crayfish, *Cherax destructor*, as revealed by comparative proteomics analysis. *Fish Shellfish Immunol.* **2023**, *142*, 109122. [CrossRef]
39. Li, Y.; Huang, Y.; Zhang, M.; Chen, Q.; Fan, W.; Zhao, Y. Effect of dietary vitamin E on growth, immunity and regulation of hepatopancreas nutrition in male oriental river prawn, *Macrobrachium nipponense*. *Aquac. Res.* **2019**, *50*, 1741–1751. [CrossRef]
40. Reiter, R.J.; Tan, D.X.; Galano, A. Melatonin: Exceeding expectations. *Physiology* **2014**, *29*, 325–333. [CrossRef]
41. Zhao, Y.; Song, X.; Zhao, P.; Li, T.; Xu, J.W.; Yu, X. Role of melatonin in regulation of lipid accumulation, autophagy and salinity-induced oxidative stress in microalga *Monoraphidium* sp. QLY-1. *Algal Res.* **2021**, *54*, 102196. [CrossRef]
42. Winiarska, K.; Fraczyk, T.; Malinska, D.; Drozak, J.; Bryla, J. Melatonin attenuates diabetes-induced oxidative stress in rabbits. *J. Pineal Res.* **2006**, *40*, 168–176. [CrossRef]
43. Marí, M.; Morales, A.; Colell, A.; García-Ruiz, C.; Fernández-Checa, J.C. Mitochondrial glutathione, a key survival antioxidant. *Antioxid. Redox Signal.* **2009**, *11*, 2685–2700. [CrossRef]
44. Pal, P.K.; Maitra, S.K. Response of gastrointestinal melatonin, antioxidants, and digestive enzymes to altered feeding conditions in carp (*Catla catla*). *Fish Physiol. Biochem.* **2018**, *44*, 1061–1073. [CrossRef] [PubMed]
45. Yang, X.; Shi, X.; Wu, M.; Pang, Y.; Song, X.; Shi, A.; Niu, C.; Cheng, Y. Effects of melatonin feed on histology and antioxidant ability of the gills and oxygen consumption of Chinese mitten crab (*Eriocheir sinensis*), exposed to acute hypoxia stress. *Aquaculture* **2021**, *544*, 737015. [CrossRef]
46. Song, Y.; Wu, M.; Pang, Y.; Song, X.; Shi, A.; Shi, X.; Niu, C.; Cheng, Y.; Yang, X. Effects of melatonin feed on the changes of hemolymph immune parameters, antioxidant capacity, and mitochondrial functions in Chinese mitten crab (*Eriocheir sinensis*) caused by acute hypoxia. *Aquaculture* **2021**, *535*, 736374. [CrossRef]
47. Coccia, E.; Varricchio, E.; Paolucci, M. Digestive enzymes in the crayfish *Cherax albidus*: Polymorphism and partial characterization. *Int. J. Zool.* **2011**, *2011*, 31037. [CrossRef]
48. Yang, X.; Shi, A.; Song, Y.; Niu, C.; Yu, X.; Shi, X.; Pang, Y.; Ma, X.; Cheng, Y. The effects of ammonia-N stress on immune parameters, antioxidant capacity, digestive function, and intestinal microflora of Chinese mitten crab, *Eriocheir sinensis*, and the protective effect of dietary supplement of melatonin. *Comp. Biochem. Physiol. C Toxicol. Pharmacol.* **2021**, *250*, 109127. [CrossRef]
49. Song, Y.; Song, X.; Wu, M.; Pang, Y.; Shi, A.; Shi, X.; Niu, C.; Cheng, Y.; Yang, X. The protective effects of melatonin on survival, immune response, digestive enzymes activities and intestinal microbiota diversity in Chinese mitten crab (*Eriocheir sinensis*) exposed to glyphosate. *Comp. Biochem. Physiol. C Toxicol. Pharmacol.* **2020**, *238*, 108845. [CrossRef]
50. Mardones, O.; Devia, E.; Labbé, B.S.; Oyarzún, R.; Vargas-Chacoff, L.; Muñoz, J.L. Effect of L-tryptophan and melatonin supplementation on the serotonin gastrointestinal content and digestive enzymatic activity for *Salmo salar* and *Oncorhynchus kisutch*. *Aquaculture* **2018**, *482*, 203–210. [CrossRef]
51. Takekida, S.; Yan, L.; Maywood, E.S.; Hastings, M.H.; Okamura, H. Differential adrenergic regulation of the circadian expression of the clock genes Period1 and Period2 in the rat pineal gland. *Eur. J. Neurosci.* **2000**, *12*, 4557–4561. [CrossRef]

Article

Effect of Dietary Copper on Growth Performance, Antioxidant Capacity, and Immunity in Juvenile Largemouth Bass (*Micropterus salmoides*)

John Cosmas Kayiira [1,2], Haifeng Mi [3], Hualiang Liang [1,2], Mingchun Ren [1,2], Dongyu Huang [1,*], Lu Zhang [3,*] and Tao Teng [3]

[1] Key Laboratory of Integrated Rice-Fish Farming Ecology, Ministry of Agriculture and Rural Affairs, Freshwater Fisheries Research Center, Chinese Academy of Fishery Sciences, Wuxi 214081, China
[2] Wuxi Fisheries College, Nanjing Agricultural University, Wuxi 214081, China
[3] Tongwei Agricultural Development Co., Ltd., Key Laboratory of Nutrition and Healthy Culture of Aquatic Livestock and Poultry, Ministry of Agriculture and Rural Affairs, Healthy Aquaculture Key Laboratory of Sichuan Province, Chengdu 610093, China
* Correspondence: huangdongyu@ffrc.cn (D.H.); zhangl21@tongwei.com (L.Z.)

Abstract: This study evaluated the optimal dietary copper (Cu) levels and their effects on growth performance, body composition, and antioxidant capacity in juvenile largemouth bass (*Micropterus salmoides*). A total of 360 fish (initial average weight (1.67 ± 0.01 g) and initial average length (2.5 ± 0.2 cm)) were randomly assigned to 18 tanks, each containing 20 fish and six dietary Cu concentrations: 2.13 (control), 3.00, 3.66, 4.58, 4.64, and 5.72 mg/kg. The results indicated that fish receiving 3.00 mg/kg of Cu exhibited the best final body weight (FBW), weight gain rate (WGR), and specific growth rate (SGR), with a significantly reduced feed conversion ratio (FCR). While body composition (moisture, protein, lipid, and ash) remained consistent across groups, plasma total protein (TP) levels increased with Cu supplementation. Elevated triglycerides (TG) and albumin (ALB) were noted at 4.64 and 5.72 mg/kg, respectively, while glucose (GLU) levels decreased with an increase in dietary Cu. Antioxidant capacity, assessed via hepatic glutathione (GSH) and the activities of catalase (CAT), and showed significant improvements at 3.00 and 3.66 mg/kg Cu, while superoxide dismutase (SOD) showed the highest activity at a dietary Cu level of 5.72 mg/kg. Additionally, the expressions of *tgf-β* and *tnf-α* genes were significantly upregulated at a dietary Cu level of 5.72 mg/kg, while *il-8* and *il-10* genes were upregulated at dietary 3.66 mg/kg. The expression of *nrf2* was significantly upregulated in response to a dietary Cu level of 3.66 mg/kg compared to the control group, and the expression of the *keap1* gene was significantly upregulated in the fish fed with 5.72 mg/kg of dietary Cu. The results indicated that appropriate dietary supplementation could promote the growth performance and improve the antioxidant status the immunity of largemouth bass, and the optimal Cu requirement for FCR and SGR were approximately 3.10 mg/kg and 3.00 mg/kg, respectively.

Keywords: largemouth bass; copper requirement; plasma biochemistry; Nrf2 signaling pathway; NF-κb signaling pathway

Key Contribution: dietary Cu enhances growth performance and immune and antioxidant capacity of largemouth bass.

Citation: Kayiira, J.C.; Mi, H.; Liang, H.; Ren, M.; Huang, D.; Zhang, L.; Teng, T. Effect of Dietary Copper on Growth Performance, Antioxidant Capacity, and Immunity in Juvenile Largemouth Bass (*Micropterus salmoides*). *Fishes* **2024**, *9*, 369. https://doi.org/10.3390/fishes9090369

Academic Editor: Aires Oliva-Teles

Received: 11 August 2024
Revised: 18 September 2024
Accepted: 20 September 2024
Published: 23 September 2024

1. Introduction

The global demand for fresh and nutrient-rich fish products has promoted the growth of aquaculture, providing food, nourishment, revenue, and improved quality of life for communities worldwide [1,2]. Minerals including copper (Cu) play crucial roles in diverse bodily processes, such as the formation of skeletal structure, maintenance of colloidal systems, generation of membrane potential, and acid–base balance control [3].

Fish require Cu, an essential trace element. Cu is a cofactor for several proteins that perform essential tasks for development and growth [4]. Dietary copper supplementation increased growth performance in genetically improved farmed tilapia (GIFT) (*Oreochromis niloticus*) [5]. For grouper (*Epinephelus malabaricus*), a diet deficient in Cu results in stunted growth and low feed efficiency (FE) [6]. Growth results from the utilization of feed, which is associated with the capacity for digestion and absorption [7]. Cu is an essential cofactor for key enzymes involved in metabolism, including tyrosinase, dopamine hydroxylase, ceruloplasmin, lysyl oxidase, superoxide dismutase (SOD), and cytochrome oxidase [8,9]. In healthy humans, the majority of Cu is either linked to proteins or coupled with enzyme prosthetic groups, with the highest proportion in the liver and brain [10,11]. However, excess levels of Cu may lead to oxidation of proteins and lipids, emphasizing the need for optimal dietary Cu intake and homeostasis to maintain cell viability and optimal fish development [12]. Enough dietary Cu can enhance the antioxidative or general immune functions of fish and shrimp, such as in large yellow croaker (*Larimichthys croceus*) [13]. But Cu as a heavy metal as well as Cu ion and its complexes can result in the production of reactive oxygen species (ROS), which can lead to oxidative stress-related harm to cells, organs, lipids, proteins, and DNA [14]. Cu ions play significant physiological roles, but when they accumulate at high tissue concentrations, they can also be toxic [15]. Excessive dietary Cu in aquatic animals frequently results in tissue damage, especially ionic and osmotic regulation disruption [16], oxidative stress, loss of cell membrane selective permeability, and enzyme inhibition [17,18].

Micronutrient consumption is closely linked to the immune system and general health of aquatic animals, highlighting the importance of supplementing fish feed with mineral nutrients [19]. Notably, insufficient Cu levels in fish-rearing water are unable to meet the fish's physiological demands [20]. Various fish species have specific dietary Cu requirements, including rainbow trout (*Oncorhynchus mykiss*) [21], carp (*Cyprinus carpio*) [21], blunt snout bream (*Megalobrama amblycephala*) [22], grouper [6], Atlantic salmon (*Salmo salar*) [23,24], abalone (*Haliotis discus hannai Ino*) [25], and channel catfish (*Ictalurus punctatus*) [26].

Largemouth bass (*Micropterus salmoides*) is a predatory freshwater species indigenous to North America, known for its fast growth rate, wide temperature tolerance, adaptability to different climatic conditions, and value as a food source [27]. Fish such as largemouth bass can experience stress in aquatic settings due to various factors, such as high temperatures, low oxygen levels, changes in social interactions, and limited range of motion [28]. The cellular defense against oxidative stress is predominantly governed by the nuclear factor erythroid 2-related factor 2 (Nrf2) and nuclear factor kappa B (NF-κB) pathways [29,30]. However, the impact of Cu on the Nrf2 and NF-κB signaling pathways in largemouth bass has not been investigated. This research endeavor aimed to investigate the effects of dietary Cu on the growth efficiency, body composition, antioxidant capability, and immune response of largemouth bass. The findings will provide supplementary materials for the development of cost-effective, environmentally friendly, and efficacious feed formulations for carnivorous fish.

2. Materials and Methods

2.1. Experimental Diet

The dietary components are presented in Table 1. We designed five addition levels of 0, 1, 2, 3, 4, 5, and 6 mg/kg in the diet with reference to the copper requirement (3.5 mg/kg) of *Lateolabrax japonicus* [31], which belongs to the order Perciformes with the largemouth bass, a closely related species. Then, six diets were formulated by supplementing copper sulfate pentahydrate ($CuSO_4 \cdot 5H_2O$) to achieve varying levels of dietary copper (Cu) (final measured values): 2.13 (control group), 3.00, 3.66, 4.58, 4.64, and 5.72 mg/kg. The ingredients were crushed through an 80-mesh sieve, weighed, and thoroughly mixed for feed formulation, followed by the addition of water (23%, 25%, mass fraction) and oil.

Concurrently, $CuSO_4 \cdot 5H_2O$ was dissolved in water and incorporated into the mixture, and then, the mixture was processed into 2 mm diameter granules using a meat grinder (SJPS56×2; Jiangsu Muyang Holdings Co., LTD, Yangzhou, China) and then air-dried and stored at $-20\ ^\circ$C until used.

Table 1. Experimental basic formula (%, dry matter).

Ingredients	Level (%)	Ingredients	Level (%)
Fish meal [1]	20	Choline chloride	0.5
Casein [1]	28	Vitamin premix [2]	1
Gelatin [1]	7	Mineral premix [3] (no copper)	1
Wheat Flour [1]	16	Calcium phosphate	4
Fish oil	4	Microcrystalline cellulose	14.45
Soybean oil	4	Vitamin C	0.05
		Component Analysis	
Crude protein (%)		46.08 ± 0.21	
Crude lipid (%)		9.97 ± 0.11	
Crude ash (%)		4.05 ± 0.24	
Crude fiber		13.69 ± 0.85	
Gross energy (KJ/g)		15.35 ± 0.28	

Note: [1] Fish meal, crude protein 67.8%, crude lipid 9.3%; casein, crude protein 90.0%; gelatin, crude protein 90.3%; wheat flour, crude protein 13.1%, crude lipid 4.0%; the above ingredients were obtained from Wuxi Tongwei feedstuffs Co., Ltd., Wuxi, China. [2] Vitamin premix (IU or mg/kg of premix, purchased by HANOVE Biotechnology Co., Ltd., Coventry, UK). [3] Mineral premix (no copper) (mg/kg of premix, purchased by HANOVE Biotechnology Co., Ltd., Coventry, UK): manganese sulfate, 31.5 mg; ferrous sulphate, 180.0 mg; zinc sulfate, 75 g; magnesium sulfate, 225.0 mg; sodium selenite, 0.8 mg; calcium iodate, 0.9 mg; cobalt chloride 0.7 mg; zeolite was used as a carrier.

2.2. Experimental Procedures

A total of 360 largemouth bass were obtained from Zhengda Aquatic Products Co., LTD (Huzhou, China) and acclimated to the control diet and indoor environment for two weeks before the feeding trial. Prior to stocking, the fish were fasted for 24 h and weighed. Fish with a uniform average weight of 1.67 ± 0.01 g and average length of 2.5 ± 0.2 cm were placed in 18 circular tanks, each containing 20 fish and a water capacity of 300 L. The tanks were randomly divided into six groups, with three tanks representing triplicates for each group. The tanks were connected to a share water reservoir, forming a closed recirculating system with biofilters to eliminate impurities and reduce ammonia concentrations. The six experimental diets were administered to the respective groups twice daily (at 7:30 and 18:00 h), and no remnants of food were observed in the tanks after each feeding session. Feed intake was monitored daily, and waste was promptly removed using a plastic pipe. The mortality rate was monitored daily, and water parameters were maintained by Octadem Multi-Parameter Water Quality Analyser (Type OCT-A) (Wuxi Octadem Biotechnology Co., LTD, Wuxi, China) within the specified range (pH 7.5–8.0, dissolved oxygen > 7.0 mg/L, temperature $30 \pm 2\ ^\circ$C, ammonia nitrogen and nitrite ranged between 0–0.2 mg/L and <0.01 mg/L, and copper ion content < 0.01 mg/L, respectively).

2.3. Sample Collection Analytical Methods

After 8 weeks, the fish were fasted for 24 h to eliminate the impact of feed on their weight and then sedated using 150 mg/L of MS-222. The total count and weight of fish in each tank were recorded to calculate the growth performance, including final body weight (FBW), weight gain rate (WGR), specific growth rate (SGR), and feed conversion ratio (FCR). The formulas for calculating the parameters are as follows:

Final body weight (FBW, g) = Final body total weight at the end (g)/number of fish

Weight gain rate (WGR, %) = 100 × (Final body weight (g) − Initial body weight (g))/Initial body weight (g)

Specific growth rate (SGR, %/day) = 100 × [ln(Final body weight) − ln(Initial body weight)]/duration (days)

Feed conversion ratio (FCR) = dry feed fed (g)/(Final body weight (g) − Initial body weight (g)

Three fish were randomly selected from each tank for blood sample collection from the caudal vessels. The blood samples were centrifuged at 3000 rpm for 10 min at 4 °C (Eppendorf 5424R centrifuge, Leipzig, Germany) to collect the plasma, which was immediately stored at −20 °C for subsequent analysis. Afterwards, 300 μL plasma of each was placed on the Mindray BS-400 Automatic Biochemical Analyser for testing. From the same fish, the liver tissues were removed and frozen in liquid nitrogen to estimate the antioxidant and immune parameters. Additionally, six fish from each tank were randomly selected and stored at −20 °C to determine the overall body composition.

2.4. Chemical Analysis

The proximate composition of the diets and fish body, including lipids, proteins, ash, and moisture, were determined in triplicate using standard methods [32]. The specific details of the analytical methods and instrumentation employed in the biochemical testing are provided in Table 2.

Table 2. Primary methodologies and analytical instruments used.

Items	Methodologies
Moisture	Dried sample in an oven at 105 °C
Crude protein	Using the Kjeldahl procedure after acid digestion (multiplied by N × 6.25)
Lipids	Analysed through ether extraction using the Soxhlet system
Ash	Examined by combusting at 550 °C for 5 h in an intelligent muffle furnace (model number XL-2A, Hangzhou, China: Zhang Chi Instruments Co., Ltd.)
Gross energy	Examined by combusting in an oxygen bomb calorimeter: IKA C6000 (IKA Works, Guangzhou, China)
Fiber	Fibercarp method by Fiber analysis system (FiberCap™ 2021, FOSS, Hilleroed, Denmark)
Plasma total protein (TP, Mindray 105-000451-00) Albumin (ALB, Mindray 105-000450-00) Total cholesterol (TC, Mindray 105-000448-00) Glucose (GLU, Mindray 105-000460-00) Triglyceride (TG, Mindray 105-000449-00) Aspartate aminotransferase (AST, Mindray, 105-000443-00) Alanine aminotransferase (ALT, Mindray 105-000442-00)	Measured using a Mindray BS-400 Automatic Biochemical Analyser (Mindray Medical International Ltd., Shenzhen, China)
Malondialdehyde (MDA, A003-1-2) Glutathione (GSH, A006-2-1) Glutathione peroxidase (GPx, A005-1-2) Superoxide dismutase (SOD, A001-3-2) Total antioxidant capacity (T-AOC, A015-2-1) Catalase (CAT, A007-1-1)	Determined using biochemical kits from Nanjing Jiancheng Bioengineering Institute, Nanjing, China

2.5. Real-Time PCR Analysis

The total RNA from the liver tissues was isolated using the RNA extraction reagent (Vazyme, Nanjing, China), followed by the qualitative and quantitative analysis using a NanoDrop 2000 spectrophotometer, and the A260/280 value of 1.8–2.0 served as a standard. The qPCR analysis employed the CFX96 Touch system (Bio-Rad, Hercules, CA, USA), and the reagents were obtained from Vazyme. GAPDH served as the housekeeping gene or standard, and mRNA levels were determined by the relative standard curve method. The PCR primers were designed according to their nucleic acid sequences and retrieved from the National Center for Biotechnology Information (NCBI; Bethesda, MD, USA) via Primer Premier 6.0 (https://primer-premier.software.informer.com/6.1/, accessed on 29 May 2024). Table 3 presents the primer details for real-time qPCR analysis.

Table 3. Primer sequences for real-time quantitative PCR analysis.

Genes		Primer Sequence (5′-3′)	Reference
gapdh	Forward	ACTGTCACTCCTCCATCTT	AZA04761.1
	Reverse	CACGGTTGCTGTATCCAA	
tgf-β	Forward	GCTCAAAGAGAGCGAGGATG	[33]
	Reverse	TCCTCTACCATTCGCAATCC	
il-8	Forward	CGTTGAACAGACTGGGAGAGATG	[34]
	Reverse	AGTGGGATGGCTTCATTATCTTGT	
il-10	Forward	CGGCACAGAAATCCCAGAGC	[34]
	Reverse	CAGCAGGCTCACAAAATAAACATCT	
nrf2	Forward	AGAGACATTCGCCGTAGA	NM_212855.2
	Reverse	TCGCAGTAGAGCAATCCT	
keap1	Forward	CGTACGTCCAGGCCTTACTC	XP_018520553.1
	Reverse	TGACGGAAATAACCCCCTGC	
tnf-α	Forward	CTTCGTCTACAGCCAGGCATCG	[33]
	Reverse	TTTGGCACACCGACCTCACC	
Cu/Zn sod	Forward	TGGCAAGAACAAGAACCACA	[33]
	Reverse	CCTCTGATTTCTCCTGTCACC	
cat	Forward	CTATGGCTCTCACACCTTC	MK614708.1
	Reverse	TCCTCTACTGGCAGATTCT	
gpx	Forward	GAAGGTGGATGTGAATGGA	MK614713.1
	Reverse	CCAACCAGGAACTTCTCAA	
nf-κb	Forward	CCACTCAGGTGTTGGAGCTT	XP_027136364.1
	Reverse	TCCAGAGCACGACACACTTC	

Note: *gapdh*, glyceraldehyde-3-phosphate dehydrogenase; *tgf-β*, transforming growth factor beta; *il-8/10*, interleukin 8/10; *nrf2*, nuclear factor erythroid 2-related factor 2; *keap1*, kelch-like ECH-associated protein l; *tnf-α*, tumor necrosis factor-α; *cat*, catalase; Cu/Zn sod, Cu/Zn superoxide dismutase; *gpx*, glutathione peroxidase; *nf-κb*, nuclear factor kappa-B.

2.6. Statistical Analysis

The data were processed and analyzed using SPSS version 23.0 software. One-way analysis of variance (ANOVA) was used to assess the significant differences between the means, and Duncan's multiple range test was applied for post hoc comparisons. The results are presented as the mean ± standard error of the mean (SEM), and p-values < 0.05 were considered statistically significant. The broken-line regression model was selected to determine the optimum dietary copper requirement by comparing the estimation coefficient (R^2) among the linear regression model (SGR, 0.191; FCR, 0.370), quadratic regression model (SGR, 0.321; FCR, 0.601), and broken-linear regression model (SGR, 0.542; FCR, 0.722).

3. Results

3.1. Growth Performance

Table 4 shows the growth performance results: Fish fed diets containing 3.00–5.72 mg/kg of Cu exhibited a significantly higher weight gain rate (WGR) compared to the control group with 2.13 mg/kg of Cu ($p < 0.05$). Additionally, a significant increase in the final body weight (FBW) and specific growth rate (SGR) was observed in the experimental groups relative to the control ($p < 0.05$). The diet containing 3.00 mg/kg of Cu displayed the highest values for both FBW and SGR. Furthermore, the experimental groups demonstrated a significantly reduced feed conversion ratio (FCR) compared to the control group ($p < 0.05$). The broken-line regression analysis based on SGR and FCR indicated the optimal dietary Cu requirement to be 3.00 and 3.10 mg/kg, respectively (Figure 1).

Table 4. Effects of dietary Cu levels on growth performance of largemouth bass.

Dietary Cu Levels (mg/kg)	Growth Parameters				
	IBW (g)	FBW (g)	FCR	WGR (%)	SGR (%/Day)
2.13	1.68 ± 0.01	17.71 ± 0.04 [a]	0.96 ± 0.01 [b]	953.92 ± 5.47 [a]	4.21 ± 0.01 [a]
3	1.68 ± 0.01	19.67 ± 0.28 [c]	0.92 ± 0.01 [a]	1075.60 ± 20.53 [c]	4.40 ± 0.03 [c]
4.58	1.67 ± 0.01	18.73 ± 0.36 [b]	0.92 ± 0.00 [a]	1022.39 ± 17.30 [b]	4.31 ± 0.03 [b]
4.64	1.67 ± 0.01	19.21 ± 0.27 [bc]	0.92 ± 0.00 [a]	1053.65 ± 17.60 [bc]	4.37 ± 0.03 [bc]
5.72	1.67 ± 0.01	19.15 ± 0.16 [bc]	0.92 ± 0.00 [a]	1046.59 ± 13.80 [bc]	4.36 ± 0.02 [bc]

Note: The data values are presented as means ± standard error mean (SEM). Significant differences between the six treatments are indicated by different letters ($p < 0.05$).

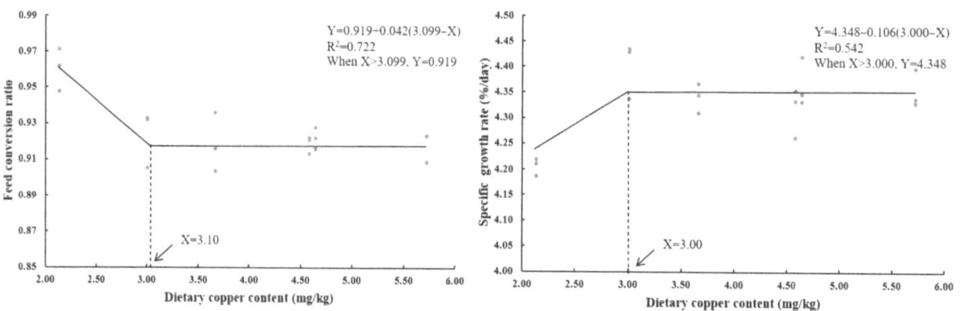

Figure 1. Broken-line regression analysis of feed conversion ratio (FCR) and specific growth rate (SGR, %/day) against graded different levels of dietary copper.

3.2. Whole Body Composition

The results of the whole body composition analysis are presented in Table 5. No statistically significant difference was observed in the moisture, crude protein, crude lipid, and ash content of the whole body ($p > 0.05$).

Table 5. Effects of dietary Cu levels on whole body composition of largemouth bass.

Dietary Cu levels (mg/kg)	Body Composition			
	Moisture (%)	Protein (%)	Lipid (%)	Ash (%)
2.13	71.99 ± 0.13	14.91 ± 0.76	6.41 ± 0.46	3.65 ± 0.07
3.00	71.63 ± 0.22	15.81 ± 0.14	7.09 ± 0.35	3.43 ± 0.14
3.66	71.45 ± 0.04	16.16 ± 0.09	6.39 ± 0.07	3.74 ± 0.11
4.58	71.92 ± 0.56	15.63 ± 0.35	7.78 ± 0.37	3.85 ± 0.13
4.64	71.37 ± 0.46	16.27 ± 0.47	6.72 ± 0.59	3.57 ± 0.06
5.72	71.78 ± 0.04	16.08 ± 0.13	6.27 ± 0.29	3.54 ± 0.19

Note: The data values are means ± standard error mean (SEM).

3.3. Plasma Biochemical Parameters

The results of the biochemical analyses of the plasma samples are presented in Table 6. Dietary treatments did not affect the AST and ALT activities ($p > 0.05$). However, the TP levels were significantly increased with higher dietary Cu levels when compared to the control group ($p < 0.05$). Conversely, the GLU levels showed a significant decrease with increasing dietary Cu concentrations compared to the control group ($p < 0.05$). Furthermore, 4.64 and 5.72 mg/kg of Cu significantly increased the ALB and TG levels compared to the control group ($p < 0.05$). Additionally, the TC levels were significantly elevated in the diets with 3.00–5.72 mg/kg of Cu ($p < 0.05$).

Table 6. Effects of dietary Cu levels on plasma biochemical indices of largemouth bass.

Parameters	Dietary Cu Levels (mg/kg)					
	2.13	3.00	3.66	4.58	4.64	5.72
ALT (U/L)	15.73 ± 3.39	12.03 ± 2.07	15.75 ± 2.98	20.88 ± 5.47	12.30 ± 2.78	24.50 ± 5.66
AST (U/L)	288.00 ± 59.11	251.80 ± 32.56	238.68 ± 34.52	297.45 ± 40.12	202.18 ± 33.62	285.55 ± 54.68
TP (g/L)	23.01 ± 2.16 [a]	30.24 ± 1.71 [b]	31.47 ± 2.04 [b]	29.06 ± 1.26 [b]	33.87 ± 1.71 [b]	34.04 ± 1.39 [b]
ALB (g/L)	2.30 ± 0.40 [a]	4.03 ± 0.58 [ab]	3.88 ± 0.74 [ab]	4.23 ± 0.45 [ab]	4.98 ± 0.68 [b]	4.88 ± 0.96 [b]
TC (mmol/L)	8.21 ± 0.90 [a]	11.20 ± 0.74 [b]	11.49 ± 0.53 [b]	10.78 ± 0.76 [b]	11.00 ± 1.35 [b]	12.73 ± 0.74 [b]
TG (mmol/L)	10.46 ± 1.24 [a]	14.42 ± 1.20 [ab]	15.02 ± 0.78 [ab]	12.53 ± 2.04 [ab]	17.08 ± 2.33 [b]	17.32 ± 2.10 [b]
GLU (mmol/L)	8.18 ± 0.68 [b]	6.93 ± 0.53 [ab]	5.85 ± 0.37 [a]	7.97 ± 0.41 [b]	6.23 ± 0.65 [a]	5.40 ± 030 [a]

Note: The data values are presented as means $\pm$ standard error mean (SEM). Significant differences between the six treatments are denoted by different letters ($p < 0.05$).

3.4. The Antioxidant Parameters of Liver

Table 7 presents the results of the liver antioxidant parameters. The levels of T-AOC, GPx, and MDA were not affected by the different dietary Cu levels ($p > 0.05$). However, the GSH level increased significantly at 3.00 mg/kg of dietary Cu relative to the control group ($p < 0.05$). Moreover, diets supplemented with 3.00–3.66 mg/kg of Cu significantly elevated the CAT levels ($p < 0.05$). Similarly, fish fed dietary Cu levels ranging from 3.00 to 4.58 mg/kg exhibited a significant difference in SOD levels ($p < 0.05$), with the highest activity observed at 5.72 mg/kg.

Table 7. Effects of dietary Cu levels on antioxidant parameters in the liver of largemouth bass.

Parameters	Dietary Cu Levels (mg/kg)					
	2.13	3	3.66	4.58	4.64	5.72
CAT (U/mgprot)	4.25 ± 2.31 [a]	13.84 ± 1.95 [b]	13.85 ± 2.85 [b]	5.99 ± 1.29 [ab]	9.17 ± 3.91 [ab]	6.18 ± 1.51 [ab]
SOD (U/mgprot)	18.30 ± 2.81 [ab]	11.30 ± 2.31 [a]	12.07 ± 2.02 [a]	13.23 ± 3.60 [a]	16.16 ± 1.99 [ab]	21.36 ± 1.72 [b]
T-AOC (mmol/gprot)	0.26 ± 0.04	0.41 ± 0.08	0.40 ± 0.04	0.39 ± 0.06	0.31 ± 0.02	0.35 ± 0.05
GSH (μmol/gprot)	19.38 ± 4.10 [ab]	39.77 ± 6.23 [c]	32.54 ± 5.24 [bc]	14.90 ± 4.46 [a]	27.21 ± 5.91 [abc]	21.92 ± 3.05 [ab]
GPx (U/mgprot)	65.36 ± 25.35	63.81 ± 23.46	39.45 ± 14.35	51.56 ± 15.72	40.54 ± 13.45	36.76 ± 10.27
MDA (nmol/mgprot)	2.44 ± 0.48	2.40 ± 0.56	3.49 ± 0.54	2.50 ± 0.43	4.56 ± 0.97	4.48 ± 1.01

Note: The data values are means $\pm$ standard error mean (SEM). Significant differences between the six treatments are indicated by different letters. Significant differences between the six treatments are indicated by different letters ($p < 0.05$).

3.5. The Gene Expressions of the NF-κB Signaling Pathway in Liver

Figure 2 showed that the gene expression pattern exhibited significant differences induced by the various dietary Cu levels. Specifically, the expressions of *tgf-β* and *tnf-α* genes were significantly upregulated at a dietary Cu level of 5.72 mg/kg ($p < 0.05$). Similarly, the expressions of *il-8* and *il-10* genes were significantly upregulated at 3.66 mg/kg of Cu ($p < 0.05$). However, no significant difference was observed in the gene expression of *nf-κb* ($p > 0.05$).

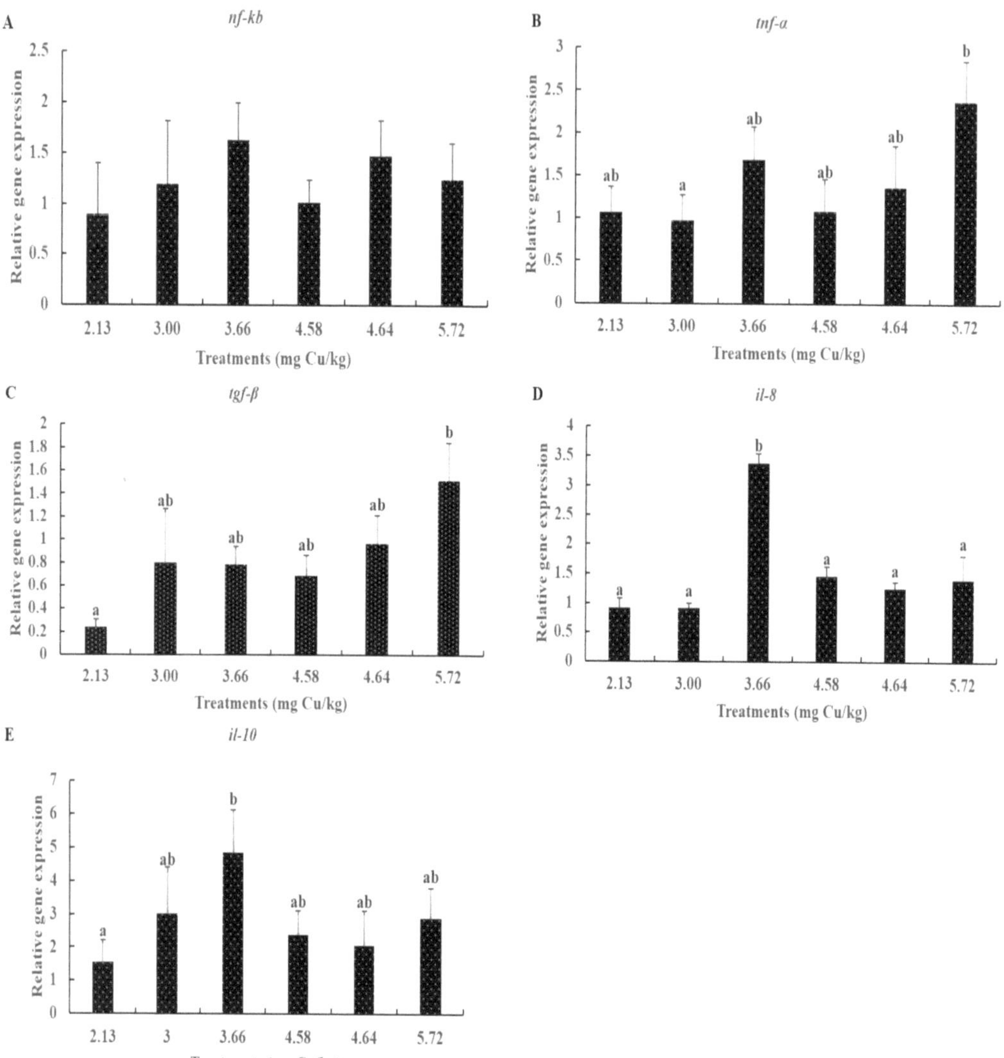

Figure 2. Gene expression of the NF-κB signaling pathway in the liver of largemouth bass. (**A**) *nf-κb*; (**B**) *tnf-α*; (**C**) *tgf-β*; (**D**) *il-8*; (**E**) *il-10*. The data values are presented as means ± standard error mean (S.E.M), and significant differences between the six treatments are denoted by different letters ($p < 0.05$).

3.6. The Core Gene Expressions of Nrf2 Signaling Pathway

Figure 3 shows that the expression of *nrf2* was significantly upregulated in response to a dietary Cu level of 3.66 mg/kg compared to the control group ($p < 0.05$); however, it declined with further increases in Cu levels. Additionally, the expression of the keap1 gene was significantly upregulated in the fish fed with 5.72 mg/kg of dietary Cu relative to the control group ($p < 0.05$). Furthermore, no significant differences were observed in the expressions of *cat*, *sod*, and *gpx* genes among all the groups ($p > 0.05$).

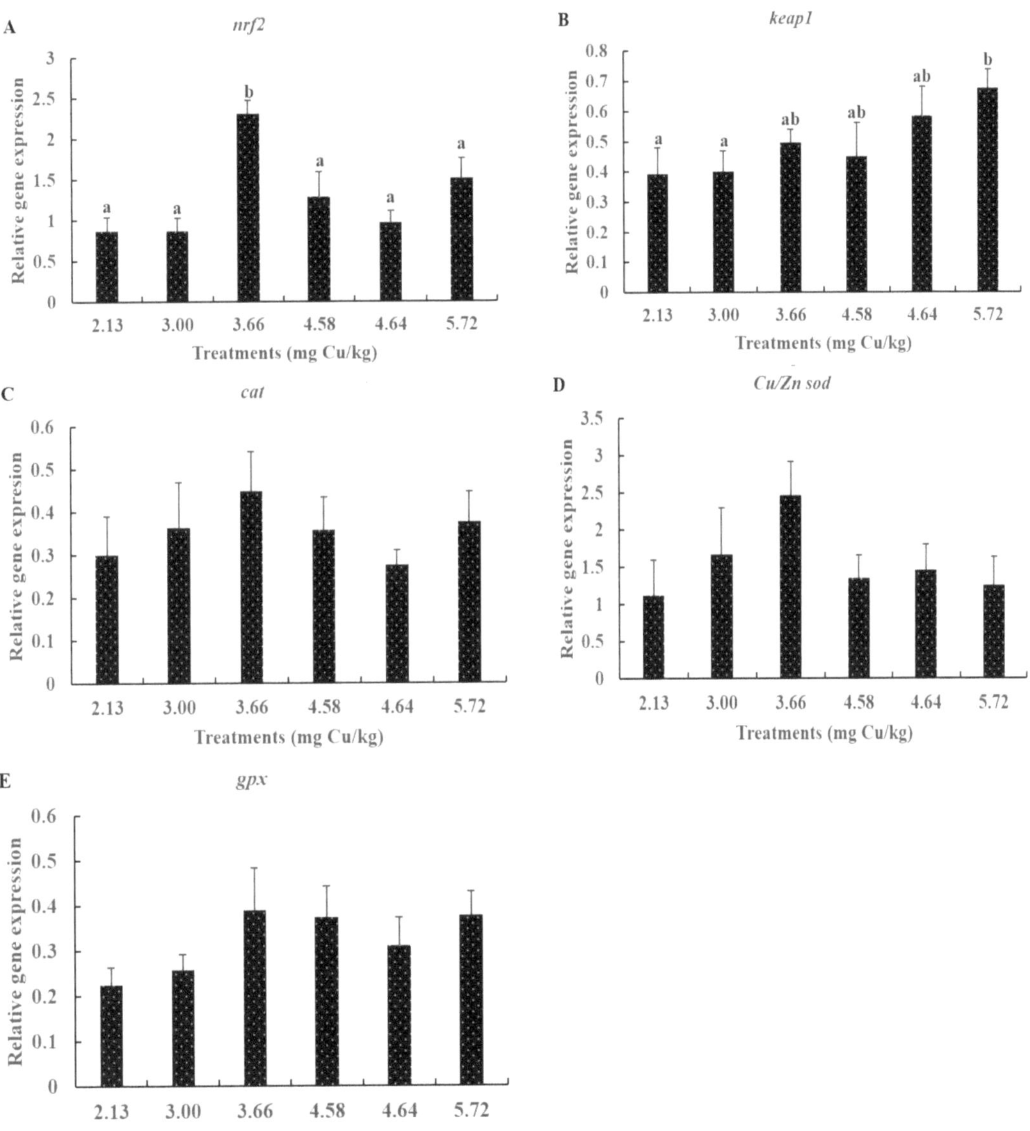

Figure 3. Nrf2 signaling pathway gene expression in the liver of largemouth bass. (**A**) *nrf2*; (**B**) *keap1*; (**C**) *cat*; (**D**) *Cu/Zn sod*; (**E**) *gpx*. The data values are means ± standard error mean (SEM). Significant differences between the six treatments are indicated by different letters ($p < 0.05$).

4. Discussion

Cu is crucial for all living organisms, including fish, as it plays a significant role in various physiological processes [35]. Promoting the growth of the aquaculture industry is essential for improving the productivity and profitability [36]. In this study, increased dietary Cu levels significantly improved WGR and SGR compared to the control group, consistent with the juvenile beluga survey (*Huso huso*) [37]. Additionally, the optimal dietary Cu levels for juvenile largemouth bass were estimated to be 3.00–3.10 mg/kg based on the broken-line regression analysis, which was supported by earlier studies on various fish species. Variations in fish species, size, feeding schedule, and experimental conditions contribute to different optimal Cu levels for different fish [38]. Additionally, the study highlighted the effect of dietary Cu on decreasing the FCR while improving the growth performance, as observed in fish fed with the optimal dietary Cu levels. Previous studies have also shown that decreased FCR was associated with improved growth performance in fish fed with optimal dietary Cu [19,37].

Blood-based biomarkers provide valuable insights into fish physiology and stress responses [39]. This experiment demonstrated that increased dietary Cu levels could significantly increase the TP and TC levels at all Cu concentrations (3.00–5.72 mg/kg) relative to the control group. Notably, 4.64 and 5.72 mg/kg of Cu could increase the levels of ALB and TG remarkably. These results are consistent with the earlier observations in snow trout (*Schizothorax zarudnyi*) [40] and red sea bream (*Pagrus major*), where Cu supplementation increased the total serum protein levels [41]. Moreover, our study also indicated that increased dietary Cu levels enhanced the innate immunity of largemouth bass. ALT is a cytosolic enzyme found in the liver, whereas AST is a mitochondrial enzyme found in the heart, skeletal muscle, kidney, and liver [42,43]. Our results found no significant differences in the activities of these enzymes, suggesting that different dietary Cu levels had no adverse effect on liver health. Another measure of fish stress is the GLU concentration in the plasma since its elevated levels often correspond to higher levels of stress in fish [44]. In the present study, increased dietary Cu levels could significantly reduce plasma GLU levels. Similar results were observed in snow trout [45], where elevated dietary Cu levels were associated with reduced plasma GLU levels and improved energy homeostasis. Furthermore, all dietary Cu treatments increased TC levels when compared with the control group, reaching their peak values at 5.72 mg/kg of dietary Cu. However, this finding was different from those observed in juvenile beluga [35], indicating that Cu may have oxidized polyunsaturated fats (PUFA), making them unsuitable for esterification [45].

Cu is a crucial trace element in living organisms, influencing biochemistry, enzyme activity, cellular respiration, free radical defense, neurotransmitter function, and tissue biosynthesis [46]. High dietary Cu levels were linked to increased SOD activities, highlighting the importance of its supplementation in improving the antioxidant mechanism in largemouth bass. SOD is responsible for converting two superoxide radicals into oxygen and hydrogen peroxide [47,48], which are then used as substrates by the CAT and GPx enzymes to remove harmful reactive oxygen species (ROS) from the cellular environment [49,50]. Our research revealed that SOD activities significantly increased as dietary Cu increased, demonstrating the importance of dietary Cu in boosting the antioxidant system in largemouth bass. Similar findings of increased SOD activities upon dietary copper supplementation have been documented in Russian sturgeon (*Acipenser gueldenstaedtii*) [51] and blunt snout bream [22]. SOD is not the same as their trend, probably because SOD is a copper-containing enzyme, and reduced dietary Cu-Zn SOD proprotein and insufficient concentrations of cofactor/chaperone proteins to saturate Cu-Zn SOD proprotein [5] may have allowed SOD activity to consistently increase with increasing dietary copper levels. Moreover, fish fed with 3.00 to 3.66 mg/kg of dietary Cu levels exhibited elevated levels of GSH and increased activities of CAT, thereby improving the antioxidant defense mechanisms. Similar results were reported in Pacific white shrimp (*Litopenaeus vannamei*) and freshwater fish *Channa punctatus* [52,53]. However, excessive copper negatively impacts these enzymes' abilities, which then suppresses the antioxidant capacity. Similar results have been documented for the crucian carp (*Carassius carassius*) [54], grass carp (*Ctenopharyngodon Idella*) [55], and Russian sturgeon [51]. MDA is an important marker of peroxidation [53], which directly affects cell membrane fluidity and integrity [56]. No significant difference in MDA levels was found in all groups in this study, suggesting that dietary Cu supplementation had no adverse effects on juvenile largemouth bass. The mRNA levels of antioxidant enzymes in fish are closely associated with their expression [57]. This experiment revealed that dietary Cu supplementation significantly increased the expression of *nrf2* and *keap1* genes in largemouth bass, suggesting that it promotes antioxidant activity in the fish. Similar findings were reported in other studies on juvenile blunt snout bream [21]. However, the gene expression of *cat*, *sod*, and *gpx* showed no significant difference in dietary Cu supplementation. These genes do not share the same significance in the expression of their associated enzyme activity, which may be due to the fact that in some cases the enzyme activity may be regulated by post-transcriptional regulation rather than being solely dependent on the level of expression of the gene; the increase in enzyme activity may also be related to changes in substrate concentration, and even if the amount of enzyme decreases, the enzyme activity may increase if the substrate concentration increases [58]. Therefore, optimal dietary Cu levels influence the Nrf2-Keap1 signaling pathway and expression of antioxidant enzyme genes regulated by it to improve the antioxidant capacity of largemouth bass.

The inflammatory response involves both pro- and anti-inflammatory cytokines [59], playing a crucial role in regulating immune responses in aquatic animals [60]. Anti-inflammatory cytokines (*tgf-β* and *il-10*) promote the production of non-inflammatory immunoglobulin isotypes IgG4 and IgA while suppressing IgE and effector cells such as mast cells, basophils, and eosinophils, which cause allergic inflammation [61]. In our study, the fish fed with a diet containing 3.66 and 5.72 mg/kg of Cu showed a significantly upregulated expression of *il-10* and *tgf-β* genes, respectively, relative to the control group. Previous studies have also shown that proper nutrient supplementation could enhance the expression of *il-10* and *tgf-β* in blunt snout bream [62]. In addition, our results showed that the dietary Cu intake led to an upregulation of *il-8* and *tnf-α* gene expression in largemouth bass, suggesting that adequate Cu intake might stimulate an immune response and support overall health. Numerous studies suggest that exposure to Cu causes inflammation, and various inflammatory cytokines regulate immune and inflammatory responses [63]. Simultaneously, inflammatory cytokines like *il-8* and *tnf-α* significantly impact the innate immunity in aquatic animals to protect against stress conditions [64].

5. Conclusions

In general, our research indicated that appropriate dietary supplementation (3.00–3.66 mg/kg) could enhance antioxidant capacity and influence relative Nrf2 and NF-κB signaling pathways, hence improving the health and immunity of largemouth bass. According to broken-line regression on SGR and FCR, the optimal dietary Cu requirement of largemouth bass was 3.00 and 3.10 mg/kg, respectively.

Author Contributions: Formal analysis, J.C.K., H.L. and H.M.; writing—original draft preparation, J.C.K.; writing—review and editing, D.H., M.R. and L.Z.; project administration, D.H. and L.Z.; methodology, H.M. and T.T.; investigation, H.L. and T.T. All authors have read and agreed to the published version of the manuscript.

Funding: This study was financially supported by the National Key R & D Program of China (2023YFD2400601), the earmarked fund for CARS (CARS-46), National Natural Science Foundation of China (32102806).

Institutional Review Board Statement: The study was approved by the Laboratory Animal Ethics Committee of the Freshwater Fisheries Research Center (LAECFFRC-2023-05-11).

Informed Consent Statement: Not applicable.

Data Availability Statement: The authors confirm that the data supporting the findings of this study are available within the manuscript, tables, and figures.

Conflicts of Interest: Authors Haifeng Mi, Tao Teng, and Lu Zhang are employed by Tongwei Agricultural Development Co., Ltd. The remaining authors declare that the research was conducted in the absence of any commercial or financial relationships that could be construed as potential conflicts of interest. Lu Zhang played a major role in the design of the study, writing—review and editing and project administration. Haifeng Mi and Tao Teng played a role in data analysis, methodology and investigation.

References

1. Boyd, C.E.; McNevin, A.A.; Davis, R.P. The contribution of fisheries and aquaculture to the global protein supply. *Food Secur.* **2022**, *14*, 805–827. [CrossRef] [PubMed]
2. Costello, C.; Cao, L.; Gelcich, S.; Cisneros-Mata, M.; Free, C.M.; Froehlich, H.E.; Golden, C.D.; Ishimura, G.; Maier, J.; Macadam-Somer, I.; et al. The future of food from the sea. *Nature* **2020**, *588*, 95–100. [CrossRef] [PubMed]
3. Araújo, J.G.; Nebo, C.; Pádua, D.M.C.; Souto, C.N.; Guimarães, I.G. Apparent Digestibility of Minerals from Several Ingredients for Tambaqui, *Colossoma macropomum*, Juveniles. *J. World Aquac. Soc.* **2018**, *49*, 1026–1038. [CrossRef]
4. Uauy, R.; Olivares, M.; Gonzalez, M. Essentiality of copper in humans. *Am. J. Clin. Nutr.* **1998**, *67* (Suppl. S5), 952S–959S. [CrossRef] [PubMed]
5. Zhong, C.C.; Ke, J.; Song, C.C.; Tan, X.Y.; Xu, Y.C.; Lv, W.H.; Song, Y.F.; Luo, Z. Effects of dietary copper (Cu) on growth performance, body composition, mineral content, hepatic histology and Cu transport of the GIFT strain of Nile tilapia (*Oreochromis niloticus*). *Aquaculture* **2023**, *574*, 739638. [CrossRef]
6. Lin, Y.H.; Shie, Y.Y.; Shiau, S.Y. Dietary copper requirements of juvenile grouper, *Epinephelus malabaricus*. *Aquaculture* **2008**, *274*, 161–165. [CrossRef]
7. Perez-Casanova, J.C.; Murray, H.M.; Gallant, J.W.; Ross, N.W.; Douglas, S.E.; Johnson, S.C. Development of the digestive capacity in larvae of haddock (*Melanogrammus aeglefinus*) and Atlantic cod (*Gadus morhua*). *Aquaculture* **2006**, *251*, 377–401. [CrossRef]
8. Malhotra, N.; Ger, T.R.; Uapipatanakul, B.; Huang, J.C.; Chen, K.H.C.; Der Hsiao, C. Review of copper and copper nanoparticle toxicity in fish. *Nanomaterials* **2020**, *10*, 1126. [CrossRef]
9. Luo, J.; Zhu, T.; Wang, X.; Cheng, X.; Yuan, Y.; Jin, M.; Betancor, M.B.; Tocher, D.R.; Zhou, Q. Toxicological mechanism of excessive copper supplementation: Effects on coloration, copper bioaccumulation and oxidation resistance in mud crab *Scylla paramamosain*. *J. Hazard. Mater.* **2020**, *395*, 122600. [CrossRef]
10. Festa, R.A.; Thiele, D.J. Copper: An essential metal in biology. *Curr. Biol.* **2011**, *21*, R877–R883. [CrossRef]
11. Gaetke, L.M.; Chow-Johnson, H.S.; Chow, C.K. Copper: Toxicological relevance and mechanisms. *Arch. Toxicol.* **2014**, *88*, 1929–1938. [CrossRef] [PubMed]
12. Tang, L.; Huang, K.; Xie, J.; Yu, D.; Sun, L.; Huang, Q.; Bi, Y. Dietary copper affects antioxidant and immune activity in hybrid tilapia (*Oreochromis niloticus* × *Oreochromis aureus*). *Aquac. Nutr.* **2017**, *23*, 1003–1015. [CrossRef]
13. Cao, J.; Miao, X.; Xu, W.; Li, J.; Zhang, W.; Mai, K. Dietary copper requirements of juvenile large yellow croaker *Larimichthys croceus*. *Aquaculture* **2014**, *432*, 346–350. [CrossRef]
14. Jomova, K.; Valko, M. Advances in metal-induced oxidative stress and human disease. *Toxicology* **2011**, *283*, 65–87. [CrossRef] [PubMed]
15. White, S.L.; Rainbow, P.S. Regulation and Accumulation of Copper, Zinc and Cadmium by the Shrimp Palaemon elegans. *Mar. Ecol.-Prog. Ser.* **1982**, *8*, 95–101. [CrossRef]
16. Craig, P.M.; Wood, C.M.; Mcclelland, G.B. Water Chemistry Alters Gene Expression and Physiological End Points of Chronic Waterborne Copper Exposure in Zebrafish, *Danio rerio*. *Environ. Sci. Technol.* **2010**, *44*, 2156. [CrossRef]
17. Machado, A.A.; Hoff, M.L.M.; Klein, R.D.; Gordeiro, G.J.; Avila, J.M.L.; Costa, P.G.; Bianchini, A. Oxidative stress and DNA damage responses to phenanthrene exposure in the estuarine guppy *Poecilia vivipar*. *Mar. Environ. Res.* **2014**, *98*, 96–105. [CrossRef]
18. Zimmer, A.M.; Fernanda, I.; Wood, C.M.; Bianchini, A. Waterborne copper exposure inhibits ammonia excretion and branchial carbonic anhydrase activity in euryhaline guppies acclimated to both fresh water and seawater. *Aquat. Toxicol.* **2012**, *122–123*, 172–180. [CrossRef]
19. Yuan, Y.; Luo, J.; Zhu, T.; Jin, M.; Jiao, L.; Sun, P.; Ward, T.L.; Ji, F.; Xu, G.; Zhou, Q. Alteration of growth performance, meat quality, antioxidant and immune capacity of juvenile *Litopenaeus vannamei* in response to different dietary dosage forms of zinc: Comparative advantages of zinc amino acid complex. *Aquaculture* **2020**, *522*, 735120. [CrossRef]
20. El Basuini, M.F.; El-Hais, A.M.; Dawood, M.A.O.; Abou-Zeid, A.E.S.; EL-Damrawy, S.Z.; Khalafalla, M.M.E.S.; Koshio, S.; Ishikawa, M.; Dossou, S. Effect of different levels of dietary copper nanoparticles and copper sulfate on growth performance, blood biochemical profiles, antioxidant status and immune response of red sea bream (*Pagrus major*). *Aquaculture* **2016**, *455*, 32–40. [CrossRef]
21. Ogino, C.; Yang, G.Y. Requirements of Carp and Rainbow Trout for Dietary Manganese and Copper. *Bull. Jpn. Soc. Sci. Fish* **1980**, *46*, 455–458. [CrossRef]
22. Liang, H.; Ji, K.; Ge, X.; Mi, H.; Xi, B.; Ren, M. Effects of dietary copper on growth, antioxidant capacity and immune responses of juvenile blunt snout bream (*Megalobrama amblycephala*) as evidenced by pathological examination. *Aquac. Rep.* **2020**, *17*, 100296. [CrossRef]
23. Lorentzen, M.; Maage, A.; Julshamn, K. Supplementing copper to a fish meal based diet fed to Atlantic salmon parr affects liver copper and selenium concentrations. *Aquac. Nutr.* **1998**, *4*, 67–72. [CrossRef]
24. Lall, S.P.; Hines, J.A. Iron and copper requirement of Atlantic salmon (*Salmo salar*) grown in sea water. In Proceedings of the International Symposium on Feeding and Nutrition of Fish, Bergen, Norway, 23–27 August 1987.
25. Wang, W.; Mai, K.; Zhang, W.; Ai, Q.; Yao, C.; Li, H.; Liufu, Z. Effects of dietary copper on survival, growth and immune response of juvenile abalone, *Haliotis discus hannai* Ino. *Aquaculture* **2009**, *297*, 122–127. [CrossRef]
26. Gatlin, D.M.; Wilson, R.P. Dietary copper requirement of fingerling channel catfish. *Aquaculture* **1986**, *54*, 277–285. [CrossRef]
27. Brown, T.G.; Runciman, B.; Pollard, S.; Grant, A.D.A. (*Micropterus salmoides*). *Lake* **2002**, *31*, 1763–1769. [CrossRef]

28. Wedemeyer, G.A. *Physiology of Fish in Intensive Culture Systems*; Chapman & Hall: New York, NY, USA, 1996. [CrossRef]

29. Karin, M.; Lin, A. NF-κB at the crossroads of life and death. *Nat. Immunol.* **2002**, *3*, 221–227. [CrossRef]

30. Mourente, G.; Diaz-Salvago, E.; Bell, J.; Tocher, D. Increased activities of hepatic antioxidant defence enzymes in juvenile gilthead sea bream. *Aquaculture* **2002**, *214*, 343–361. [CrossRef]

31. Wang, L.M.; Wang, J.; Bharadwaj, A.S.; Xue, M.; Qin, Y.C. Effects of dietary copper sources on growth, tissue copper accumulation and physiological responses of Japanese sea bass (*Lateolabrax japonicus*) (Cuvier, 1828) fed semipurified or practical diets. *Aquac. Res.* **2015**, *46*, 1619–1627. [CrossRef]

32. Association of Official Analytical Chemists (AOAC). *Official Methods of Analysis of the Association of Official Analytical Chemists*, 15th ed.; Association of Official Analytical Chemists Inc.: Arlington, VA, USA, 2003; Volume 1.

33. Gu, J.Z.; Liang, H.L.; Ge, X.P.; Xia, D.; Pan, L.K.; Mi, H.F.; Ren, M.C. A study of the potential effect of yellow mealworm (Tenebrio molitor) substitution for fish meal on growth, immune and antioxidant capacity in juvenile largemouth bass (Micropterus salmoides). *Fish & Shellfish Immunology* **2022**, *120*, 214–221. [CrossRef]

34. Yang, P.; Wang, W.Q.; Chi, S.Y.; Mai, K.S.; Song, F.; Wang, L. Effects of dietary lysine on regulating GH-IGF system, intermediate metabolism and immune response in largemouth bass (*Micropterus salmoides*). *Aquaculture Reports* **2020**, *17*, 100323. [CrossRef]

35. Watanabe, T.; Kiron, V.; Satoh, S. Trace minerals in fish nutrition. *Aquaculture* **1997**, *151*, 185–207. [CrossRef]

36. Wang, J.L.; Meng, X.L.; Lu, R.H.; Wu, C.; Luo, Y.T.; Yan, X.; Li, X.J.; Kong, X.H.; Nie, G.X. Effects of *Rehmannia glutinosa* on growth performance, immunological parameters and disease resistance to *Aeromonas hydrophila* in common carp (*Cyprinus carpio* L.). *Aquaculture* **2015**, *435*, 293–300. [CrossRef]

37. Mohseni, M.; Pourkazemi, M.; Bai, S.C. Effects of dietary inorganic copper on growth performance and immune responses of juvenile beluga, *Huso huso*. *Aquac. Nutr.* **2014**, *20*, 547–556. [CrossRef]

38. Song, J.; Li, L.Y.; Chen, B.B.; Shan, L.L.; Yuan, S.Y.; Yu, H.R. Dietary copper requirements of postlarval coho salmon (*Oncorhynchus kisutch*). *Aquac. Nutr.* **2021**, *27*, 2084–2092. [CrossRef]

39. Shahjahan, M.; Islam, M.J.; Hossain, M.T.; Mishu, M.A.; Hasan, J.; Brown, C. Blood biomarkers as diagnostic tools: An overview of climate-driven stress responses in fish. *Sci. Total Environ.* **2022**, *843*, 156910. [CrossRef]

40. Afshari, A.; Sourinejad, I.; Gharaei, A.; Johari, S.A.; Ghasemi, Z. The effects of diet supplementation with inorganic and nanoparticulate iron and copper on growth performance, blood biochemical parameters, antioxidant response and immune function of snow trout *Schizothorax zarudnyi* (Nikolskii, 1897). *Aquaculture* **2021**, *539*, 736638. [CrossRef]

41. Minganti, V.; Drava, G.; De Pellegrini, R.; Siccardi, C. Trace elements in farmed and wild gilthead seabream, *Sparus aurata*. *Mar. Pollut. Bull.* **2010**, *60*, 2022–2025. [CrossRef]

42. Yanpallewar, S.U.; Sen, S.; Tapas, S.; Kumar, M.; Raju, S.S.; Acharya, S.B. Effect of *Azadirachta indica* on paracetamol-induced hepatic damage in albino rats. *Phytomedicine* **2003**, *10*, 391–396. [CrossRef]

43. Chellappan, D.K.; Ganasen, S.; Batumalai, S.; Candasamy, M.; Krishnappa, P.; Dua, K.; Chellian, J.; Gupta, G. The Protective Action of the Aqueous Extract of Auricularia polytricha in Paracetamol Induced Hepatotoxicity in Rats. *Recent Pat. Drug Deliv. Formul.* **2016**, *10*, 72–76. [CrossRef]

44. Eslamloo, K.; Falahatkar, B.; Yokoyama, S. Effects of dietary bovine lactoferrin on growth, physiological performance, iron metabolism and non-specific immune responses of Siberian sturgeon *Acipenser baeri*. *Fish Shellfish Immunol.* **2012**, *32*, 976–985. [CrossRef] [PubMed]

45. Hahnel, D.; Huber, T.; Kurze, V.; Beyer, K.; Engelmann, B. Contribution of copper binding to the inhibition of lipid oxidation by plasmalogen phospholipids. *Biochem. J.* **1999**, *340*, 377–383. [CrossRef] [PubMed]

46. Hefnawy, A.; Khaiat, H. The Importance of Copper and the Effects of Its Deficiency and Toxicity in Animal Health. *Int. J. Livest. Res.* **2015**, *5*, 1–20. [CrossRef]

47. Sevcikova, M.; Modra, H.; Blahova, J.; Dobsikova, R.; Plhalova, L.; Zitka, O.; Hynek, D.; Kizek, R.; Skoric, M.; Svobodova, Z. Biochemical, haematological and oxidative stress responses of common carp (*Cyprinus carpio* L.) after sub-chronic exposure to copper. *Vet. Med.* **2016**, *61*, 35–50. [CrossRef]

48. Saffari, S.; Keyvanshokooh, S.; Zakeri, M.; Johari, S.A.; Pasha-Zanoosi, H. Effects of different dietary selenium sources (sodium selenite, selenomethionine and nanoselenium) on growth performance, muscle composition, blood enzymes and antioxidant status of common carp (*Cyprinus carpio*). *Aquac. Nutr.* **2017**, *23*, 611–617. [CrossRef]

49. Uyisenga, A.; Liang, H.; Ren, M.; Huang, D.; Xue, C.; Yin, H.; Mi, H. The Effects of Replacing Fish Meal with Enzymatic Soybean Meal on the Growth Performance, Whole-Body Composition, and Health of Juvenile Gibel Carp (*Carassius auratus gibelio*). *Fishes* **2023**, *8*, 423. [CrossRef]

50. Musharraf, M.; Khan, M.A. Estimation of dietary copper requirement of fingerling Indian major carp, *Labeo rohita* (Hamilton). *Aquaculture* **2022**, *549*, 737742. [CrossRef]

51. Wang, H.; Li, E.; Zhu, H.; Du, Z.; Qin, J.; Chen, L. Dietary copper requirement of juvenile Russian sturgeon *Acipenser gueldenstaedtii*. *Aquaculture* **2016**, *454*, 118–124. [CrossRef]

52. Shi, B.; Lu, J.; Hu, X.; Betancor, M.B.; Zhao, M.; Tocher, D.R.; Zhou, Q.; Jiao, L.; Xu, F.; Jin, M. Dietary copper improves growth and regulates energy generation by mediating lipolysis and autophagy in hepatopancreas of Pacific white shrimp (*Litopenaeus vannamei*). *Aquaculture* **2021**, *537*, 736505. [CrossRef]

53. Parvez, S.; Sayeed, I.; Pandey, S.; Ahmad, A.; Bin-Hafeez, B.; Haque, R.; Ahmad, I.; Raisuddin, S. Modulatory effect of copper on nonenzymatic antioxidants in freshwater fish *Channa punctatus* (Bloch.). *Biol. Trace Elem. Res.* **2003**, *93*, 237–248. [CrossRef]

54. Jiang, H.; Kong, X.; Wang, S.; Guo, H. Effect of Copper on Growth, Digestive and Antioxidant Enzyme Activities of Juvenile Qihe Crucian Carp, *Carassius carassius*, during Exposure and Recovery. *Bull. Environ. Contam. Toxicol.* **2016**, *96*, 333–340. [CrossRef] [PubMed]

55. Tang, Q.Q.; Feng, L.; Jiang, W.D.; Liu, Y.; Jiang, Y.; Li, S.H.; Kuang, S.Y.; Tang, L.; Zhou, X.Q. Effects of dietary copper on growth, digestive, and brush border enzyme activities and antioxidant defense of hepatopancreas and intestine for young grass carp (*Ctenopharyngodon Idella*). *Biol. Trace Elem. Res.* **2013**, *155*, 370–380. [CrossRef] [PubMed]

56. Dzoyem, J.P.; Kuete, V.; Eloff, J.N. Biochemical Parameters in Toxicological Studies in Africa: Significance, Principle of Methods, Data Interpretation, and Use in Plant Screenings. In *Toxicological Survey of African Medicinal Plants*; Elsevier Inc.: London, UK, 2014. [CrossRef]

57. Fontagné-Dicharry, S.; Lataillade, E.; Surget, A.; Larroquet, L.; Cluzeaud, M.; Kaushik, S. Antioxidant defense system is altered by dietary oxidized lipid in first-feeding rainbow trout (*Oncorhynchus mykiss*). *Aquaculture* **2014**, *424–425*, 220–227. [CrossRef]

58. Li, X.; Egervari, G.; Wang, Y.; Berger, S.L.; Lu, Z. Regulation of chromatin and gene expression by metabolic enzymes and metabolites. *Nat. Rev. Mol. Cell Biol.* **2018**, *19*, 563–578. [CrossRef] [PubMed]

59. Liang, H.; Mokrani, A.; Ji, K.; Ge, X.; Ren, M.; Pan, L.; Sun, A. Effects of dietary arginine on intestinal antioxidant status and immunity involved in Nrf2 and NF-κB signaling pathway in juvenile blunt snout bream, *Megalobrama amblycephala*. *Fish Shellfish Immunol.* **2018**, *82*, 243–249. [CrossRef]

60. Han, J.; Ulevitch, R.J. Limiting inflammatory responses during activation of innate immunity. *Nat. Immunol.* **2005**, *6*, 1198–1205. [CrossRef]

61. Taylor, A.; Verhagen, J.; Blaser, K.; Akdis, M.; Akdis, C.A. Mechanisms of immune suppression by interleukin-10 and transforming growth factor-β: The role of T regulatory cells. *Immunology* **2006**, *117*, 433–442. [CrossRef]

62. Ji, K.; Liang, H.; Ren, M.; Ge, X.; Liu, B.; Xi, B.; Pan, L.; Yu, H. Effects of dietary tryptophan levels on antioxidant status and immunity for juvenile blunt snout bream (*Megalobrama amblycephala*) involved in Nrf2 and TOR signaling pathway. *Fish Shellfish Immunol.* **2019**, *93*, 474–483. [CrossRef]

63. Zhao, L.; Yuan, B.D.; Zhao, J.L.; Jiang, N.; Zhang, A.Z.; Wang, G.Q.; Li, M.Y. Amelioration of hexavalent chromium-induced bioaccumulation, oxidative stress, tight junction proteins and immune-related signaling factors by *Allium mongolicum Regel* flavonoids in *Ctenopharyngodon idella*. *Fish Shellfish Immunol.* **2020**, *106*, 993–1003. [CrossRef]

64. Ali, Z.; Khan, I.; Iqbal, M.S.; Zhang, Q.; Ai, X.; Shi, H.; Ding, L.; Hong, M. Toxicological effects of copper on bioaccumulation and mRNA expression of antioxidant, immune, and apoptosis-related genes in Chinese striped-necked turtle (*Mauremys sinensis*). *Front. Physiol.* **2023**, *14*, 1296259. [CrossRef]

Article

Effects of Dietary Protein Levels on Growth Performance, Plasma Parameters, and Digestive Enzyme Activities in Different Intestinal Segments of *Megalobrama amblycephala* at Two Growth Stages

Wuxiao Zhang [1,2], Silei Xia [1], Bo Liu [3], Hongyan Tian [1], Fei Liu [1], Wenping Yang [1], Yebing Yu [1], Caiyuan Zhao [4], Naresh Kumar Dewangan [4], Aimin Wang [1,*] and Tao Teng [3,5,*]

[1] College of Marine and Biology Engineering, Yancheng Institute of Technology, Yancheng 224051, China; zhangwx257258@163.com (W.Z.); susanxia1990323@126.com (S.X.); tianhyy@outlook.com (H.T.); liufei@ycit.edu.en (F.L.); yangwenping@ycit.edu.cn (W.Y.); yuyebing2005@126.com (Y.Y.)

[2] Key Laboratory of Agricultural Environmental Microbiology, Ministry of Agriculture and Rural Affairs, College of Life Sciences, Nanjing Agricultural University, Nanjing 210095, China

[3] Key Laboratory of Freshwater Fisheries and Germplasm Resources Utilization, Ministry of Agriculture, Freshwater Fisheries Research Center, Chinese Academy of Fishery Sciences, Wuxi 214081, China; liub@ffrc.cn

[4] College of Animal Science and Technology, Henan University of Animal Husbandry and Economy, Zhengzhou 450046, China; zhaocaiyuan326@foxmail.com (C.Z.); nareshbiotech@yahoo.in (N.K.D.)

[5] Tongwei Agricultural Development Co., Ltd., Key Laboratory of Nutrition and Healthy Culture of Aquatic, Livestock and Poultry, Ministry of Agriculture and Rural Affairs, Healthy Aquaculture Key Laboratory of Sichuan Province, Chengdu 610093, China

* Correspondence: blueseawam@ycit.cn (A.W.); tengt@tongwei.com (T.T.)

Academic Editor: Seyed Hossein Hoseinifar

Received: 31 December 2024
Revised: 28 January 2025
Accepted: 29 January 2025
Published: 31 January 2025

Citation: Zhang, W.; Xia, S.; Liu, B.; Tian, H.; Liu, F.; Yang, W.; Yu, Y.; Zhao, C.; Dewangan, N.K.; Wang, A.; et al. Effects of Dietary Protein Levels on Growth Performance, Plasma Parameters, and Digestive Enzyme Activities in Different Intestinal Segments of *Megalobrama amblycephala* at Two Growth Stages. *Fishes* 2025, 10, 60. https://doi.org/10.3390/fishes10020060

Abstract: An 8-week rearing trial was designed to estimate the dietary protein requirement and evaluate the effects of dietary protein on growth performance, plasma parameters, and digestive enzyme activities of blunt snout bream at two growth stages. Six practical diets were prepared to feed two sizes of fish (larger fish: initial weight of 153.69 ± 0.85 g; smaller fish: initial weight of 40.89 ± 0.28 g) with graded protein levels (26%, 28%, 30%, 32%, 34%, and 36%). Our results show that the final weight, weight gain (WG), and specific growth rate (SGR) of the fish initially rose to peak values and then declined as the dietary protein levels increased. The higher WG and SGR were recorded in the larger fish fed diets containing 30%, 32%, and 34% protein, and in the smaller fish fed a 30% protein diet, all significantly higher than those in the control group ($p < 0.05$). No significant differences were observed in the feed conversion ratio (FCR), viscerosomatic ratio (VR), hepatosomatic index (HSI), condition factor (CF), or survival rate among the treatments at both growth stages ($p > 0.05$). The plasma total protein (TP) content was highest at both growth stages in fish fed a 30% protein diet ($p < 0.05$). As the dietary protein level increased, the plasma urea content of the larger fish increased, peaked in the 34% protein group ($p < 0.05$), and then remained stable. In contrast, no significant difference in the plasma urea content was seen among the treatment groups of the smaller fish ($p > 0.05$). Protease activity in the fish foregut at both growth stages peaked in the 32% protein group ($p < 0.05$). In the midgut of the larger fish, protease activity was higher in the control group, while in the smaller fish, it was higher in the 36% protein group ($p < 0.05$). In the larger fish, hindgut protease activity was higher in the 34% protein group ($p < 0.05$), while in the smaller fish, there was no significant difference in the hindgut protease activity among all groups ($p > 0.05$). The dietary protein levels had no significant effect on lipase activity in the foregut, midgut, or hindgut, or on amylase activity in the foregut or midgut of the fish at the two growth stages ($p > 0.05$). However, hindgut amylase activity was highest in the control group of the smaller fish ($p < 0.05$). Based on regression analysis, the optimal dietary protein levels for the larger and smaller fish were 30.45% and 29.95%, respectively. Overall, appropriate dietary protein levels (30%) could improve the growth performance, immune function, and health status of fish at two growth stages and promote the adaptive response of their digestive system, especially the spatial regulation of protease activity in different gastrointestinal regions.

Keywords: protein requirement; *Megalobrama amblycephala*; growth performance; plasma parameters; digestive enzyme activities; different growth stages

Key Contribution: The optimal protein requirements for two sizes of *Megalobrama amblycephala* were determined, and the effects of dietary protein levels on plasma protein metabolism and digestive enzyme activities in different parts of the intestinal tract were investigated.

1. Introduction

Protein is regarded as the main nutrient for fish, providing energy and essential amino acids [1,2]. It is an important macronutrient in nutritional studies, as it forms the costly part of commercial fish feed and is crucial for growth and development, also key for maintaining fish optimal growth and reproduction [3–5]. Insufficient dietary protein can lead to stunted growth or even cessation of growth, as the body must divert protein from less critical tissues to sustain the function of more vital organs and systems [6]. An excessive intake of dietary protein can be equally detrimental. The surplus protein is metabolized into energy, which not only increases feed costs but also leads to higher levels of ammonia nitrogen in the water, potentially harming the aquatic environment [7]. Therefore, determining the optimal protein requirements for fish is crucial for promoting their best growth and efficient diet utilization.

Megalobrama amblycephala, also referred to as the blunt snout bream, is a herbivorous freshwater fish that is indigenous to China. In recent years, the aquaculture industry has experienced rapid growth in the farming of this species, primarily due to its rapid growth rate, high feed efficiency, tender flesh, and strong disease resistance [8]. As a result, the total production of this fish in China reached approximately 0.74 million tons in 2023 [9]. The natural diet of the blunt snout bream consists mainly of aquatic plants, such as zooplankton. Consequently, its nutritional requirements are relatively simple. However, the blunt snout bream can adapt to formulated feeds, and understanding its nutritional requirements is essential for developing appropriate feed formulations that support optimal growth and health. In the context of fish nutrition, the demand for dietary protein is a critical factor that can significantly influence the overall health and productivity of the fish. The blunt snout bream, with its herbivorous tendencies, is no exception to this rule. In China, diets for blunt snout bream fingerlings were recommended to have at least 30% protein [10]. A previous study found that blunt snout bream requires 27–30% protein for optimal growth at about 20 °C water temperature, and the optimum protein requirement ranges from 25.6% to 41.4% when the water temperature is from 25 °C to 30 °C [11]. The specific protein requirements for this species can vary depending on its size, life stage, and environmental conditions [12]. Therefore, research into the most effective protein levels for different sizes of *Megalobrama amblycephala* is essential to optimize aquaculture practices.

The plasma biochemical parameters of fish provide valuable insights into their metabolic status and overall health [13]. For instance, levels of enzymes, such as alanine aminotransferase (ALT) and aspartate aminotransferase (AST), can indicate liver function, while urea and creatinine levels can reflect kidney function [6,14]. Monitoring these parameters in relation to dietary protein levels can help in assessing whether the protein intake is appropriate or excessive, potentially leading to stress or metabolic disorders [15].

The activity of digestive enzymes in fish depends on their diet, and different diets can affect it [16]. It was reported that the utilization of feed protein is affected by the activity of digestive enzymes [17]. Also, protease activity changes in response to different levels of protein in the diet [18]. The activity of digestive enzymes in the intestine is a critical factor in determining the efficiency with which dietary protein is utilized. Enzymes such as proteases, amylases, and lipases play a vital role in breaking down complex nutrients into simpler forms that can be absorbed by the fish [19]. Understanding how different protein levels affect these enzymes can help in designing feeds that maximize nutrient absorption and minimize waste.

Therefore, this study aimed to carry out a comprehensive investigation into the effects of dietary protein levels on the growth performance, plasma protein metabolism, and intestinal digestive enzyme activity of *Megalobrama amblycephala* at two sizes. Through the determination of optimal protein levels tailored to various sizes of this fish, researchers can make contributions to the advancement of more efficient and ecologically sustainable aquaculture feed formulations.

2. Materials and Methods

2.1. Experimental Diets

Based on previous research on the protein requirements of blunt snout bream [10,11], six isonitrogenous and isoenergetic diets were formulated. Using fish meal, casein, and gelatin as protein sources and soybean oil as a lipid source, they contained graded levels of protein (26%, 28%, 30%, 32%, 34%, and 36%) (Table 1). All the ingredients were ground into a powder and thoroughly mixed with the soybean oil and water. Then, they were forced through a pelletizer (Made in South China University of Technology, Guangzhou, China) and air-dried at 20 °C for 24 h, and then packaged in airtight plastic bags and stored at −20 °C.

Table 1. Formulation and chemical composition of experimental diets (% dry diet).

Ingredients	Group					
	26	28	30	32	34	36
Casein [a]	16.86	18.86	19.5	20.8	22	23.5
Gelatin [a]	4.25	4.25	4.85	5.5	5.71	5.86
White fish meal [a]	8.95	9.95	10.7	11.5	12.26	13
Corn starch	33.75	30.75	28.5	25.2	22.85	20
Dextrin	10	10	10	10	10	10
Microcrystalline cellulose	5.84	5.84	6.2	6.9	7.23	7.84
Carboxylmethyl cellulose	10	10	10	10	10	10
Soybean Oil	5.5	5.5	5.45	5.4	5.35	5.3
Vitamin mix and mineral mix [b]	1	1	1	1	1	1
Soybean lecithin	1	1	1	1	1	1
Calcium dihydrogen phosphate	2.65	2.65	2.6	2.5	2.4	2.3
Chlorinated choline	0.15	0.15	0.15	0.15	0.15	0.15
Ethoxyquin	0.05	0.05	0.05	0.05	0.05	0.05
Methionine + cystine	0.47	0.51	0.53	0.56	0.59	0.63
Lysine	2.09	2.16	2.17	2.20	2.24	2.28
Analyzed nutrient compositions (% of dry diet)						
Crude protein	25.89	28.46	30.07	32.37	34.18	36.20
Crude lipid	6.56	6.62	6.61	6.60	6.60	6.59
Gross energy (KJ/g)	19.48	19.61	19.60	19.56	19.59	19.56

[a] Casein, obtained from Hua'an Biological Products Lit. (Linxia, China), 90.2% crude protein; gelatin, obtained from Zhanyun Chemical Lit. (Shanghai, China), 91.3% crude protein; white fish meal, obtained from Copeinca (Lima, Peru), 67.4% crude protein and 9.3% crude lipid. [b] Vitamin premix (IU or mg/kg of diet): vitamin A, 25,000 IU; vitamin D3, 20,000 IU; vitamin E, 200 mg; vitamin K3, 20 mg; thiamin, 40 mg; riboflavin, 50 mg; calcium pantothenate, 100 mg; pyridoxine HCl, 40 mg; cyanocobalamin, 0.2 mg; biotin, 6 mg; folic acid, 20 mg; niacin, 200 mg; inositol, 1000 mg; vitamin C, 2000 mg; choline, 2000 mg. Cellulose was used as a carrier. Mineral premix (g/kg of diet): calcium biphosphate, 20 g; sodium chloride, 2.6; potassium chloride, 5 g; magnesium sulphate, 2 g; ferrous sulphate, 0.9 g; zinc sulphate, 0.06 g; cupric sulphate, 0.02; manganese sulphate, 0.03 g; sodium selenate, 0.02 g; cobalt chloride, 0.05 g; potassium iodide, 0.004. Zeolite was used as a carrier.

2.2. Experimental Fish and Feeding Trial

Experimental fish at two sizes were obtained from the Freshwater Fisheries Research Center of the Chinese Academy of Fishery Sciences (Wuxi, China). Prior to the feeding trial, these two sizes fish were reared in 200 L and 750 L tanks, respectively, for 2 weeks to acclimate to the experimental conditions while fed a commercial diet containing 32% protein and 5% lipid (1.5 mm in diameter, Wuxi Tongwei Feedstuffs, Wuxi, China). After fasting for 24 h, the healthy, similarly sized fish with a mean body weight of 40.89 ± 0.28 g were randomly sorted into eighteen 200 L tanks with 20 fish per tank (3 replicates per group). The same was carried out for another size of blunt snout bream, where the healthy, similarly sized fish with a mean body weight of 153.69 ± 0.85 g were randomly sorted into eighteen 750 L tanks with 25 fish per tank (3 replicates per group).

During the eight-week feeding trial, the fish were hand-fed the experimental diets to satiation three times daily (8:00, 12:00, and 16:00 h). Feed consumption and the number and weight of dead fish were recorded daily. The water temperature was controlled at 26–28 °C, pH 7.2–7.8. The dissolved oxygen concentration was higher than 5 mg/L, and the ammonia nitrogen concentration was <0.1 mg/L. The concentrations of ammonia-N were determined using the Nesslerization method, as described by Zhang et al. [20]. The temperature, pH, and dissolved oxygen were measured using a water quality instrument (YSI Inc., Yellow Springs, OH, USA).

The calculation formulas used for growth performance were as follows:

Specific growth rate, (SGR) = 100 × (Ln finial individual weight − Ln initial individual weight)/number of days;

Weight gain, (WG) = 100 × (final weight − initial weight)/initial weight;

Feed conversation ratio, (FCR) = (wet weight gain, g)/(dry feed weight, g);

Viscerosomatic ratio, (VR) = 100 × (viscera weight, g)/(body weight, g);

Hepatosomatic index, (HSI) = 100 × (liver weight, g)/(body weight, g);

Condition factor, (CF) = $100 \times W/L^3$, where W is the weight (g), and L is the length (cm);

Survival rate (%) = 100 × (final number of fish)/(initial number of fish).

2.3. Sample Collection

At the end of the experiment, the fish were fasted for 24 h and individually counted and weighed from each tank. Three fish were chosen from each tank and anesthetized with MS-222 (100 mg/L, Sigma Chemical Company, St. Louis, MO, USA), and then blood samples were collected immediately from the caudal vein using heparinized syringes. Following centrifugation at 3000 g for 10 min at 4 °C, the plasma was separated and stored at −80 °C. After the blood samples were collected, the intestine was excised and stored at −80 °C.

2.4. Chemical Analysis

The crude protein and crude lipid contents in the diets were determined according to the established methods of (AOAC, 2003) [21]: the crude protein content (N × 6.25) was determined by the Kjeldahl method using the semi-automatic Kjeldahl system (1030 Auto-analyzer, Tecator, Hoganos, Sweden) after acid digestion; the crude lipid content was measured by the ether extraction method using the Soxhlet system HT6 (Soxtec System, Tecator, Sweden); and the gross energy content was measured using the IKA C2000 basic bomb calorimeter (IKA Works Inc., Wilmington, NC, USA).

The plasma levels of alkaline phosphatase (ALP), total protein (TP), albumin (ALB), urea, and creatinine (Crea) were measured with an automatic biochemical analyzer (Mindray BS-400, Mindray Bio-Medical Electronics Co., Ltd., Shenzhen, China) using the AMP buffer method, biuret, bromocresol green, UV method, and sarcosine oxidase method, respectively. Determination of the low-density lipoprotein (LDL) and high-density lipoprotein (HDL) was carried out following the methods described by Mozanzadeh et al. [22]. The total globulin (GLB) content was estimated by subtracting the albumin from the total protein [22].

Intestinal digestive enzyme activity was determined according to the method described by Bowyer et al. [23]. The fish gut was divided into three sections (foregut, midgut, and hindgut). The intestines were homogenized in 10 volumes (w/v) of $4000\times g$ for 20 min at 4 °C, and the supernatant was used as the enzyme source. The activities of the protein concentration, lipase, and amylase in the intestines of blunt snout bream were determined by spectrophotometry using kits from Nanjing Jiancheng Bioengineering Institute (Nanjing, China). The activity of protease was assayed following the Forint phenol reagent method in 0.01 M Tris-HCL (pH 7.4) buffer using 2% casein as a substrate. The reaction was carried out for 10 min at 30 °C, stopped with 0.1 M trichloroacetic acid (TCA), and then centrifuged at $3000\times g$ for 5 min at 4 °C. Then, 0.5 mL of the supernatant was added to 2.5 mL of 0.4 M $NaHCO_3$ and 0.5 mL of 50% Folin phenol reagent, and the optical density was read at 680 nm using tyrosine as a standard. The specific activity of protease was expressed as 1 μmol of hydrolyzed tyrosine per minute per milligram of protein (U/mg). The specific activity of amylase was expressed as 1 mol of reducing sugars per min per mg of protein (U/mg). The specific activity of lipase was expressed as 1 μmol of hydrolyzed substrate per minute per gram of protein (U/g).

2.5. Statistical Analysis

All data were presented as the means $\pm$ SEMs (standard errors of the mean). The data were subjected to one-way analysis of variance (ANOVA) using SPSS 27.0 software for Windows (International Business Machines Corporation, Armonk, NY, USA). Significant differences in the means among the dietary treatments were evaluated by Tukey's multiple range test. Mean differences were considered significant at a p value of less than 0.05.

3. Results

3.1. Growth Performance

As the dietary protein levels increased, the final weight, weight gain (WG), and specific growth rate (SGR) of the blunt snout bream at two sizes initially rose to their peak values, and then subsequently exhibited a downward trend (as shown in Table 2). A higher final weight was observed in the larger fish (initial weight of 153.69 g) fed 30% and 32% protein, significantly higher than that in the control group ($p < 0.05$). Moreover, higher WG and SGR were recorded in the larger fish fed diets with 30%, 32%, and 34% protein, also significantly higher than those in the control group ($p < 0.05$). For the smaller fish (with an initial weight of 40.89 g), the highest final weight, WG, and SGR were achieved with a 30% protein diet, which were also significantly higher than those in the 28% protein group and the control group ($p < 0.05$). Broken-line regression analysis showed, based on the WG, that the optimal dietary protein levels in the larger fish and smaller fish were estimated to be 30.45% and 29.95% (Figure 1), respectively. The feed conversion ratio (FCR), viscerosomatic ratio (VR), hepatosomatic index (HSI), condition factor (CF), and survival rate of the fish at two growth stages were not significantly different among the treatments ($p > 0.05$).

Table 2. Effects of dietary protein levels on growth performance in blunt snout bream at two growth stages.

Item	Protein Level (% of Dry Diet)						*p* Value
	26	28	30	32	34	36	
Trial 1 (initial body weight of 153.69 g)							
Initial weight (g)	152.86 ± 0.29	153.76 ± 0.64	154.27 ± 0.27	153.72 ± 0.87	153.52 ± 0.36	153.67 ± 0.65	0.659
Final weight (g)	278.98 ± 1.41 [a]	285.80 ± 1.79 [ab]	297.24 ± 0.12 [b]	294.72 ± 5.05 [b]	294.36 ± 1.08 [ab]	289.48 ± 4.60 [ab]	0.018
SGR (% day^{-1})	1.07 ± 0.01 [a]	1.11 ± 0.01 [ab]	1.17 ± 0.00 [b]	1.16 ± 0.02 [b]	1.16 ± 0.01 [b]	1.13 ± 0.02 [ab]	0.009
WG	82.50 ± 0.61 [a]	85.88 ± 1.21 [ah]	93.02 ± 0.08 [b]	91.70 ± 2.21 [b]	91.74 ± 1.15 [b]	88.37 ± 2.49 [ab]	0.007
FCR	2.75 ± 0.02	2.45 ± 0.10	2.21 ± 0.20	1.89 ± 0.04	1.98 ± 0.02	2.17 ± 0.21	0.281
VR	13.91 ± 3.54	15.49 ± 1.38	11.16 ± 0.72	15.24 ± 1.16	13.43 ± 1.49	11.88 ± 1.66	0.38
HSI	1.68 ± 0.32	1.63 ± 0.13	1.56 ± 0.15	1.82 ± 0.10	1.58 ± 0.17	1.76 ± 0.09	0.873
CF	2.12 ± 0.11	2.15 ± 0.03	2.02 ± 0.02	2.14 ± 0.03	2.18 ± 0.02	2.08 ± 0.02	0.204
Survival (%)	92.00 ± 2.30	98.67 ± 1.33	96.00 ± 4.00	96.00 ± 2.31	98.67 ± 1.33	94.67 ± 1.33	0.227
Trial 2 (initial body weight of 40.89 g)							
Initial weight (g)	40.78 ± 0.04	40.81 ± 0.02	41.02 ± 0.09	40.92 ± 0.09	40.93 ± 0.04	40.87 ± 0.04	1.53
Final weight (g)	80.97 ± 1.05 [a]	81.81 ± 0.67 [a]	90.62 ± 1.85 [b]	85.32 ± 0.26 [ab]	84.34 ± 1.66 [ab]	84.27 ± 0.17 [ab]	0.004
SGR (% day^{-1})	1.23 ± 0.02 [a]	1.25 ± 0.02 [a]	1.41 ± 0.04 [b]	1.31 ± 0.01 [ab]	1.29 ± 0.03 [ab]	1.30 ± 0.01 [ab]	0.005
WG	98.54 ± 2.59 [a]	100.50 ± 1.65 [a]	120.92 ± 4.39 [b]	108.51 ± 0.71 [ab]	106.04 ± 3.91 [ab]	106.04 ± 0.67 [ab]	0.005
FCR	1.99 ± 0.10	2.12 ± 0.08	1.72 ± 0.08	1.85 ± 0.01	2.26 ± 0.33	2.13 ± 0.24	0.348
VR	16.17 ± 3.70	14.05 ± 0.60	15.71 ± 0.96	14.02 ± 2.01	18.70 ± 2.54	16.25 ± 0.66	0.637
HSI	1.72 ± 0.10	1.70 ± 0.17	1.28 ± 0.09	1.59 ± 0.20	1.54 ± 0.09	1.29 ± 0.11	0.143
CF	2.09 ± 0.14	1.97 ± 0.01	2.02 ± 0.11	1.97 ± 0.02	2.07 ± 0.05	1.92 ± 0.01	0.55
Survival (%)	73.33 ± 13.02	98.33 ± 1.67	100	98.33 ± 1.67	86.67 ± 8.33	91.39 ± 3.25	0.120

Note: Values are presented as means $\pm$ SEM (n = 3). Values with different superscripts in the same row are significantly ($p < 0.05$) different. SGR, specific growth rate; WG, weight gain; FCR, feed conversation ratio; VR, viscerosomatic ratio; HSI, hepatosomatic index; CF, condition factor.

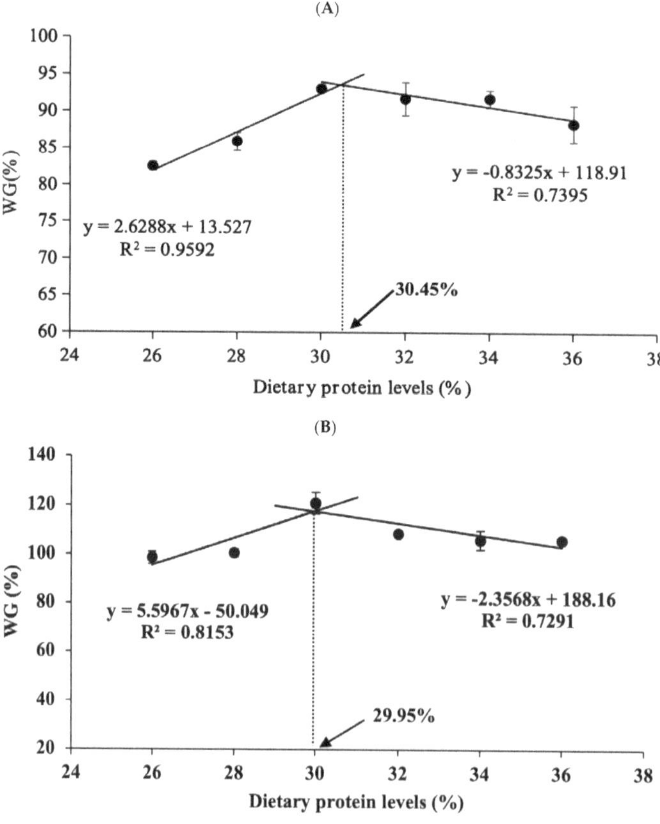

Figure 1. Regression analysis of weight gain (WG, %) against different graded levels of dietary protein. (**A**) Larger fish, initial body weight of 153.69 g; (**B**) smaller fish, initial body weight of 40.89 g.

3.2. Plasma Parameters

In all treatment groups, no significant differences were observed in the plasma levels of alkaline phosphatase (ALP), albumin (ALB), globulin (GLB), low-density lipoprotein (LDL), high-density lipoprotein (HDL), or creatinine (Crea) at either fish growth stage ($p > 0.05$, Table 3). However, when fed a diet containing 30% protein, the plasma total protein (TP) content at both growth stages of blunt snout bream was highest compared to that of the other groups ($p < 0.05$). As the dietary protein level increased, the plasma urea content of the larger fish (with an initial weight of 153.69 g) increased accordingly, peaking in the 34% protein group. This value was significantly higher than that of the control group ($p < 0.05$), and subsequently remained stable. Conversely, no significant difference in the plasma urea content was observed among all treatment groups of the smaller fish (with an initial weight of 40.89 g) ($p > 0.05$). Furthermore, the plasma levels of urea and creatinine, particularly creatinine, were substantially higher in the larger fish compared to the smaller fish.

Table 3. Effects of dietary protein levels on plasma parameters in blunt snout bream at two growth stages.

Item	Protein Level (% of Dry Diet)						*p* Value
	26	28	30	32	34	36	
Trial 1 (initial body weight of 153.69 g)							
ALP	39.28 ± 8.46	44.93 ± 6.91	44.99 ± 4.46	45.57 ± 5.47	35.12 ± 3.71	43.68 ± 4.51	0.783
TP	37.27 ± 0.30 [a]	37.68 ± 2.55 [a]	49.87 ± 1.55 [b]	38.48 ± 0.29 [a]	38.93 ± 1.39 [a]	42.77 ± 1.74 [a]	0.002
ALB	17.80 ± 1.33	16.65 ± 1.55	19.26 ± 0.83	17.93 ± 1.70	17.40 ± 1.17	18.78 ± 1.10	0.744
GLB	25.62 ± 1.95	23.88 ± 2.04	26.83 ± 0.61	26.03 ± 1.75	25.06 ± 1.24	27.78 ± 1.02	0.554
Urea	0.67 ± 0.05 [a]	0.82 ± 0.09 [ab]	0.85 ± 0.04 [ab]	0.90 ± 0.03 [ab]	1.13 ± 0.19 [b]	0.95 ± 0.07 [ab]	0.037
LDL	1.13 ± 0.18	1.55 ± 0.26	1.62 ± 0.16	1.91 ± 0.29	1.82 ± 0.31	2.01 ± 0.15	0.093
HDL	4.36 ± 0.16	4.20 ± 0.24	4.79 ± 0.20	4.77 ± 0.36	4.64 ± 0.32	5.00 ± 0.36	0.396
Crea	1008.54 ± 107.99	981.98 ± 45.56	1028.69 ± 50.62	964.60 ± 58.11	1066.96 ± 44.51	1116.52 ± 47.46	0.509
Trial 2 (initial body weight of 40.89 g)							
ALP	62.34 ± 8.98	50.23 ± 6.71	55.06 ± 5.97	61.54 ± 8.15	63.74 ± 7.36	47.81 ± 4.56	0.460
TP	26.44 ± 1.00 [a]	28.52 ± 1.20 [a]	33.83 ± 0.11 [b]	30.15 ± 1.16 [a]	28.33 ± 0.79 [a]	26.93 ± 1.75 [a]	0.009
ALB	4.17 ± 0.52	1.85 ± 0.10	2.96 ± 0.70	4.34 ± 0.64	3.28 ± 0.46	4.10 ± 0.36	0.087
GLB	27.59 ± 0.68	26.50 ± 1.51	26.50 ± 1.32	25.58 ± 1.39	25.33 ± 0.67	24.33 ± 2.73	0.611
Urea	0.44 ± 0.05	0.31 ± 0.01	0.45 ± 0.04	0.43 ± 0.05	0.44 ± 0.03	0.45 ± 0.03	0.123
LDL	1.21 ± 0.18	1.05 ± 0.07	1.04 ± 0.12	1.29 ± 0.14	1.02 ± 0.11	1.02 ± 0.15	0.611
HDL	3.01 ± 0.16	2.97 ± 0.26	2.79 ± 0.14	2.98 ± 0.10	2.77 ± 0.07	2.56 ± 0.18	0.356
Crea	38.79 ± 1.48	33.13 ± 0.90	38.47 ± 2.22	37.16 ± 1.37	34.87 ± 1.86	37.66 ± 1.15	0.113

Note: Values are presented as means ± SEM (n = 3). Values with different superscripts in the same row are significantly ($p < 0.05$) different.

3.3. Intestinal Digestive Enzyme Activity

As the dietary protein levels increased, the protease activity in the foregut of blunt snout bream at both growth stages exhibited an initial increase followed by a decrease (Table 4). Notably, the protease activity in the foregut at both fish growth stages peaked in the 32% protein group, which was significantly higher than that in the control group ($p < 0.05$). In the midgut of the larger fish, protease activity was highest in the control group, significantly higher than that of the 30% protein group ($p < 0.05$). Conversely, for the smaller fish, protease activity was highest in the 36% protein group, which was significantly higher than that in all other groups except the 32% protein group ($p < 0.05$). In the larger fish, hindgut protease activity was highest in the 34% protein group, significantly higher than that in the 30% protein group ($p < 0.05$), and not significantly different from other groups ($p > 0.05$). In smaller fish, no significant difference in hindgut protease activity was observed among all groups ($p > 0.05$).

Table 4. Effects of dietary protein levels on digestive enzyme activity in blunt snout bream at two growth stages.

Item	Protein Level (% of Dry Diet)						*p* Value
	26	28	30	32	34	36	
Trial 1 (initial body weight of 153.69 g)							
Protease activity							
Foregut	1.13 ± 0.17 [a]	1.35 ± 0.11 [a]	2.28 ± 0.06 [bc]	2.84 ± 0.06 [c]	1.72 ± 0.14 [ab]	1.49 ± 0.14 [ab]	0.006
Midgut	2.22 ± 0.30 [b]	1.42 ± 0.21 [ab]	0.81 ± 0.20 [a]	1.29 ± 0.31 [ab]	1.66 ± 0.24 [ab]	1.75 ± 0.20 [ab]	0.022
Hindgut	$1.66 + 0.10$ [ab]	1.63 ± 0.33 [ab]	0.71 ± 0.19 [a]	1.47 ± 0.20 [ab]	1.86 ± 0.16 [b]	1.56 ± 0.25 [ab]	0.033
Lipase activity							
Foregut	4.40 ± 0.52	4.00 ± 0.26	3.33 ± 0.32	4.12 ± 0.40	4.74 ± 0.44	4.98 ± 0.51	0.939
Midgut	5.44 ± 0.77	3.06 ± 0.24	4.87 ± 0.43	3.51 ± 0.48	3.20 ± 0.29	5.74 ± 0.52	0.109
Hindgut	4.90 ± 0.52	2.65 ± 0.37	3.00 ± 0.43	3.26 ± 0.36	4.41 ± 0.40	5.98 ± 0.63	0.251
Amylase activity							
Foregut	38.81 ± 3.85	37.41 ± 2.16	40.90 ± 4.38	52.46 ± 14.10	53.93 ± 14.20	47.84 ± 9.31	0.081
Midgut	47.24 ± 12.58	33.06 ± 5.76	74.50 ± 0.90	51.86 ± 9.56	37.72 ± 8.00	61.43 ± 11.33	0.067
Hindgut	45.41 ± 6.16	50.64 ± 6.66	42.41 ± 7.73	47.72 ± 10.67	56.75 ± 10.61	54.39 ± 9.19	0.842
Trial 1 (initial body weight of 40.89 g)							
Protease activity							
Foregut	1.27 ± 0.11 [a]	1.20 ± 0.19 [a]	1.57 ± 0.20 [ab]	1.91 ± 0.09 [b]	1.71 ± 0.28 [ab]	1.47 ± 0.10 [ab]	0.02
Midgut	1.29 ± 0.13 [a]	1.00 ± 0.23 [a]	1.36 ± 0.20 [a]	1.50 ± 0.23 [ab]	1.35 ± 0.29 [a]	2.74 ± 0.53 [b]	0.005
Hindgut	1.86 ± 0.27	1.55 ± 0.18	1.47 ± 0.08	1.55 ± 0.13	1.75 ± 0.10	1.78 ± 0.18	0.588
Lipase activity							
Foregut	9.19 ± 0.94	10.19 ± 1.12	12.26 ± 1.20	10.11 ± 1.12	9.46 ± 1.02	10.00 ± 1.09	0.727
Midgut	8.62 ± 1.01	7.70 ± 0.78	9.24 ± 1.04	8.73 ± 1.02	11.31 ± 1.25	11.26 ± 1.05	0.195
Hindgut	7.43 ± 0.77	5.17 ± 0.49	5.93 ± 0.68	5.09 ± 0.64	5.39 ± 0.71	6.48 ± 0.65	0.697
Amylase activity							
Foregut	48.35 ± 4.38	49.13 ± 4.23	61.18 ± 7.67	58.71 ± 4.33	54.10 ± 7.52	40.58 ± 2.57	0.137
Midgut	56.36 ± 6.43	47.07 ± 6.03	61.02 ± 3.62	47.37 ± 3.93	51.21 ± 6.54	67.19 ± 6.45	0.098
Hindgut	78.48 ± 6.56 [b]	72.40 ± 7.26 [ab]	38.97 ± 5.52 [a]	60.94 ± 12.44 [ab]	60.58 ± 9.72 [ab]	55.12 ± 5.53 [ab]	0.043

Note: Values are presented as means $\pm$ SEM (n = 3). Values with different superscripts in the same row are significantly ($p < 0.05$) different.

The dietary protein levels had no significant difference on the lipase activity in the foregut, midgut, or hindgut at either growth stage of fish ($p > 0.05$). Similarly, the dietary protein levels did not significantly affect the amylase activity in the foregut or midgut at either growth stage of fish ($p > 0.05$). However, in the hindgut of the smaller fish, the amylase activity was the highest in the control group, significantly higher than that in the 30% protein group ($p < 0.05$), with no significant difference compared to the other groups ($p > 0.05$). In contrast, no significant difference in the hindgut amylase activity was observed among the groups of larger fish ($p > 0.05$).

4. Discussion

The results of this study indicate that the dietary protein levels had a significant impact on the growth performance of blunt snout bream, as evidenced by the changes in the final weight, WG, and SGR. The observed trend of an initial increase followed by a subsequent decrease in these growth parameters suggests an optimal dietary protein range for both larger and smaller fish. This pattern is consistent with previous research in aquaculture, which has identified protein requirement plateaus for various fish species [24]. For the larger fish, the peak values of WG and SGR were achieved with diets containing 30%, 32%, and 34% protein, with the highest values observed at 30%. This suggests that the protein requirement for optimal growth in larger fish is around 30%, and that increasing protein levels beyond this point does not provide additional benefits. This finding is similar to those of studies that have reported optimal protein levels for growth in other fish species, such as *Oreochromis niloticus* [25] and *Lepomis macrochirus* [24]. In the smaller fish, the highest WG and SGR were also observed with a 30% protein diet. This indicates that smaller fish have a similar optimal protein requirement, also around 30%. The similar response of the growth parameters to the dietary protein levels between the larger and smaller fish is consistent with previous research that has reported similar protein requirements for different life stages of the same fish species. Specifically, no differences were found in the protein requirement for Dicentrarchus labrax with an initial body weight (IBW) of 99 or 160 g [26], humpback grouper *Cromileptes altivelis* (Valenciennes) with an IBW of 136, 175, or 225 g [27], and Scophthalmus maximus with an IBW of 4, 59, or 209 g [28]. In addition, according to broken-line regression analysis of the WG against dietary protein, the optimal dietary protein contents for the larger fish and smaller fish were determined to be 30.45% and 29.95%, respectively. These results further support the notion that the optimal protein requirement for growth is relatively consistent across different life stages and sizes within the same fish species. The broken-line regression analysis provided a more precise estimation of the optimal protein level, and the slight difference between the larger and smaller fish may be due to variations in their metabolic rate, digestive efficiency, or other physiological factors. The overall trend is clear: an intermediate level of dietary protein is beneficial for optimal growth in both larger and smaller fish, while excessive protein does not confer additional advantages. This has important implications for aquaculture practices, as it suggests that feed formulations can be optimized to meet the protein requirements of fish without unnecessary over-supplementation. Such optimizations can lead to more cost-effective production, reduced environmental impact, and improved animal welfare.

Interestingly, the dietary protein levels did not affect the feed conversion ratio (FCR), viscerosomatic ratio (VR), hepatosomatic index (HSI), or condition factor (CF) of the two sizes of fish. Our results suggest that while protein levels affected WG, they did not significantly alter the efficiency of feed utilization or the overall health status of the fish, as indicated by the VR, HSI, and CF. The absence of significant differences in these parameters across treatments imply that the fish were able to efficiently utilize the protein provided in the diets, and that the protein levels tested were not so high as to cause adverse effects on the fish's health or metabolism. This finding is supported by previous research that has reported similar results in terms of feed utilization and health status in fish fed diets with different protein levels [27,29,30].

The consistent plasma levels of alkaline phosphatase (ALP), albumin (ALB), globulin (GLB), low-density lipoprotein (LDL), and high-density lipoprotein (HDL) across all the various treatments administered indicate that the protein levels tested did not have any adverse effects on the metabolic processes associated with these biochemical parameters. This suggests that the tested protein levels maintained normal metabolic functions without causing any significant disruptions [31]. However, the total protein (TP) content was highest in both sizes of fish when fed a diet containing 30% protein. This suggests that a 30% protein diet may be optimal for promoting protein synthesis and maintaining plasma TP levels in blunt snout bream, irrespective of fish size. TP in the blood is a critical indicator of overall health and nutritional status in fish, as it reflects the balance between protein synthesis and degradation [32]. The observed increase in TP levels with a 30% protein diet aligns with findings in other species, where dietary protein levels have been shown to directly influence plasma TP concentrations [33,34]. Moreover, the relationship between dietary protein intake and TP levels is crucial for understanding the metabolic demands of fish during growth and development [15,35]. In juvenile fish, higher protein diets have been associated with enhanced growth rates and improved feed conversion efficiency, which are correlated with increased TP levels [36]. The maintenance of normal TP levels is also indicative of a well-balanced diet that supports the immune system and the overall health of the fish. Therefore, the observed increase in TP levels with a 30% protein diet not only reflects enhanced protein synthesis but also suggests improved immune function and health status.

In the present study, the observation of elevated plasma urea content in the larger fish that were fed diets with higher protein levels indicates that protein metabolism might be more efficient and perhaps more optimal at lower protein levels for this particular size group of fish. This is evidenced by the fact that the urea levels did not escalate further in the fish fed diets containing up to 34% protein, suggesting a plateau or an upper limit to the efficiency of protein metabolism at these higher levels [31]. Additionally, the higher plasma creatinine levels observed in the larger fish could be attributed to their greater muscle mass and overall metabolic activity when compared to their smaller counterparts. The increased muscle mass in larger fish likely results in higher metabolic rates and, consequently, higher creatinine production [37].

The observed trends in protease activity within the gastrointestinal tract of blunt snout bream in response to varying dietary protein levels provide intriguing insights into the digestive physiology of this species. The initial increase, followed by a subsequent decrease in protease activity in the foregut of both the small and large fish, suggests an adaptive response to the protein content in the diet [38]. The peak protease activity at the 32% protein level in the foregut indicates an optimal dietary protein concentration for protease induction in blunt snout bream, which aligns with the notion that protein intake is a critical regulator of proteolytic enzyme production [38]. Interestingly, the midgut protease activity in the larger fish peaked in the control group, which was significantly higher than that of the 30% protein group. This finding implies that lower protein levels could potentially stimulate higher protease activity in the midgut of larger fish, possibly as a compensatory mechanism to enhance protein digestion [39]. Conversely, in the smaller fish, the 36% protein level resulted in the highest protease activity, suggesting that higher protein diets could also stimulate protease production, albeit with a different optimal concentration than observed in the foregut.

In the present study, we observed that the hindgut protease activity was highest in the larger fish fed a diet containing 34% protein, which was significantly higher than that in the group receiving 30% protein. This finding suggests that the dietary protein level had a significant impact on the protease activity in the hindgut of the larger fish. The increased protease activity in the higher protein group could be attributed to the greater demand for protein digestion, as higher dietary protein levels require more efficient enzymatic breakdown to meet the nutritional requirements of the fish [40]. However, no significant differences in the hindgut protease activity were observed among the groups of smaller fish. Our result suggests that the protein level may have a less pronounced effect on protease activity in the hindgut of smaller individuals. Similar results were also found in *Haliotis laevigata* [41]. Moreover, the optimal protein level for protease induction in the hindgut may differ from that in the foregut and midgut. This difference could be due to the distinct physiological roles and digestive capacities of these gut segments. It is plausible that the hindgut may require a higher protein level to stimulate protease production, particularly in larger fish, to ensure the efficient digestion and absorption of dietary proteins. The differences in protease activity responses between small and large fish further highlight the complexity of dietary protein regulation in fish digestion. The adaptive mechanisms observed in the foregut, midgut, and hindgut suggest that the digestive system of blunt snout bream is capable of fine-tuning protease production in response to dietary protein levels. This suggests that the digestive system of blunt snout bream is not only responsive to dietary protein levels but also exhibits spatial regulation of protease activity within different gastrointestinal regions. In a study of yellowtail kingfish (*Seriola lalandi*), it was also found that under the action of nutrients, the protease activity of different intestinal segments also showed different adaptive changes [23].

The absence of significant differences in the lipase activity across different dietary protein levels in both sizes of fish suggests that the dietary protein concentration does not significantly influence lipase production or activity in blunt snout bream. Similar results were also found in studies of *Apostichopus japonicus* [42], *Rhamdia quelen* [43], *Eriocheir sinensis* [44], and *Macrobrachium nipponense* [45]. Similarly, the lack of significant effects on the amylase activity in the foregut and midgut of both sizes of fish indicates that dietary protein levels do not substantially impact amylase production or activity in these gut regions. However, in the hindgut of the smaller fish, the control group exhibited the highest amylase activity, which was significantly higher than that in the 30% protein group. Our results suggest that lower protein diets may support amylase activity in the hindgut of smaller fish, potentially due to a compensatory mechanism to utilize carbohydrates more efficiently when protein intake is limited [36].

Therefore, the differential responses observed among the foregut, midgut, and hindgut, as well as between the different sizes of fish, underscore the importance of considering the digestive physiology of specific gut regions and life stages when formulating diets for optimal nutrient utilization.

5. Conclusions

In summary, the results show that dietary protein levels affect the growth performance of blunt snout bream at two growth stages. Based on broken-line regression analysis, the optimal dietary protein levels for larger and smaller fish are recommended to be 30.45% and 29.95% of the diet, respectively. The dietary protein levels maintained normal metabolic functions, and a 30% protein diet might be optimal for maintaining plasma TP levels. The digestive system had adaptive responses to the protein levels. They impacted the protease activity in the gastrointestinal tract, with the optimal protein concentrations for protease induction differing among the gut regions. The larger fish had significant differences in the protease activity with higher-protein diets, especially in the foregut and hindgut, while the smaller fish showed the same in the foregut but less pronounced effects in the hindgut. The dietary protein levels had no significant impact on the lipase activity and minimally influenced the amylase activity in the foregut and midgut, but lower-protein diets might support amylase activity in the hindgut of smaller fish.

Author Contributions: Conceptualization, W.Z.; methodology, S.X. and B.L.; validation, W.Z. and A.W.; formal analysis, H.T.; investigation, Y.Y. and W.Y.; resources N.K.D. and C.Z.; data curation, F.L.; writing—original draft preparation, W.Z.; writing—review and editing, W.Z.; visualization, T.T.; supervision, T.T.; project administration, A.W.; funding acquisition, A.W. All authors have read and agreed to the published version of the manuscript.

Funding: This research was funded by the National Natural Science Foundation of China (32202913), the China Postdoctoral Science Foundation (2024M761452), and school-level research projects of the Yancheng Institute of Technology (xjr2022043).

Institutional Review Board Statement: The experimental procedures were conducted in accordance with the standards for scientific breeding and the utilization of aquatic animal established by the Animal Care and Use Committee of the Committee on the Ethics of Animal Experiments of the Freshwater Fisheries Research Center (Approval Code: LAECFFRC-2019-08-27).

Informed Consent Statement: Not applicable.

Data Availability Statement: The data are contained within the article.

Conflicts of Interest: Author T.T. was employed by Tongwei Agricultural Development Co., Ltd. The remaining authors declare that the research was conducted in the absence of any commercial or financial relationships that could be construed as a potential conflict of interest.

References

1. Dong, M.; Zhang, L.; Wu, P.; Feng, L.; Jiang, W.D.; Liu, Y.; Kuang, S.Y.; Li, S.W.; Mi, H.F.; Tang, L.; et al. Dietary protein levels changed the hardness of muscle by acting on muscle fiber growth and the metabolism of collagen in sub-adult grass carp (*Ctenopharyngodon idella*). *J. Anim. Sci. Biotechnol.* **2022**, *13*, 109. [CrossRef]

2. Ma, B.H.; Wang, L.G.; Lou, B.; Tan, P.; Xu, D.D.; Chen, R.Y. Dietary protein and lipid levels affect the growth performance, intestinal digestive enzyme activities and related genes expression of juvenile small yellow croaker (*Larimichthys polyactis*). *Aquac. Rep.* **2020**, *17*, 100403. [CrossRef]

3. Fuertes, J.B.; Celada, J.D.; Carral, J.M.; Sáez-Royuela, M.; González-Rodríguez, Á. Effects of dietary protein and different levels of replacement of fish meal by soybean meal in practical diets for juvenile crayfish (*Pacifastacus leniusculus*, Astacidae) from the onset of exogenous feeding. *Aquaculture* **2012**, *364*, 338–344. [CrossRef]

4. Martínez-Palacios, C.A.; Ríos-Durán, M.G.; Ambriz-Cervantes, L.; Jauncey, K.J.; Ross, L.G. Dietary protein requirement of juvenile Mexican silverside (*Menidia estor* Jordan 1879), a stomachless zooplanktophagous fish. *Aquac. Nutr.* **2007**, *13*, 304–310. [CrossRef]

5. Deng, D.F.; Ju, Z.Y.; Dominy, W.; Murashige, R.; Wilson, R.P. Optimal dietary protein levels for juvenile pacific threadfin (*Polydactylus sexfilis*) fed diets with two levels of lipid. *Aquaculture* **2011**, *316*, 25–30. [CrossRef]

6. Mirzakhani, M.K.; Ebrahimzadeh, S.M.; Samandaki, B.E.; Vardastzadeh, H.; Vafadar, A.; Abdel-Tawwab, M. Effects of dietary protein levels on growth performance, hemato–biochemical, and immune parameters in ship sturgeon (*Acipenser nudiventris*) juveniles. *Anim. Feed Sci. Technol.* **2024**, *316*, 116069. [CrossRef]

7. Wu, X.Y.; Gatlin, D.M., III. Effects of altering dietary protein content in morning and evening feedings on growth and ammonia excretion of red drum (*Sciaenops ocellatus*). *Aquaculture* **2014**, *434*, 33–37. [CrossRef]

8. Ren, M.C.; Tsion, H.M.; Liu, B.; Miao, L.H.; Ge, X.P.; Xie, J.; Liang, H.L.; Zhou, Q.L.; Pan, L.K. Dietary leucine level affects growth performance, whole body composition, plasma parameters and relative expression of TOR and TNF-α in juvenile blunt snout bream, *Megalobrama amblycephala*. *Aquaculture* **2015**, *448*, 162–168. [CrossRef]

9. Ministry of Agriculture of the People's Republic of China. *Chinese Fisheries Yearbook*; Chinese Agricultural Press: Beijing, China, 2024.

10. Ministry of Agriculture of the People's Republic of China. *Formula Feed for Blunt Snout Bream (Megalobrama amblycephala) (SC/T 1074-2004)*; Chinese Agricultural Press: Beijing, China, 2005.

11. Shi, W.L.; Shan, J.; Liu, M.Z.; Yan, H.; Huang, F.Q.; Zhou, X.W.; Shen, L. *A Study of the Optimum Demand of Protein by Blunt Snout Bream (Megalobrama amblycephala)*; FAO Library: Rome, Italy, 1988; Report No: FAO-FI–RAS/86/047; FAO-FI–NACA/WP/88/68; Accession No: 289611.

12. Tibbetts, S.M.; Lall, S.P.; Milley, J.E. Effects of dietary protein and lipid levels and DP DE^{-1} ratio on growth, feed utilization and hepatosomatic index of juvenile haddock, *Melanogrammus aeglefinus* L. *Aquac. Nutr.* **2005**, *11*, 67–75. [CrossRef]

13. Zhao, S.Y.; Lin, H.Z.; Huang, Z.; Zhou, C.P.; Wang, J.; Wang, Y.; Qi, C.L. Effect of dietary protein level on growth performance, plasma biochemical indices and flesh quality of grouper (*Epinephelus lanceolatus* × *E. fuscoguttatus*) at two growth stages. *South China Fish. Sci.* **2017**, *13*, 87–96.

14. El-Dahhar, A.A.; Elhetawy, A.I.G.; Shawky, W.A.; El-Zaeem, S.Y.; Abdel-Rahim, M.M. Diverse carbon sources impact the biofloc system in brackish groundwater altering water quality, fish performance, immune status, antioxidants, plasma biochemistry, pathogenic bacterial load and organ histomorphology in Florida red tilapia. *Aquac. Int.* **2024**, *32*, 9225–9252. [CrossRef]

15. Wang, J.T.; Jiang, Y.D.; Li, X.Y.; Han, T.; Yang, Y.X.; Hu, S.X.; Yang, M. Dietary protein requirement of juvenile red spotted grouper (*Epinephelus akaara*). *Aquaculture* **2016**, *450*, 289–294. [CrossRef]

16. Wang, L.G.; Hu, S.Y.; Lou, B.; Chen, D.X.; Zhan, W.; Chen, R.Y.; Liu, F.; Xu, D.D. Effect of different dietary protein and lipid levels on the growth, body composition, and intestinal digestive enzyme activities of juvenile yellow drum *Nibea albiflora* (Richardson). *J. Ocean. Univ. China* **2018**, *17*, 1261–1267. [CrossRef]

17. Jayant, M.; Muralidhar, A.P.; Sahu, N.P.; Jain, K.K.; Pal, A.K.; Srivastava, P.P. Protein requirement of juvenile striped catfish, *Pangasianodon hypophthalmus*. *Aquac. Int.* **2018**, *26*, 375–389. [CrossRef]

18. Pavasovic, A.; Richardson, N.A.; Mather, P.B.; Anderson, A.J. Influence of insoluble dietary cellulose on digestive enzyme activity, feed digestibility and survival in the red claw crayfish, *Cherax quadricarinatus* (von Martens). *Aquac. Res.* **2006**, *37*, 25–32. [CrossRef]

19. Habte-Tsion, H.M.; Liu, B.; Ge, X.P.; Xie, J.; Xu, P.; Ren, M.C.; Zhou, Q.L.; Pan, L.K.; Chen, R.L. Effects of Dietary Protein Level on Growth Performance, Muscle Composition, Blood Composition, and Digestive Enzyme Activity of Wuchang Bream (*Megalobrama amblycephala*) Fry. *Isr. J. Aquac. Bamidgeh* **2013**, *65*, 1–9. [CrossRef]

20. Zhang, W.X.; Xia, S.L.; Zhu, J.; Miao, L.H.; Ren, M.C.; Lin, Y.; Ge, X.P.; Sun, S.M. Growth performance, physiological response and histology changes of juvenile blunt snout bream, *Megalobrama amblycephala* exposed to chronic ammonia. *Aquaculture* **2019**, *506*, 424–436. [CrossRef]

21. AOAC. *Official Methods of Analysis of the Association of Official Analytical Chemists*, 15th ed.; Association of Official Analytical Chemists Inc.: Arlington, VA, USA, 2003.

22. Mozanzadeh, M.T.; Marammazi, J.G.; Yavari, V.; Agh, N.; Mohammadian, T.; Gisbert, E. Dietary n−3 LC-PUFA requirements in silvery-black porgy juveniles (*Sparidentex hasta*). *Aquaculture* **2015**, *448*, 151–161. [CrossRef]

23. Bowyer, J.N.; Qin, J.G.; Adams, L.R.; Thomson, M.J.S.; Stone, D.A.J. The response of digestive enzyme activities and gut histology in yellowtail kingfish (*Seriola lalandi*) to dietary fish oil substitution at different temperatures. *Aquaculture* **2012**, *368*, 19–28. [CrossRef]

24. Yang, M.; Wang, J.T.; Han, T.; Yang, Y.X.; Li, X.Y.; Jiang, Y.D. Dietary protein requirement of juvenile bluegill sunfish (*Lepomis macrochirus*). *Aquaculture* **2016**, *459*, 191–197. [CrossRef]

25. Kpundeh, M.D.; Qiang, J.; He, J.; Yang, H.; Xu, P. Effects of dietary protein levels on growth performance and haemato-immunological parameters of juvenile genetically improved farmed tilapia (GIFT), *Oreochromis niloticus*. *Aquac. Int.* **2015**, *23*, 1189–1201. [CrossRef]

26. Dias, J.; Arzel, J.; Aguirre, P.; Corraze, G.; Kaushik, S. Growth and hepatic acetylcoenzyme-A carboxylase activity are affected by dietary protein level in European seabass (*Dicentrarchus labrax*). *Comp. Biochem. Physiol. B* **2003**, *135*, 183–196. [CrossRef] [PubMed]

27. Usman, R.; Laining, A.; Ahmad, T.; Williams, K.C. Optimum dietary protein and lipid specifications for grow-out of humpback grouper *Cromileptes altivelis* (Valenciennes). *Aquac. Res.* **2005**, *36*, 1285–1292. [CrossRef]

28. Liu, X.W.; Mai, K.S.; Liufu, Z.G.; Ai, Q.H. Effects of dietary protein and lipid levels on growth, nutrient utilization, and the whole-body composition of turbot, *Scophthalmus maximus*, Linnaeus 1758, at different growth stages. *J. World Aquac. Soc.* **2014**, *45*, 355–366. [CrossRef]

29. Chen, W.; Han, D.; Yang, Y.X.; Zhang, Z.M.; Jin, J.Y.; Liu, H.K.; Zhu, X.M.; Xie, S.Q. Effects of dietary protein levels with cottonseed protein concentrate inclusion on growth, feed utilization, liver health and intestinal microbiota of juvenile largemouth bass (*Micropterus salmoides*). *Aquac. Rep.* **2024**, *39*, 102461. [CrossRef]

30. Yang, M.; Wang, J.T.; Han, T.; Yang, Y.X.; Li, X.Y.; Tian, H.L.; Zheng, P.Q. Dietary protein requirement of juvenile triangular bream *Megalobrama terminalis* (Richardson, 1846). *J. Appl. Ichthyol.* **2017**, *33*, 971–977. [CrossRef]

31. Fan, Z.; Wu, D.; Li, J.N.; Zhang, Y.Y.; Xu, Q.Y.; Wang, L.S. Dietary protein requirement for large-size Songpu mirror carp (*Cyprinus carpio Songpu*). *Aquac. Nutr.* **2020**, *26*, 1748–1759. [CrossRef]

32. Tang, L.; Xu, Q.Y.; Wang, C.A.; Yin, J.S. Effects of dietary protein levels on blood biochemical parameters in mirror common carp (*Cyprinus specularis*) at different temperatures. *J. Dalian Fish. Univ.* **2011**, *26*, 41–46.

33. Wang, L.; Zhang, W.; Gladstone, S.; Ng, W.K.; Zhang, J.; Shao, Q. Effects of isoenergetic diets with varying protein and lipid levels on the growth, feed utilization, metabolic enzymes activities, antioxidative status and serum biochemical parameters of black sea bream (*Acanthopagrus schlegelii*). *Aquaculture* **2019**, *513*, 734397. [CrossRef]

34. Wang, C.; Liu, E.; Zhang, H.; Shi, H.; Qiu, G.; Lu, S.; Han, S.; Jiang, H.; Liu, H. Dietary protein optimization for growth and immune enhancement in juvenile hybrid sturgeon (*Acipenser baerii* × *A. schrenckii*): Balancing growth performance, serum biochemistry, and expression of immune-related genes. *Biology* **2024**, *13*, 324. [CrossRef]

35. Zhang, J.; Zhou, F.; Wang, L.; Shao, Q.; Xu, Z.; Xu, J. Dietary protein requirement of juvenile black sea bream, *Sparus macrocephalus*. *J. World Aquac. Soc.* **2010**, *41*, 151–164. [CrossRef]

36. Yan, X.B.; Yang, J.J.; Dong, X.H.; Tan, B.P.; Zhang, S.; Chi, S.Y.; Yang, Q.H.; Liu, H.Y.; Yang, Y.Z. The optimal dietary protein level of large-size grouper *Epinephelus coioides*. *Aquac. Nutr.* **2020**, *26*, 705–714. [CrossRef]

37. Braun, J.P.; Lefebvre, H.P. Kidney function and damage. *Clin. Biochem. Domest. Anim.* **2008**, *6*, 485–528.

38. Zeng, B.H.; Wang, W.L.; Dong, Y.W. Dietary protein requirement of the high altitude's representative teleost juveniles *Schizopygopsis younghusbandi* (Cypriniformes: Cyprinidae). *Aquac. Res.* **2020**, *51*, 2852–2862. [CrossRef]

39. Liu, Q.D.; Wen, B.; Li, X.D.; Jiang, Y.S.; Liang, Z.P.; Zuo, R.T. An investigation on the effects of dietary protein level in juvenile Chinese mitten crab (*Eriocheir sinensis*) reared at three salinities: Survival, growth performance, digestive enzyme activities, antioxidant capacity and body composition. *Aquac. Res.* **2021**, *52*, 2580–2592. [CrossRef]

40. Hu, B. Effects of Pelleted and Extruded Diets of Different Protein Levels on Growth, Digestive Enzyme Activity, Antioxidant Capacity, Nonspecific Immunity, and Ammonia-N Stress Tolerance of Postlarval Pacific White Shrimp. *N. Am. J. Aquac.* **2022**, *84*, 239–248. [CrossRef]

41. Bansemer, M.S.; Qin, J.G.; Harris, J.O.; Schaefer, E.N.; Wang, H.R.; Mercer, G.J.; Howarth, G.S.; Stone, D.A.J. Age-dependent response of digestive enzyme activities to dietary protein level and water temperature in greenlip abalone (*Haliotis laevigata*). *Aquaculture* **2016**, *451*, 451–456. [CrossRef]

42. Xia, S.D.; Zhao, W.; Li, M.; Zhang, L.B.; Sun, L.N.; Liu, S.L.; Yang, H.S. Effects of dietary protein levels on the activity of the digestive enzyme of albino and normal *Apostichopus japonicus* (Selenka). *Aquac. Res.* **2018**, *49*, 1302–1309. [CrossRef]

43. Moro, G.V.; Camilo, R.Y.; Moraes, G.; Fracalossi, D.M. Dietary non-protein energy sources: Growth, digestive enzyme activities and nutrient utilization by the catfish jundia, *Rhamdia quelen*. *Aquac. Res.* **2010**, *41*, 394–400. [CrossRef]
44. Cui, Y.Y.; Ma, Q.Q.; Limbu, S.M.; Du, Z.Y.; Zhang, N.N.; Li, E.C.; Chen, L.Q. Effects of dietary protein to energy ratios on growth, body composition and digestive enzyme activities in Chinese mitten-handed crab, *Eriocheir sinensis*. *Aquac. Res.* **2017**, *48*, 2243–2252. [CrossRef]
45. Wang, W.L.; Li, L.Q.; Huang, X.X.; Zhu, Y.M.; Kuang, X.X.; Yi, G.F. Effects of dietary protein levels on the growth, digestive enzyme activity and fecundity in the oriental river prawn, *Macrobrachium nipponense*. *Aquac. Res.* **2022**, *53*, 2886–2894. [CrossRef]

MDPI AG
Grosspeteranlage 5
4052 Basel
Switzerland
Tel.: +41 61 683 77 34

Fishes Editorial Office
E-mail: fishes@mdpi.com
www.mdpi.com/journal/fishes